Focus on Technology

Alf Yarwood
and
David Heywood

Hodder & Stoughton

LONDON SYDNEY AUCKLAND TORONTO

Acknowledgements

The authors wish to express their appreciation of the help given by representatives of organisations who have permitted them to reproduce copyright photographs in this book. These organisations are:

The Central Electricity Generating Board;
Lego (UK) Ltd (Educational Division);
The Ford Motor Company;
Ferranti Electronics Limited;
Commotion Technology Supplies.

The following G.C.S.E. Examination Boards have kindly given permission for questions from examination papers to be reproduced within the pages of this book:

London and East Anglian Group (*LEAG*)
Midland Examining Group (*MEG*)
Northern Examining Association (*NEA*)
Southern Examining Group (*SEG*)
Welsh Joint Education Committee (*WJEC*)

We would like to thank Mr. G.R. Skirton, head teacher of the Arnewood School, New Milton, for the support he has given us in producing this book.

The authors also wish to record their gratitude to Mr. A. H. Orme, co-author with A. Yarwood of the book *Design and Technology* (published by Hodder and Stoughton Educational) for allowing the specimen projects illustrated here on pages 138 to 147 to be reproduced from that earlier book.

Finally we would like to express our appreciation of the work of those pupils from the Arnewood School, New Milton and of the Thomas Alleyne's High School, Uttoxeter, whose projects we have reproduced here. Without such work, a book of this nature would not have been possible.

A. Yarwood
D. Heywood

British Library Cataloguing in Publication Data
Yarwood, A. (Alfred), 1917–
  Focus on technology.
  1. Design & technology
  I. Title II. Heywood, D.
  600
  ISBN 0 340 41486 3

First published 1990

Typeset by Gecko Ltd, Bicester, Oxon.
Printed in Great Britain for Hodder and Stoughton Educational, a division of Hodder and Stoughton Ltd, Mill Road, Dunton Green, Sevenoaks, Kent by Butler and Tanner Ltd, Frome.

Contents

Preface

This book is designed as a textbook for pupils and students in schools and colleges who are preparing for an examination in the GCSE subject *CDT: Technology*. It is the third in a series of books for the GCSE subjects covered by CDT syllabuses. The other two books are *Design and Craft* and *Design and Communication*.

The Design Process adopted by all the examination boards in their CDT syllabuses follows the sequence *situation*; *design brief*; *investigation*; *ideas for solutions from which an appropriate solution is selected*; *the making of a model*; *realisation*; *evaluation*. This Design Process is followed in all three books in the series.

CDT syllabuses for GCSE are intended for examinations following a five-year course. This book is designed for use during the final two years of the course (which will be Key Stage 4 of the National Curriculum).

CDT examinations consist of two or three 'written' papers plus the submission of project work to moderators appointed by the examination boards. Project work is an important part of all GCSE syllabuses. For *CDT: Technology*, examination boards usually require that as much as 50 per cent of available marks are awarded for projects submitted by candidates. An important feature of this book is the very large number of suggestions for project work which it contains. A number of specimen projects are also included. This does not mean that the theoretical side of school technology has been neglected. The contents of this book are designed to cover both theoretical and project aspects of GCSE Technology.

A. Yarwood
D. Heywood
1989

1 Introduction

CDT: Technology Examinations

The examinations are in three parts. All Examining Groups examine in the following manner:

I. CDT: Technology Core Syllabuses for GCSE

Core examination papers require:
- skill in designing and making and in graphics;
- knowledge of safety problems when designing technology projects and when using tools and equipment;
- general knowledge about energy resources, materials, electronics, mechanisms and structures;
- knowledge of how technology can affect the environment and the society we live in.

2. CDT: Technology Module Syllabuses for GCSE

Module examination papers require a more specialised knowledge of at least two of the following modules:
- Energy & Power
- Electrics
- Electronics
- Pneumatics
- Computers
- Mechanisms
- Structures
- Materials

3. CDT: Technology Projects

All candidates are required to design projects, which will be marked by moderators from an Examining Group. Projects must show knowledge and practice in:
- several modules;
- designing;
- graphic skills;
- making skills.

Note: Project work is very important because 50% of the total marks for the examination are based upon moderators' marks for your project(s).

How to use this book
Core

This introduction includes information you will need for the core syllabus, as follows:
1 discussion ideas for the Technology and Society part of the syllabus;
2 notes on the Design Process common to all CDT: Technology activity. A later section also deals with the Design Process. (pages 119 to 124);
3 notes on safety in project work and in the use of tools and equipment. In addition to this information, the reader, with the guidance of a teacher or lecturer, can select from each module chapter the general information needed for the CDT: Technology core. (Each module chapter also covers the specialist knowledge needed if you follow it as an option.)

A number of core exercises are given at the end of each of the module chapters. (These require the use of information from the modules which is relevant to the core.)

Modules

Knowledge and practical applications of modular knowledge are given in chapters 2–9.

Questions taken from recent module examination papers are included in chapter 14.

Projects

1 A large number of project suggestions are given in chapter 13 'Ideas for Projects'.
2 Methods of graphics and some specimen projects are described in chapter 11, 'Presentation'.

Technology and Society

When it is first introduced, a new technology may seem to offer benefits for all. Problems may only appear after some time. When you are asked to discuss a technology, try not to make quick judgements.

1 Learn as much about it as you can.
2 Read about it in newspapers, magazines, books.
3 Keep an open mind about the technology until you have sufficient knowledge about it and its effects.
4 Try not to make judgements from watching a single television programme or from reading a single newspaper or magazine article.

People have always invented technologies to make life easier and more comfortable. However, new technologies rarely completely replace old ones. New and old technologies often exist side by side. Some examples are given below.

1 Stone Age tools were replaced by those of the Bronze Age when it was discovered that copper and tin produced a metal which could be sharpened to hold an edge. Tools of bronze brought benefits to Bronze Age people, but they caused the decline of those people who went on using stone tools. In turn, the Bronze Age was succeeded by the Iron Age. Yet groups of peoples were still using primitive tools in modern times. Examples could be found in Australia, in South America and in New Guinea, where Stone Age cultures were still existing in the 20th century.
2 The technologies involved in hunting and collecting were overtaken by those of herding and cultivation to provide food. As a result human populations began to increase.
3 In more recent times forms of land transport using horses to power them have been superseded by the technologies of vehicles driven by engines resulting in increased travel and trade.
4 The technologies of sailing ships have been overtaken by the technologies of ships driven by engines of various types. This has again resulted in vastly increased trading throughout the world.
5 At present new technologies are being developed at an increasing rate. Travelling has become much easier. People's health has been improved by advances in medicine, as well as by advances in methods of reducing air pollution, by improved methods of sanitation, by the development of water resources. Means of communication have been improved, e.g. by the use of computers and satellites and by the use of glass fibre optics.
Note 1: A technology may benefit the country which uses it, but it could be harmful to other countries.
Note 2: Once a technology has been invented, it cannot be dis-invented. For example, nuclear bombs have been made. Although their use can be controlled by agreement and they can also be destroyed, the knowledge of how to make them nevertheless exists.

Discussion

Further on there are some discussion topics for you to consider, but first let's consider what is needed for a constructive discussion.

With discussion groups, small *is* beautiful. An ideal size is three, four or five people. Larger groups may lead to some not having a chance to speak or the group being dominated by one member. A leader should be appointed for each group, to see that the discussion continues along relevant lines. When you are asked to join in a discussion group, spend some time making notes or lists of the points which you would like to discuss or which you would like to talk about. Think about what you wish to include in the discussion before it starts.

Discussions can be started off by asking for suggestions – including some which could be controversial.

Attempt to make a report of the results of a discussion: either a spoken report to the group, or to your whole class, or a written report. Below are some discussion topics.

Topics for discussion

1 New and old technologies exist side by side. Note and discuss examples.
2 One important modern technology is computing. One example of its application is computer-controlled robots and machines which have resulted in many boring and monotonous jobs, previously performed by people, being carried out by machines.
  Make lists of:
- other examples of modern technologies;
- advantages of the examples you select;
- disadvantages of the examples.

3 ■ In 1983, an engineering firm was producing valves used by water companies throughout the world. The directors decided to change from man-operated to computer-aided machining. The company installed the following throughout their factory:

■ a computer-aided design (CAD) system so that the engineers could design at computer terminals;

■ a computer-aided machining (CAM) system by which all machine operations were automated. These computer systems produced valves which were cheaper, more accurately made and more efficient. Sales of the company's products increased.

However, about 60% of the work force became redundant.

■ The French telephone service offers free computers linked to a system known as Mintel. This allows users to 'call up' any telephone number on their screens. Mintel also allows companies to run services such as conversation links between people, the leaving of messages, computer-linked mail and other facilities. The introduction of Mintel has created a large number of new jobs.

With these two examples in mind, discuss the effect of computing on job opportunities.

4 In 1987/88 biologists at an American university invented a system for inserting genes into the eggs of cows. The resulting calves grew into cows with higher milk yields than their mothers. The 'inventors' of these genes wanted to take out a patent on their discovery, which would allow them to demand payment for any cow possessing the gene responsible for the higher milk yields. Discuss the ethics of this situation.

5 Air travel has become cheap, because of advances in the technology of flying. The result is as if the world had shrunk in size. We can now travel easily on holidays. Business people can arrange meetings in one part of the world to which people from other countries can travel with ease.

Discuss the implications of travel by aeroplane.

6 By adding nitrate fertilisers to arable crops, farmers have been able to increase the yields of cereals grown on their land. During rainfall some of the nitrates run off the land into rivers. This nitrate run-off can pollute water supplies and it is known to cause illness in young children who drink nitrate-polluted water. Discuss the ethics of using such fertilisers, bearing in mind the ever-increasing need to produce more food.

7 Many radio and television satellites are in orbit. These have reduced the cost of communication between most parts of the earth. On the other hand, governments have used satellites to control military weapons and to spy on other countries.

Discuss the implications of satellite technology for future generations of people.

8 The increase in the use of motor cars has improved the life of many people. It also causes problems – atmospheric pollution, noise pollution, large numbers of people being killed and maimed as a result of road accidents.

Discuss what you think can be done to reduce these harmful results.

9 A company wanted to mine good quality limestone in a coastal beauty spot. Local people objected on the grounds that such a beautiful stretch of land should not be destroyed for purely financial gain. The area covered 50 hectares and at least 30 people would be employed in the mining operation.

Which is more important – the countryside or the financial considerations of the company?

10 In 1988 it became known that a fire at a nuclear plant in 1957 had caused nuclear emissions into the atmosphere. The government at that time did not report the accident, because of the alarm it would have caused and because it might have prevented them from building further nuclear power stations.

Discuss this situation.

11 Many technological developments have changed the running of homes, e.g. cheap, efficient refrigerators and electrical washing machines.

Make a list of other items which are labour-saving in the running of a home. Discuss how such devices have helped in the emancipation of women in this country.

12 Many houses are centrally heated by burning oil. Oil is a finite material, which will eventually run out. It is a valuable raw material from which useful chemicals and drugs can be obtained.

Discuss the ethics of using irreplaceable oil to heat houses.

13 The production of electrical power and of manufactured goods often requires huge amounts of energy, usually produced by burning fuels such as coal and oil. Without electricity, many of the amenities of life at present would be lost. Yet without sufficient control, the burning of fuels produces emissions into the atmosphere which can be harmful.

Discuss this problem.

14 Do you know of any modern developments which have given rise to noise, atmospheric or visual pollution? Make a list of these and discuss ways in which you think the problems caused by these types of pollution could be reduced or eliminated.

15 A great deal of information about people can be stored on computer files. This has great benefits, e.g. it can speed up payments to sick and unemployed people, help doctors to have quick access to their patients' files, allow banks to give credit to those people who require it faster.

Despite these advantages, a Data Protection Act was passed by Parliament in 1987 which allows anyone to have details of any information about them that is held on computer files.

Do these two details conflict? Make notes of your own comments on these two items of information.

A Design Process

Fig. 1 is a flow chart showing the stages involved in designing.

1 Problem: there must have been some problem for which the design is to be made.

2 Design Brief: a written statement defining what is to be designed, and any constraints on the design.

3 Ideas for Solutions: a number of sketches with notes of ideas for making the design.

4 Appropriate Solution: the best solution chosen from the ideas for solutions.

5 Model: you may need to make a model of the appropriate solution.

6 Realisation: the design is made.

7 Evaluation: the realised design is evaluated against the original design brief.

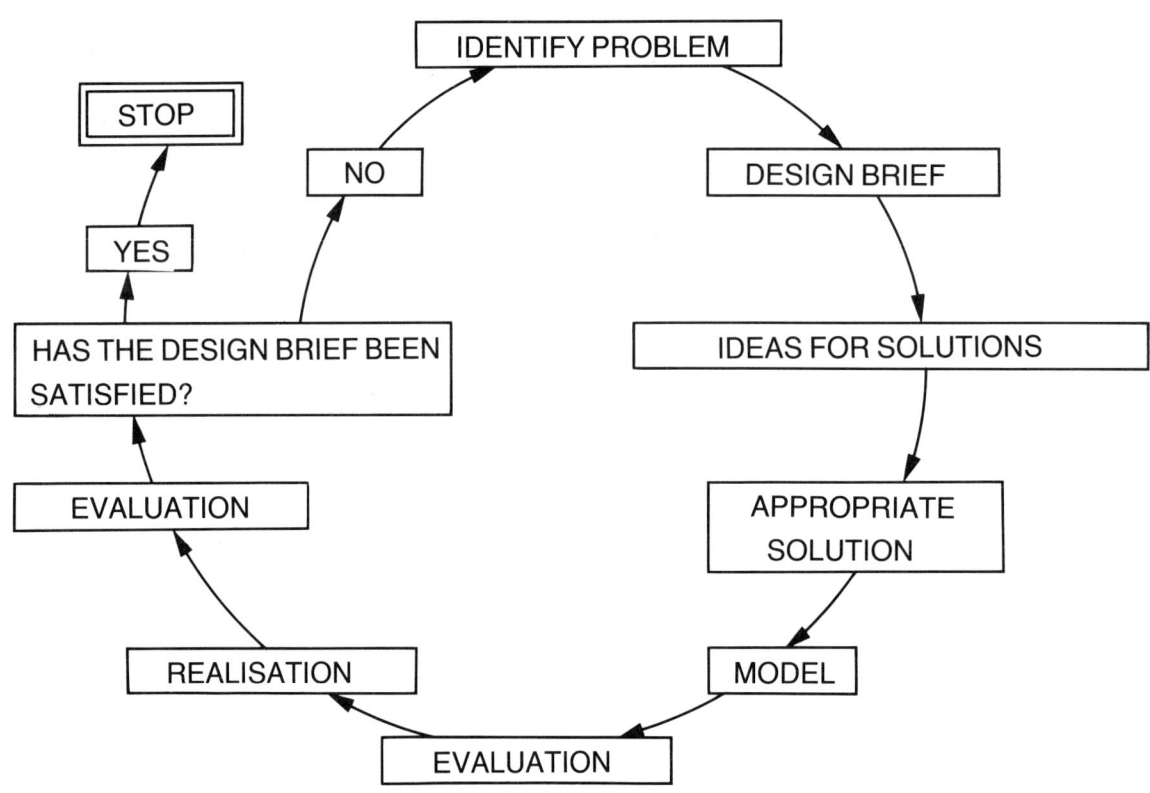

Figure 1 Flow chart showing the Design Process

Evaluation

Each stage in the design process should include evaluation. Each solution idea should be evaluated – to see if it could satisfy the demands of the design brief. The appropriate solution must be evaluated. Models need to be evaluated. A finished design must be evaluated to check whether it meets the requirements of the design brief to which it has been made.

Designing could be thought of as a continuous process of the evaluation of ideas.

When is a design completed?

In Fig. 1 the question is asked:

'Has the design brief been satisfied?'

1 If the finished design satisfies the design brief, then the answer is YES.
2 If the realised design does not satisfy the design brief, then NO – start again by looking at the problem from which the design brief came.
3 If the design is unsatisfactory, several answers are possible:

- Alterations to the completed design may produce a satisfactory result.
- Start all over again.
- After studying all the notes and drawings of the ideas for solutions and the evaluations of them, a new and more satisfactory solution may arise.

A design process mnemonic

The flow chart, Fig. 1, can be reduced to several main stages:

DESIGN BRIEF

IDEAS FOR SOLUTIONS

APPROPRIATE SOLUTION

MODIFICATIONS

REALISATION

EVALUATION

You may find it easy to remember this as a word made up from the initials of each stage, giving the code name DIAMRE. Or – if you would sooner learn a phrase – how about 'Design In Any Material Requires Effort', in which the initials of each word

are the initials of the stages in the design process. Such words or phrases which can help your memory are known as *mnemonics*. The design process is described in pictorial form on pages 5 and 6.

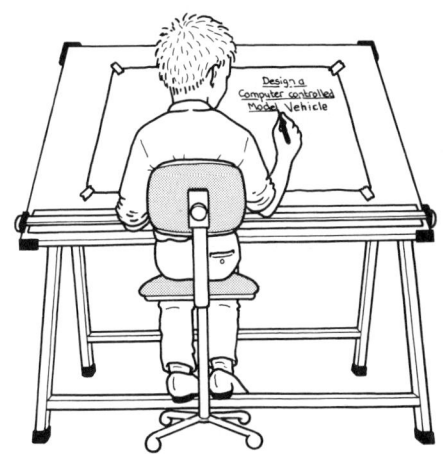

D is for Design Brief

I is for Investigation

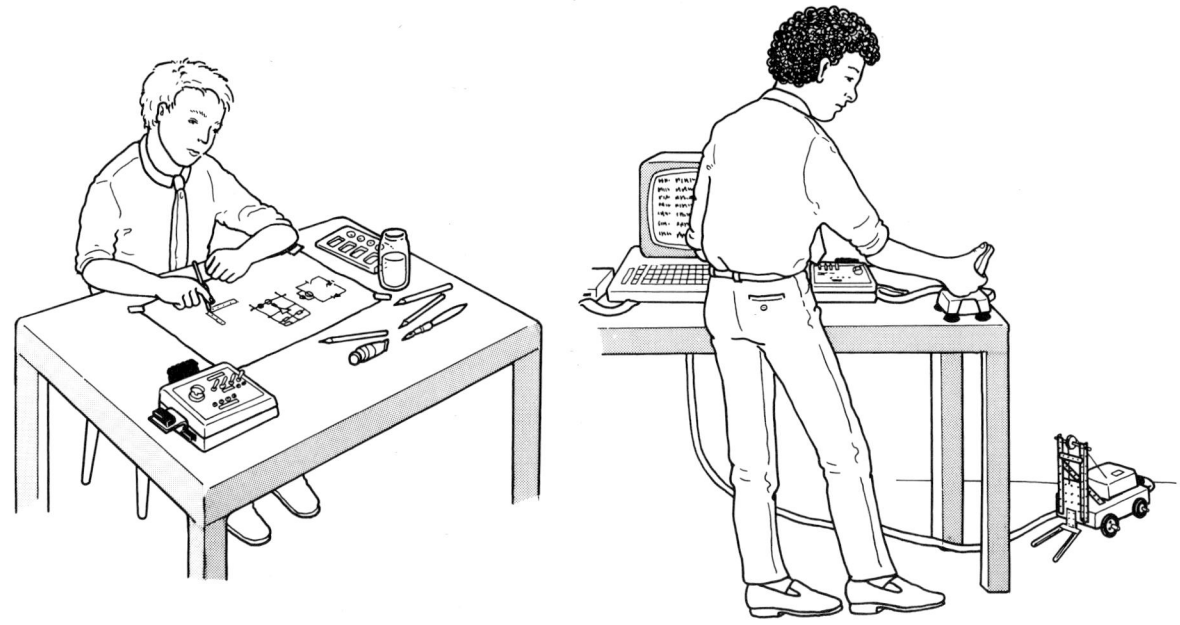

R is for Realisation

A is for Appropriate Solution

E is for Evaluation

M is for Model

Safety

Safety is an important consideration, particularly if your design includes electrical, electronic or pneumatic circuits. Problems of safety also arise when working with tools and machines in workshops. Here are some points to check as you are working.

Control systems

Is your design controlled by a mechanical, electrical, electronic or pneumatic system? Does it work as you expected? Check, re-check and check again until you are satisfied that it does work correctly. Control systems should be designed to be foolproof – if they fail for any reason, either another system should take over control, or the design should stop working.

Mechanical systems

1 Does the mechanism work as you thought it should? Test it and then test it again.
2 If a mechanism is powered by electricity, test if it works BEFORE switching on.
3 Slipping belts are a safety hazard.
4 Badly-designed gear work is unsafe.
5 Check moving parts – can they move freely without hitting people or parts of a design?

Electrical and electronic systems

1 Are all parts fully insulated?
2 If connected to the mains supply:
■ are circuits earthed?
■ would a transformer have safety advantages?
3 Check:
■ can parts become damp and so cause 'shorts'?
■ are circuits fully disconnected when OFF?
4 Disconnect the power supply when making changes to a circuit.
5 Large capacitors can carry large voltages and can be dangerous even when a circuit is OFF.
6 Short circuits may be caused by:
■ poor insulation;
■ parts being too close to each other;
■ wrong materials used for containers holding electrical parts.
7 Power: is the wattage too high? (Watts = amps × volts.)

Pneumatic systems

1 A loose compressed air line, or someone playing with an air line, can cause damage to soft human tissues.
2 Check pipework in a circuit before switching compressed air ON.

3 Check the positions of components in circuits before switching ON, to avoid the possibility of injury to those operating the circuit.

Structures

1 Are the structures in a design safe?
2 Will they take the loads they were designed to take?
3 Is a safety margin included in the design?

Materials

When choosing materials look at details such as:
1 Their strength: are they strong enough to carry the loads and stresses in the design?
2 Are they flexible (or rigid) enough to resist the stresses and strains the design must undergo?
3 Will they deteriorate under conditions of damp, heat, vibration?
4 Are they corrosion-resistant if the design is in a situation liable to corrosion?
5 Lead, asbestos, some oils, inflammable foams, and some sprays should be avoided, because of the grave health risks associated with their use.

Fire

1 If possible, use only non-combustible materials.
2 Make sure that there is sufficient protection if the output from a design generates heat.
3 If a gas is used to power a design, check that adequate safety measures are being taken.

Speed

Speed is dangerous unless it is controlled. The faster a machine works, the greater is the danger.
Vehicles which have been designed for speed must be tested at low speeds and at the fastest expected speed before you can be satisfied that the design is safe.

Instructions

Instructions issued with a design should be correct. Such instructions should be tested to ensure that the wording is accurate. Instructions on controls and switches must be correct.

Noise

Noise can be a health hazard. Ask of your design 'Is it generating too much noise?'. If the answer is 'Yes', then amend the design accordingly. Noise can also annoy people.

Sharp edges and corners

Avoid sharp edges and corners because they can cause injury. Smooth or round them off or pad them so that they are safe to those who may use or be near a design.

Some workshop safety rules

1 Be properly dressed – aprons or overalls, hair tied up, no loose clothing, sleeves rolled.
2 Hand working:
- Materials being worked must be firmly held.
- Tools must not cover your bench top.
- Badly maintained hand tools are dangerous – they should be clean and sharp.
- Keep sharp edges away from your own and other people's bodies.
3 Working with machines:
- Understand how to use the machine before switching ON.
- Tools and materials must be held firmly in the machine. Check before switching ON.
- Wear goggles.
- Switch OFF when changing tools or material held in the machine.
- Learn where emergency stops are, but DO NOT use them except in an emergency.
4 When casting in metal:
- Goggles, heavy leather gloves, a leather apron and shoe/boot guards MUST be worn.
- The floor must be dry.
- Good ventilation is essential.
5 Working with plastics:
- Working areas MUST be well ventilated. Dust and fumes from some plastics can cause eye and throat irritation.
- When sawing, filing or machining, wear face masks and goggles.
- When working with polyester resins for GRP work, use barrier cream on hands or wear gloves.
- Be careful where catalysed polyester resin waste is placed – it can generate enough heat to cause fires.

Modules

The **areas of knowledge** needed for technology design are described in GCSE syllabuses for examinations in the subject CDT: Technology as **Modules**. These modules or areas of knowledge are:

1 Energy & Power – chapter 2;
2 Electrics – chapter 3;
3 Electronics – chapter 4;
4 Pneumatics – chapter 5;
5 Computers – chapter 6;
6 Mechanisms – chapter 7;
7 Structures – chapter 8;
8 Materials – chapter 9.

This book includes some information and practical work for each of the eight modules. However, there is only room for elementary information and examples of simple practical projects. Some people will spend their working lives studying and practising in just *one* of these areas. So it is no wonder that a book of this size cannot fully cover every module.

The work in this book needs to be supplemented by further study:
1 from other books;
2 by more experimenting;
3 by the use of films, television, magazines;
4 by noting how technology is being applied to all aspects of modern life;
5 by attempting to understand how modern technological designs work.

There are further areas of knowledge which will have to be studied if other areas of technology are to be applied. Examples are *aerodynamics, hydraulics, acoustics, optics*.

The questions from examination papers set by the GCSE Examination Groups in chapter 14 include many module questions. Some of these could be used as project ideas.

Structure of technology design

Fig. 2 shows the way in which modules, graphics and design are linked. The chart shows:
1 *energy and power* are needed to power the systems used in projects;

2 *control* can be achieved using a variety of systems – *electrics*, *electronics*, *mechanisms*, *pneumatics*, *structures* or *computing*;
3 *energy and power* are needed to produce *materials*;
4 *graphics and notes* are necessary when designing;
5 *computers* can be used as design tools;
6 *graphics* and *notes* link work in the eight modules with *designing* for project work.

Control

Note the word *control*. Circuits and systems based on six of the modules – electrics, electronics, mechanisms, pneumatics, structures and computing – can be used to control the way in which the projects you design will function.

SI units of measurement

All units of measurement in technology work should be those of the international standard SI system, together with standard SI abbreviations. All SI units are metric. Appendix 1 on page 164 lists the more commonly used SI units and the prefixes used with SI units.

The newton

The SI unit of force is the newton (N). The exact weight of a 1 kg mass varies slightly from place to place, but is approximately 9.81 newtons. For the purposes of the experiments and minor projects suggested in this book the weight of a 1 kg mass is taken to be 10 newtons.

Newton's Laws

In 1687 Isaac Newton published his book *Principia Mathematica*, in which he included his *Three Laws of Motion*. These laws still apply to *structural* and *mechanical* phenomena. Newton's three laws only need corrections when astronomical speeds or distances are involved. Then Einstein's *Law of Relativity* comes into operation. We are concerned here with two of Newton's laws.

☐ **Newton's First Law:** If no total force acts on an object it either stays at rest or, if it is moving, it carries on moving at the same steady speed in the same straight line.

☐ **Newton's Third Law:** To every action there is an equal and opposite reaction.

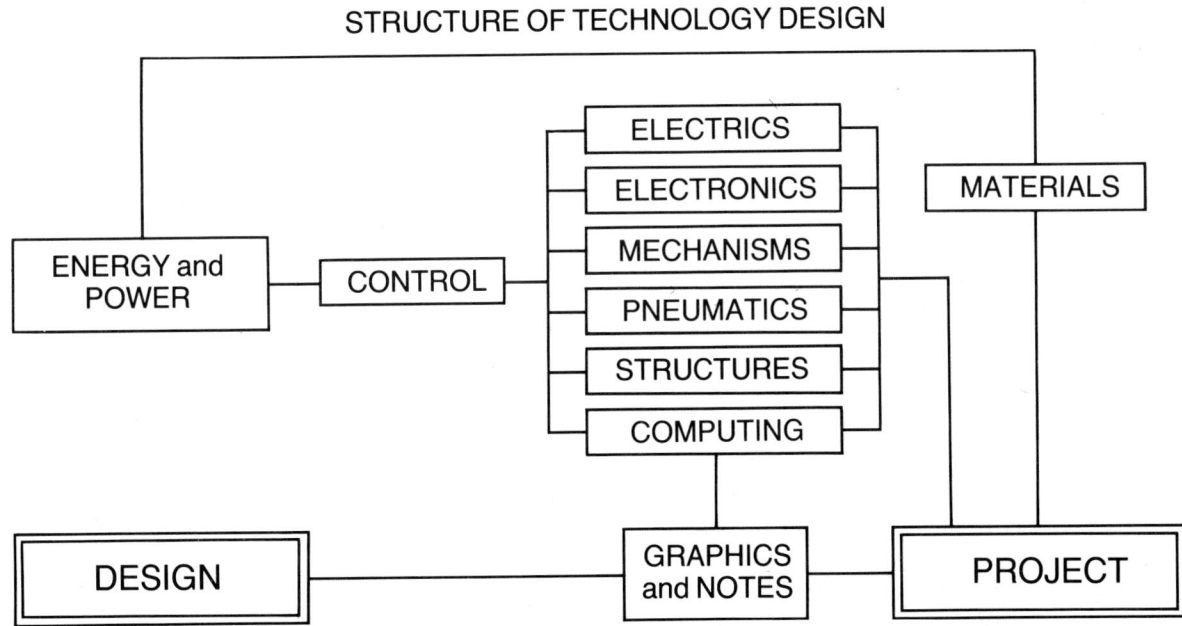

Figure 2 Structure of Technology Design

2 Energy & Power

Introduction

The sun is the original source of all energy. Without the sun, there would be no life on earth. *Chlorophyll*, the green matter of plant life, uses light energy from the sun in the chemical action known as **photosynthesis**, to produce carbohydrates from carbon dioxide and water in the atmosphere and in the soil. A by-product of this chemical action is oxygen. The well-being of the earth and its atmosphere depends completely on photosynthesis.

Forms of energy

Energy is available in many different forms – chemical, wind, heat, electrical, mechanical, sound, light, magnetic. Any one form of energy can be converted to another. For example, wind energy turns the blades of an aerogenerator; the mechanical energy of the aerogenerator creates electrical energy in a generator; electrical energy from the generator is conducted to houses; heat energy from electric fires gives warmth; light energy is obtained from the glowing bars.

This example shows the principle of **conservation of energy** – energy cannot be created or destroyed, it only changes into a different form.

Finite and Infinite Sources of Energy

Those sources of energy which are used up as time progresses, such as coal, oil and gas in the earth's crust are said to be **finite** or **non-renewable** sources of energy.

Those sources of energy which cannot be used up, such as, energy from the sun, from the wind, from the action of waves and tides, are said to be **infinite** or **renewable** sources.

Finite sources	*Infinite sources*
coal	the sun
oil	the wind
natural gas	tidal energy
wood (to a limited extent)	wave energy
	geothermal energy
nuclear energy (to a limited extent)	nuclear energy (to a limited extent)
	wind energy
	water energy
	thermal energy from the ocean
	energy from hydrogen
	heat exchange systems

Primary and secondary sources of energy

☐ **Primary sources:** e.g. coal, natural gas, the sun, water and wind.

☐ **Secondary sources:** energy sources resulting from **processing** primary sources in some way. – e.g. electricity, petrol, heating oils, smokeless fuels from coal.

Finite sources of energy

Coal

Coal was formed many millions of years ago from the remains of plants. The energy in coal comes from the photosynthesis carried out by those plants when they were alive.

Coal-fired power station

Fig. 1 shows a typical coal-fired power station. A forced draught of air carries coal (which has been ground into small pieces) into the boiler, where it catches fire. The heat produced turns water into

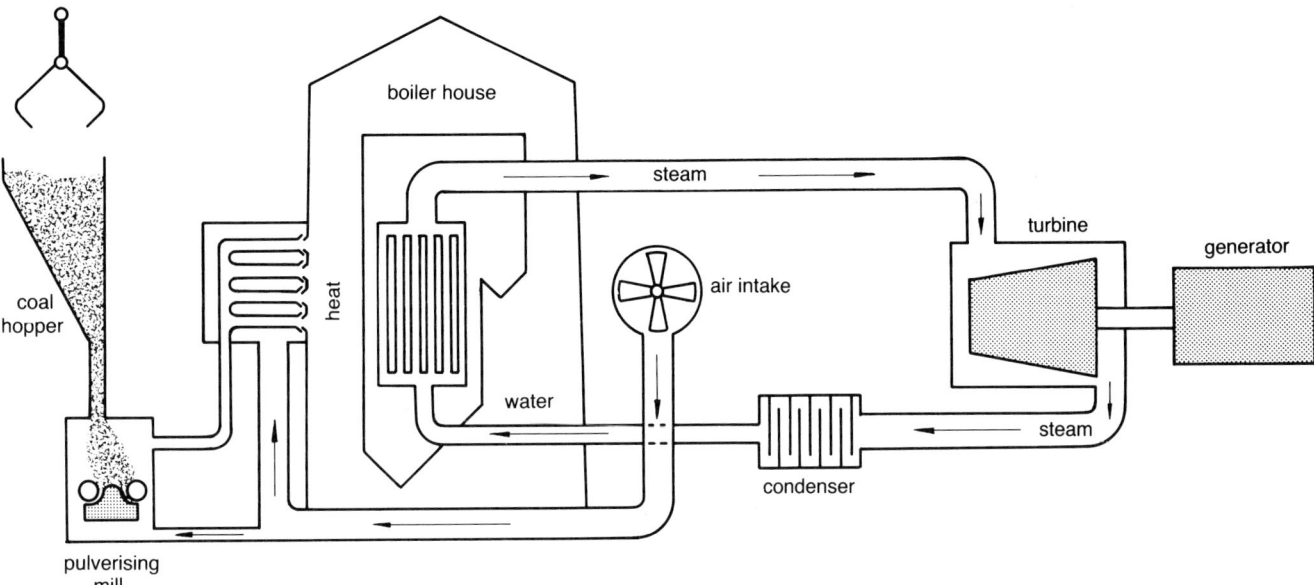

Figure 1 Coal-fired power station

steam, to drive a turbine. The turbine drives a generator which produces electricity.

In 24 hours a 1000 unit MW coal-fired power station will consume some 11 000 tonnes of coal and 100 000 tonnes of air. About 23 000 tonnes of carbon dioxide (CO_2) and 3600 tonnes of water vapour are expelled into the atmosphere. 3000 tonnes of ash needs removing. Some 7.5 tonnes of ash dust is ejected into the atmosphere.

Oil and natural gas
Refining of crude oil

The process of refining crude oil follows this pattern:
1 The oil is heated and passed into a still.
2 The oil rises through fractionating columns.

3 A separation occurs at various levels within the fractionating columns resulting in fractions as shown in Fig. 2.
4 In the cases of petrol, additions are added to:
■ ensure the petrol burns cleanly;
■ prevent gumming inside the engine cylinders.
One additive – lead – prevents the knocking noise known as *pinking*. Motorists are now being persuaded to run their vehicles on lead-free petrol because of the danger involved with lead pollution of the atmosphere.

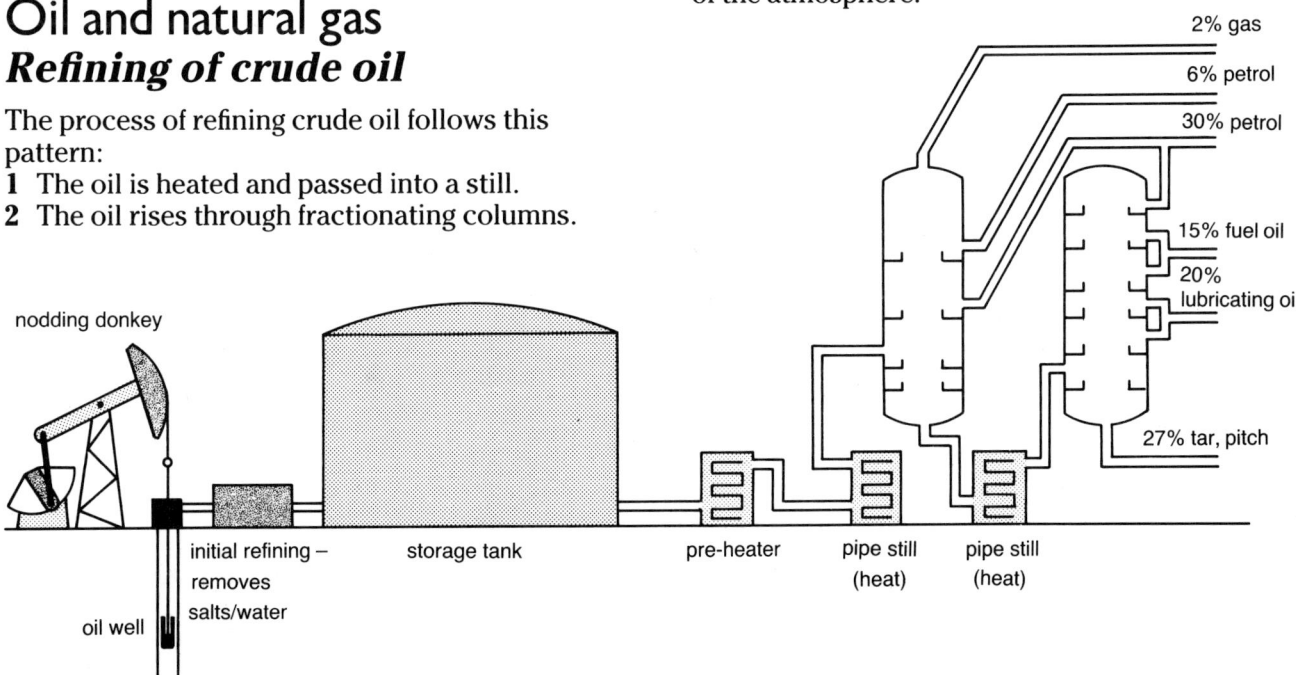

Figure 2 The refining of crude oil

Nuclear power

The following paragraphs explain how energy is released in a nuclear reactor.

The nucleus of any atom consists of protons and neutrons. Some nuclei are described as unstable – which means that they tend to break up, releasing radioactivity. Uranium–235 is one example of a radioactive substance – its nuclei are unstable and for this reason it is used in fission nuclear reactors.

If an atom of uranium–235 is bombarded with neutrons, then it will split into two fragments (fission fragments). When an atom splits in this way, more neutrons start what is called a chain reaction by bombarding any other nearby uranium–235 atoms which split into more fission fragments (as above), releasing more neutrons and yet more energy.

This process is the basis of energy production in a nuclear reactor. In a reactor, pellets of uranium are loaded into metal tubes which are sealed securely. A number of these metal tubes (nuclear fuel elements) are grouped together, with spaces between them for the heat which is generated to escape. A robot is used to load the rods into the reactor. The nuclear reaction starts spontaneously, when there is enough uranium fuel present (this can be controlled via the robot). The energy from the splitting atoms makes the fuel rods glow red hot, although there are no flames.

Heat produced in the reactor is conducted away by a coolant, such as carbon dioxide under pressure, which circulates through the core, where it absorbs heat. The coolant is directed to a heat exchanger. This consists of many rows of water-filled tubes where water absorbs the heat from the coolant and is turned to steam. This steam operates turbine generators, as in a coal-fired power station.

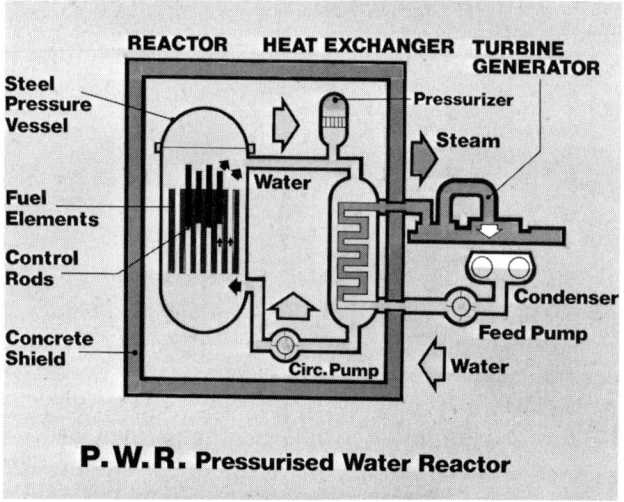

Figure 3 PWR – Pressurised Water Reactor (by courtesy of the Central Electricity Generating Board)

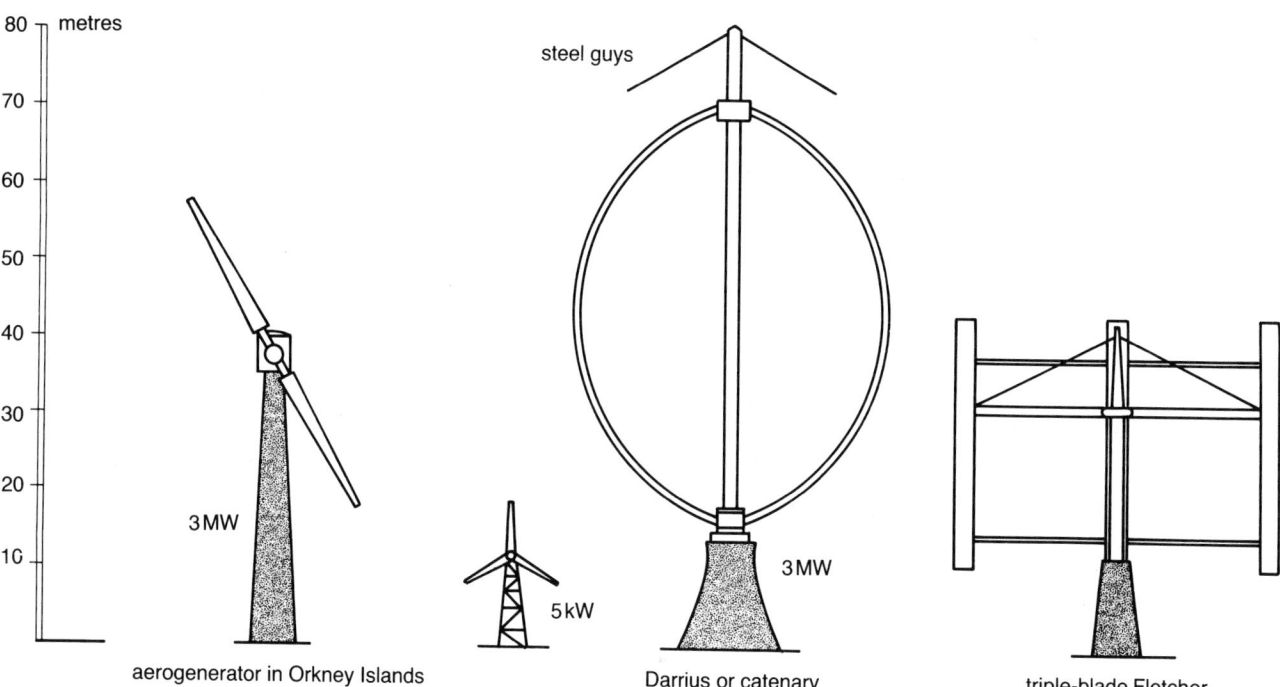

Figure 4 Wind energy machines

Infinite sources of energy

Wind energy

Great Britain is situated in an area where winds are common. So we can make good use of aerogenerators which use wind energy to produce electricity.

The drawings in Figure 4 include two windmills with blades set to produce horizontal rotation.

The Savonius Rotor

This is a cylinder cut in half vertically. One half is fixed across the other (Fig. 5). The two halves are welded onto two plates, one on each end. A central shaft allows free rotation.

Wind is caught by the open face of the device. The closed, convex face has less resistance to wind. There is a free passage of air through. Thus wind produces more pressure on one 'rotor' than on the other, making the whole device spin round on its axis.

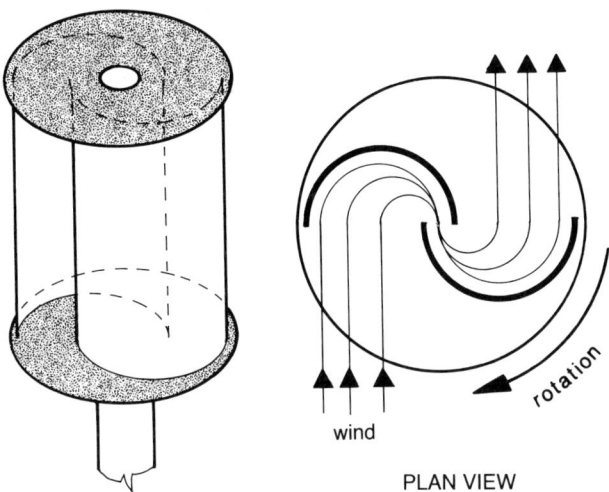

PLAN VIEW

Figure 5 Savonius rotor

Energy from the ocean

Two types of wave are suitable for the generation of electricity from energy from the oceans:
1 waves at sea – produced by winds;
2 tidal waves, such as the Severn Bore.

Waves at sea

The electricity is developed from the bobbing movement of large *pontoons* along a central spine.

The torque (turning motion) caused by the waves is converted by generators into electricity.

Tidal waves

As the tide runs up an estuary, it can travel considerable distances (an example of this is the Severn Bore). By building a barrage across the estuary, the energy in a tidal wave can be used to produce electricity as follows. The barrage allows water to flow up from the sea when the tide is coming in and then stops the water from returning down to the sea, except through man-made pipes. The pipes direct the water to drive turbines which change the energy from the wave to electricity, Fig. 6. See page 14 for more information about hydroelectric energy.

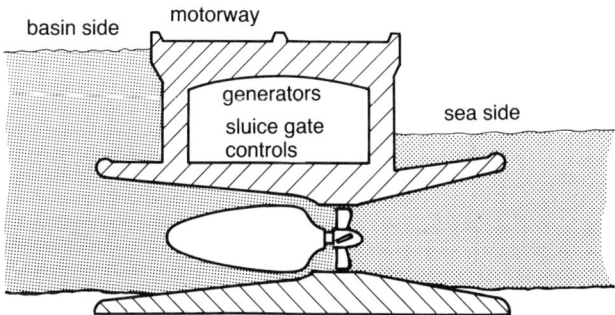

Figure 6 Tidal power station

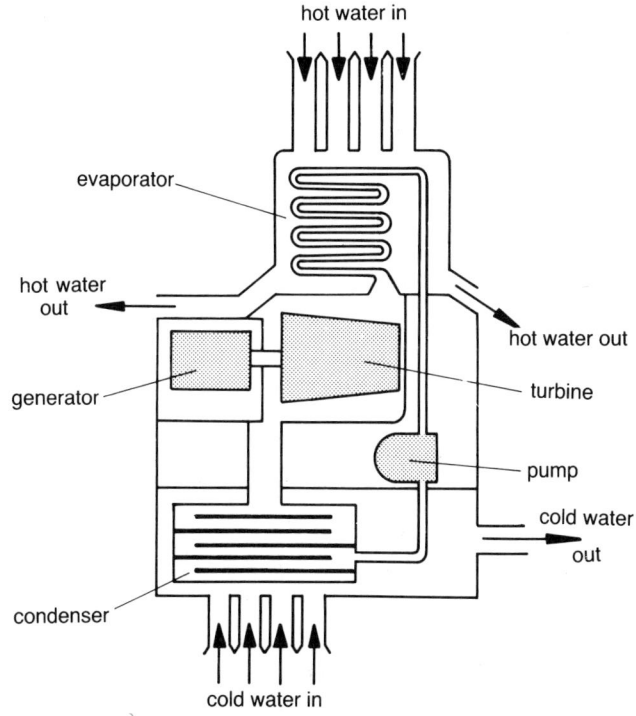

Figure 7 Thermal energy from the ocean

Thermal energy from the ocean

In the tropics, the temperature at the ocean surface is a steady 25 to 30°C, whilst at depths of 600 metres the temperature is less than 3°C. A heat engine (Fig. 7) operates in a manner similar to the way in which a heat pump operates (see page 18).

Solar energy

Solar Cells

Solar cells 'capture' energy from the sun. This energy can then be used to generate electricity or other forms of energy. One solar cell is the photo-voltaic cell. When a photo-voltaic cell is exposed to bright light, a circuit attached to the cell will produce a small current. In space, panels of solar cells are used to power satellites in orbit.

Solar energy satellite

There are several different designs under development for solar energy satellites. One design has two arms, each of which carries massive arrays of solar cells, 16 km² in area. These are surrounded by larger areas of mirrors. The mirrors are focused so as to intensify the sun's rays on the arms. The energy from the solar cells in the arms would be beamed down to earth, to be collected at receiving bowls, 7 km in diameter. One estimate of the possible power output from such bowls is 10 000 MW.

Solar panels

These are panels painted black and covered by glass. Short-wave radiation from the sun passes through the glass. This is absorbed by the black panels. Long-wave radiation (heat) from the panel cannot pass back through the glass and so it is trapped in the panel. Water passing through pipes within the panels is heated and fed into insulated storage tanks.

Energy towers

In an energy tower, the sun's rays are focused onto small mirror collectors. The sun is followed across the sky to maximise the amount of energy collected. Temperatures of several thousands of degrees can be achieved.

Generating Electricity

Hydroelectric energy

Water collected behind a dam is fed through large pipes down to a power station. Here the kinetic energy in the water is converted into rotary motion as follows. A jet of water, travelling at high speed is directed onto water turbines such as those shown in Fig. 9. Either turbine can be linked to a generator, from which electricity is fed to the national grid.

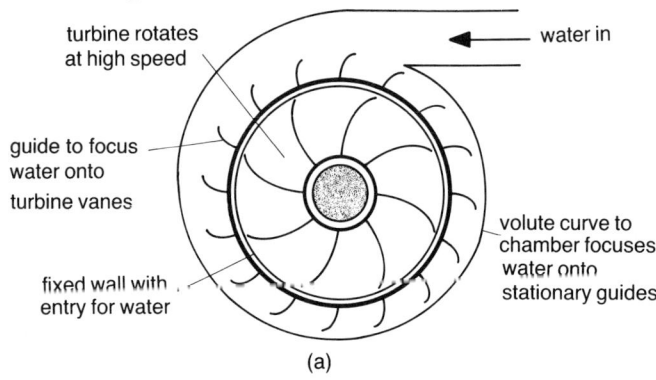

(a)

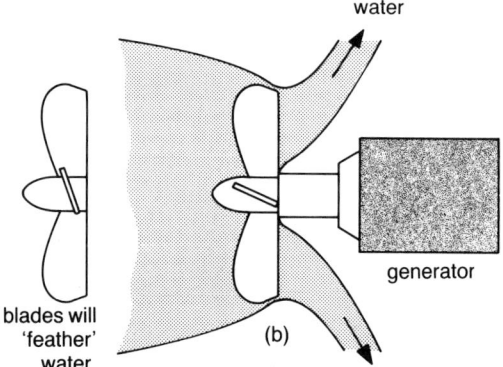

(b)

Figure 9 Two turbines

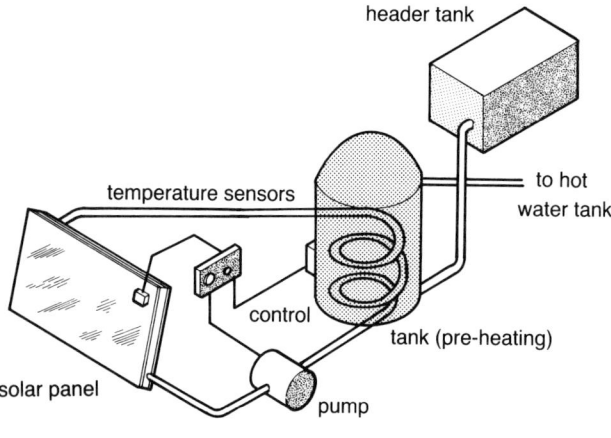

Figure 8 Water heating system by solar panel

Electric motors

An electric current can be produced by moving a wire up and down between the poles of a horseshoe magnet. Similar results are obtained by moving a magnet inside a coil of wire.

To increase the strength of the current produced:
1 the magnet or coil can be moved more quickly;
2 a stronger magnet can be used;
3 more turns can be added to the coil.

Project 1

Use a centre-zero ammeter. How does its needle move when:
1 the wire is moved up and then down – Fig. 10 (a)?
2 the magnet moves in and out of the coil – Fig. 10 (b)?

To obtain a constant supply of current, continuous movement of the wire or of the magnet is necessary.

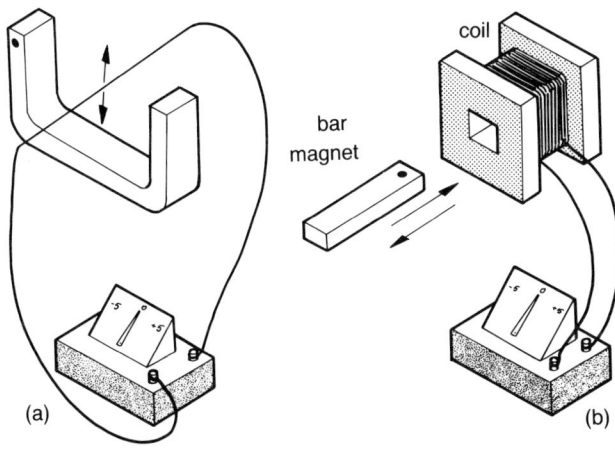

Figure 10 Project 1

Project 2

Make a turntable or use a Danum-Trent one. Place a horseshoe magnet on it. Wind a coil of wire on a pencil and transfer it to a paper tube.

Connect the two ends of the wire to an ammeter and observe the movement of the needle. What is the effect of:
1 increasing the speed of rotation of the turntable?
2 increasing the number of turns on the coil?

The movement of the needle indicates the generation of alternating current (a.c.).

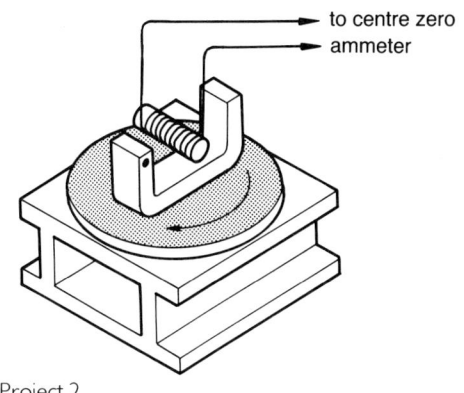

Figure 11 Project 2

☐ **Direct current (d.c.):** The electric current from a battery is direct current. Direct current may vary in strength but it always flows in one direction.

☐ **Alternating current (a.c.):** The electric current supplied to our homes (mains electricity) is a.c. This means that the current flows first in one direction and then in the other. It changes direction regularly (alternates).

A.C. generator

A simple a.c. generator is shown in Fig. 12. A rectangular coil rotates between the arms of a horseshoe magnet. **Slip rings** rotate with the coil. The carbon brushes do not move, but rub against the slip rings and produce a.c. current in the wire. This principle is used (on a bigger scale) in power stations.

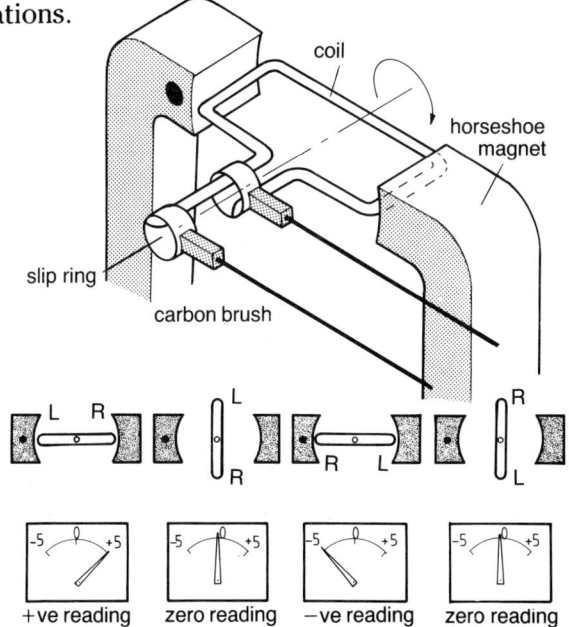

Figure 12 Generation of a.c. current

The frequency of an a.c. current depends on the number of revolutions that the coil of the generator makes per second.
1 cycle per second = 1 hertz (Hz);
Mains supply is 50 Hz.

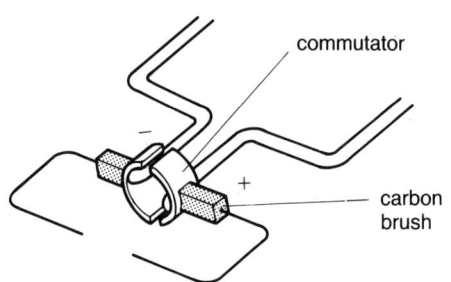

Figure 13 Generation of d.c. current

If the slip rings in Fig. 12 are replaced with a **commutator**, the arrangement shown in Fig. 13 will generate d.c.

Generators in power stations

In power station generators, the coils and their iron core remain stationary and the magnets rotate inside them. The coils and iron core are called the **stator**. The magnets, or **rotor** are electromagnets. They are supplied with electricity from a small d.c. generator (the exciter). A turbine drives the rotor at high speed inside the stator. The stator supplies a.c. current to the national grid.

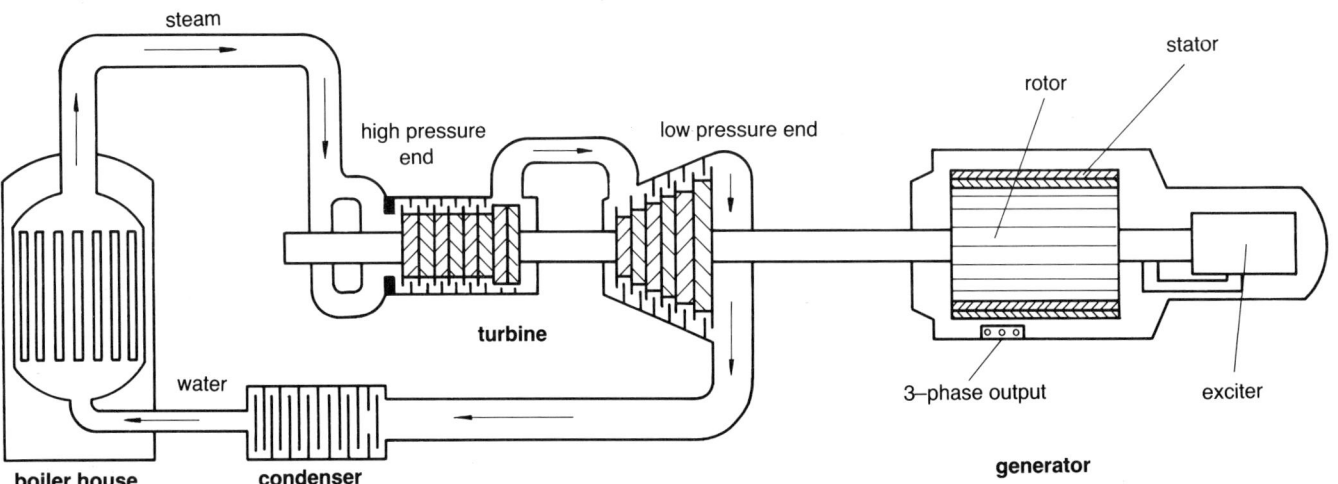

Figure 14 Conventional power station

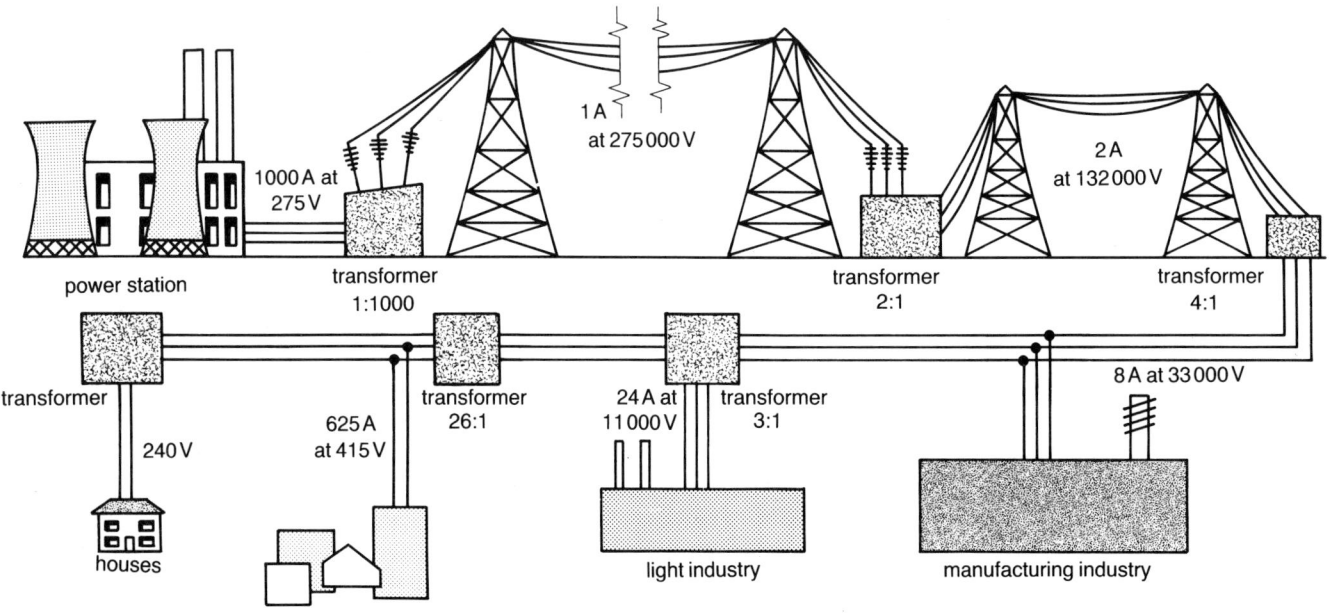

Figure 15 Transmission of electricity

The transmission of electricity

Turbo-generators convert heat energy into electricity with 90% efficiency. To maintain this high rate of efficiency, electricity is transmitted around the country at 275 000 volts. However, the voltage at which electricity is supplied to factories is usually 415 volts, and to homes 240 volts. To step the voltage up to 275 000 volts at the power station and down to 415 volts or 240 volts at the other end of the transmission line, transformers are used.

Transformers

Project 3

Place two coils end-to-end – Fig. 16 (a). Complete two circuits, with a low voltage power supply and a switch connected to one coil, and the centre-zero ammeter connected to the other.

Observe what happens when the switch:

1 is closed;
2 is opened and closed continuously.

Now place an iron core so that it passes through both coils – Fig. 16 (b). Observe the readings on the ammeter as you repeat steps 1 and 2.

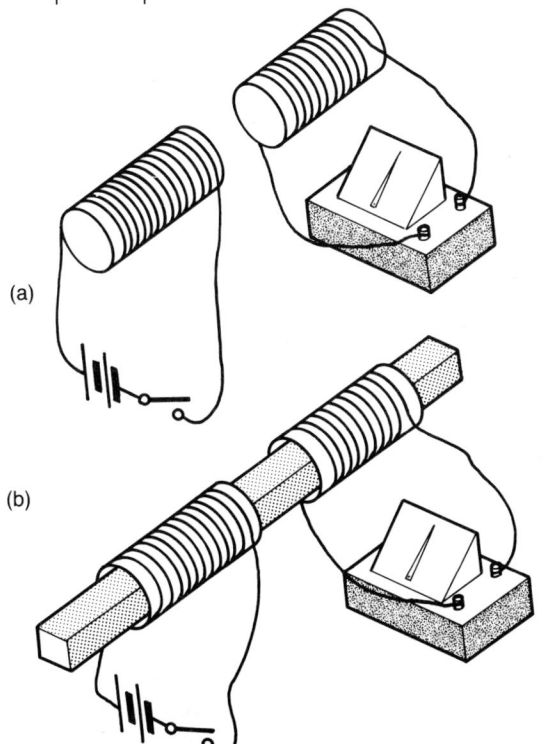

Figure 16 Project 3

The current supplied by the low-power voltage supply was d.c. If a.c. had been supplied to the first coil, what do you think would be the effect on the second coil?

What you have made is a transformer.

The iron core of a transformer consists of a number of plates joined together (laminated) with two sets of wires wound around them. At a power station, the **primary** coil of a large transformer is connected to the a.c. supply from the generator. The **secondary** coil is connected to the national grid. As the a.c. fluctuates in the primary coil, an **induced a.c.** is produced in the secondary coil.

To act as a *step-up* transformer, the secondary coil must have more turns than the primary one. This means that a higher voltage is produced in the secondary coil than is supplied to the primary coil. For *stepping down* (i.e. reducing the voltage produced in the secondary coil), the reverse would be true.

In the national grid, transformers are used for 'stepping up' the voltage of a.c. at the power station in order to reduce power losses during transmission (see above). When a.c. is to be used in homes and factories the voltage must be stepped down again.

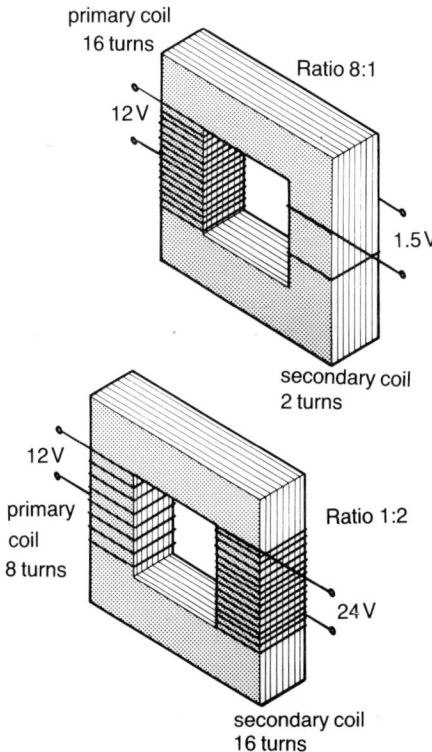

Figure 17 Step-up and step-down transformers

Geothermal energy

In places where rock formations conduct heat close to the earth's surface, water in the ground absorbs the heat. In some places this hot water gushes or bubbles to the surface (e.g. as a geyser). At other places it can be tapped by bore holes (Fig. 18(a)).

Where the hot rocks are deeper, holes can be bored down to them and cold water pumped into the rock. The heat in the rock turns the water into steam, which is brought to the surface through a second bore hole. The returned steam can be used to drive turbines for the production of electricity (Fig. 18(b)).

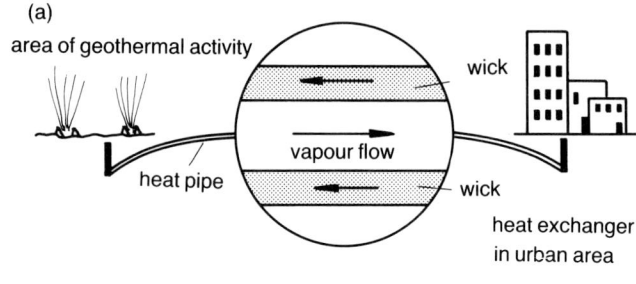

heat exchanger
in urban area

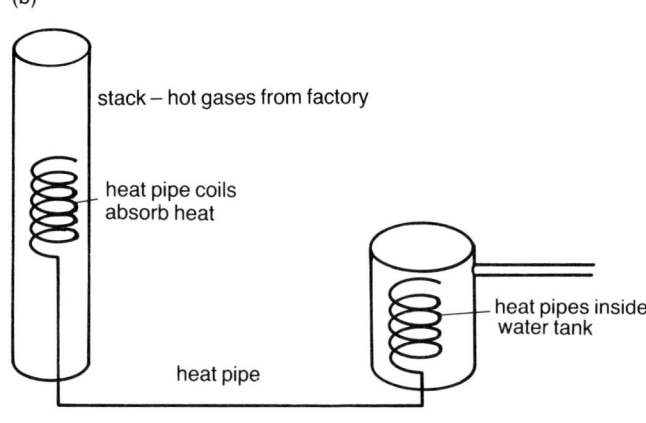

Figure 19 A heat pipe

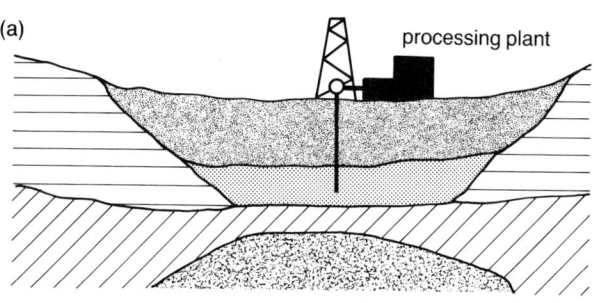

(a)

processing plant

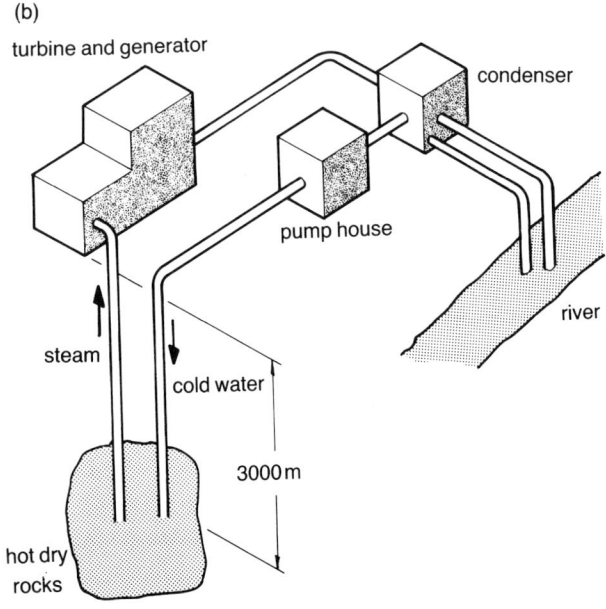

Figure 18 Geothermal energy

Heat pipes

Heat pipes can transport hot water over long distances. They work on the same principles as heat pumps (see below). Figure 19 (a) shows a section through a heat pipe. Two of the spaces in the pipe contain a liquid conductor or 'wick'. The centre space conducts vapour. One end of the pipe is placed in a source of heat, and the liquid turns into vapour. The other end is led to a heat sink. As the temperature here is low, the vapour condenses, and cools releasing its heat. The liquid so formed is transferred to the wick which transports it back to the original heat source by capillary action.

Heat pumps

Heat pumps use liquids that have the property of giving up heat easily when under pressure. These liquids can be used to carry heat from one place to another. Ammonia, ethyl chloride, and Freon are suitable liquids.

When a liquid is under high pressure it will evaporate at a much higher temperature than when it is at low pressure. This principle is used in a heat pump. Heat to raise the temperature of the liquid is obtained from low-grade heat sources such as buried pipes, rivers or geothermal sources. The liquid, under low pressure, is fed by a pipe (the evaporator) through the heat source. It immediately evaporates to form a vapour and, as it does so, it absorbs heat from its surroundings. The vapour is then put under high pressure (compressed) and passed to a condenser, where the vapour condenses (becomes a liquid again) and releases its heat. The condenser is surrounded by a water jacket, which heats up as heat passes into it from the condensing vapour. This is called heat exchange. The liquid is then passed back to the heat source.

Heat pumps can also be used to *cool* buildings – the evaporator is placed in the building and the condenser outside.

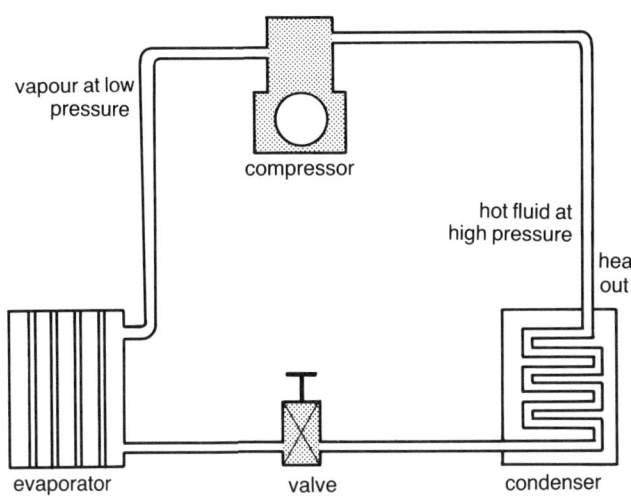

Figure 20 Heat pump

Engines

Steam engines

The Stirling boiler (Fig. 21) is a typical modern steam engine boiler. Water is pre-heated by the hot flue which goes up the chimney from the fire, and is then fed into the top header tank. Pipes carry the warmed water from the top header tank to the bottom tank, in which the water is further heated by the fire. The water in the pipes leaving the bottom tank is heated still more by the high temperature flue gases. Here the water is turned into steam. The steam produced is fed to a piston, turbine or other work unit. In a steam engine, steam from the boiler is fed into the cylinder where it expands, forcing a piston to move.

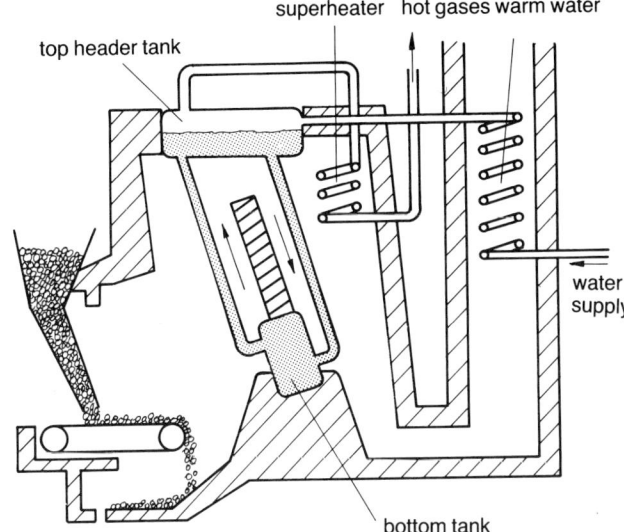

Figure 21 Stirling boiler with mechanical stoker

The internal combustion engine

The four-stroke cycle

☐ **Stroke 1. The induction stroke**: Inlet valve open; exhaust valve closed. As the piston descends petrol/air mixture is drawn into the cylinder. The inlet valve is then closed.

☐ **Stroke 2. The compression stroke**: Inlet and exhaust ports closed. The piston rises to compress the petrol/air mix. Under compression the mixture becomes hot, causing the petrol to vaporise.

☐ **Stroke 3. The power stroke**: Both valves closed. A spark from the spark plug ignites the gases causing them to expand. This expansion forces the piston down. The exhaust valve opens.

☐ **Stroke 4. The exhaust stroke**: Inlet valve closed; exhaust valve open. The piston rises, expelling burnt gases.

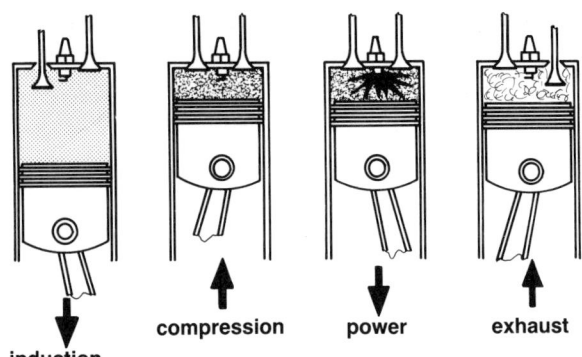

compression

power

exhaust

induction

Figure 22 The 4-stroke cycle

The two-stroke cycle

□ **Stroke 1. The power and exhaust stroke**:
The down stroke. The mixture above the piston
ignites, forcing the piston down. As the piston
descends it covers the inlet port and compresses
the air/oil/petrol mixture in the crankcase. As the
piston drops lower it uncovers the exhaust port
and the burnt gases are expelled. As the exhaust
gases leave the combustion chamber, the
compressed petrol/air mixture leaves the
crankcase and begins to enter the combustion
chamber.

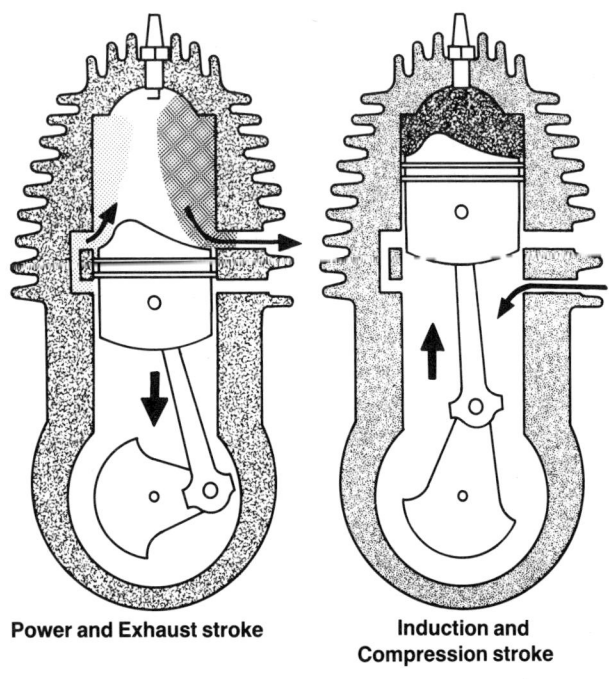

Power and Exhaust stroke

piston on down stroke

**Induction and
Compression stroke**

piston on up stroke

Figure 23 The 2-stroke cycle

□ **Stroke 2. Induction and compression
stroke**: The up stroke. The combustion chamber
is now full of petrol/oil/air mixture. The piston
rises, covering the exhaust port and compressing
the mixture. The piston rises higher, creating a
partial vacuum in the crankcase, and uncovering
the inlet port. The next charge of petrol/oil/air
mixture is drawn into the crankcase. The mixture
in the combustion chamber is now fully
compressed and the spark is about to ignite it,
beginning another cycle.

The rotary or Wankel engine

The rotary engine has fewer moving parts than a
piston engine. Instead of a piston the engine has a
triangular-shaped rotor running inside a housing.
The housing is very slightly constricted in the
middle. As the rotor turns, the seals at each corner
remain in contact with the walls of the housing.
The three spaces formed between the rotor, and
the housing increase and decrease in size as the
rotor turns. These changes in volume are used for
induction, compression, power and exhaust.
There are three ignitions for each revolution of the
rotor.

   The only two moving parts are the rotor and the
drive shaft. The gear teeth on the inside of the
rotor engage with gear teeth on the outside of the
drive shaft. The drive shaft is part of the oval

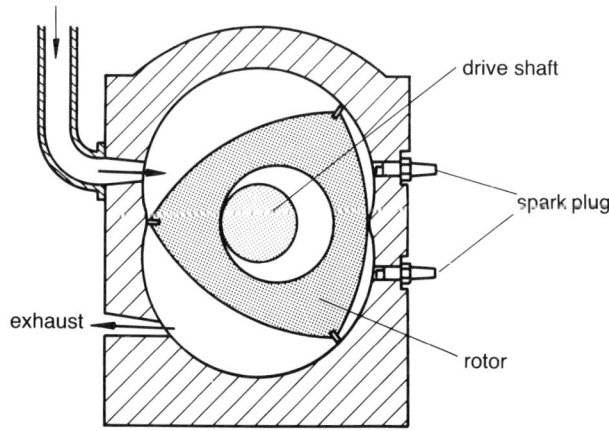

fuel/air intake

drive shaft

spark plug

exhaust

rotor

Figure 24 Diagrammatic section through a Wankel engine

housing and does *not* rotate. It is the rotor which
runs round the shaft. A system of gears transmits
the motion of a rotor to an output shaft.

□ **1. Induction:** The rotor uncovers the inlet
port and the petrol/air mixture is drawn in. The
inlet port is covered by the rotor.

☐ **2. Compression:** The space between rotor and wall becomes smaller and the mixture is compressed. The heat generated vaporises the petrol.

☐ **3. Power:** The spark plug ignites the gases and turns the rotor.

☐ **4. Exhaust:** As the rotor turns it uncovers the exhaust port and the burnt gases are expelled.
  The other two sides of the rotor follow the same sequence, but 120° behind each other.

Diesel engine

Diesel engines do not need spark plugs. They run on diesel oil. When diesel oil is compressed to a ratio of about 22:1 it will ignite spontaneously.

The diesel 4-stroke cycle

☐ **Stroke 1. Induction stroke:** Inlet valve open; exhaust valve closed. Air enters the cylinder. Inlet valve then closes.

☐ **Stroke 2. Compression stroke:** With inlet and exhaust ports closed, the piston rises and compresses the air. Diesel oil under pressure is injected into the cylinder where it vaporises.

☐ **Stroke 3. The power stroke:** Both valves remain closed. Maximum compression is reached and the vapour ignites, expands and the piston is forced down.

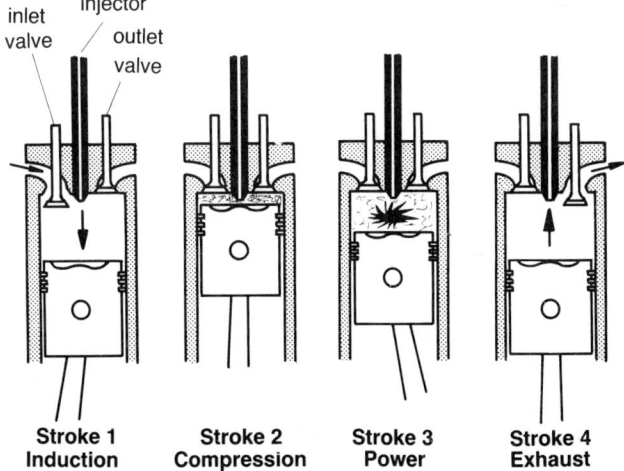

Stroke 1 **Stroke 2** **Stroke 3** **Stroke 4**
Induction **Compression** **Power** **Exhaust**

Figure 25 The diesel 4-stroke cycle

☐ **Stroke 4. The exhaust stroke:** Inlet valve remains closed; exhaust valve opens as the piston rises to expel burnt gases. The inlet valve then opens and the exhaust valve closes ready for the next cycle to begin.

Exercises

1 ■ State two examples of fossil fuels.
■ What is meant by the phrase 'renewable sources of energy'?
2 ■ State two methods for reducing energy consumption for each of the following examples:
a heating water for a central heating and hot water system;
b propelling a motor vehicle.
■ Name *three* alternative energy sources and state *two* advantages and *two* disadvantages of each.
3 Why are bottle banks popular with people who are concerned about the environment?
4 Complete the following sentences:
■ A solar cell converts . . . . energy to . . . . energy.
■ A torch converts . . . . energy to . . . . energy.
5 List all the changes in energy states that must occur before energy from ocean waves can warm a snack in a microwave oven.
6 For each of the following examples, name a machine which changes:
■ chemical energy to mechanical energy;
■ sound energy to electrical energy;
■ electrical energy to heat energy;
■ potential energy to mechanical energy.
7 Briefly explain the differences between potential energy and kinetic energy.
8 A baby's bottle warmer is rated at 240 volts, 360 watts.
■ Calculate the power required in kilowatts.
■ How much current is taken?
■ The bottle warmer is left on for eight hours. How many joules of energy are consumed?
9 Let 1 kg = 10 newtons.
■ A mother lifts a child weighing 15 kg from the floor to a high chair 1.5 m higher. Calculate the work done in joules.
10 A girl climbs a tower 30 m high in 45 seconds. If she weighs 48 kg:
■ how much work has she done?
■ at what rate has she done this work (i.e. average power)?

3 Electrics

Introduction

Electricity is the flow of electric current through a conductor around a circuit. Electric current will flow between two points only if there is a **potential difference** (p.d.) between them (just as water will flow between two places if there is a difference in height between them). The potential difference in a circuit can be provided by an **electromotive force** (e.m.f.) e.g. from a battery, dynamo, solar cell or the mains. Potential difference is measured in **volts** (V).

By convention we say that electric current flows from a positive point to a negative point (e.g. from the positive side of a battery to the negative side) – although in fact, the electrons which carry the charge in a current move from negative to positive. An electric current will flow only when a circuit is complete. Electric current is measured in **amps** (A).

When a current flows through a conductor (such as a wire), the amount of current flowing depends on the **resistance** of the conductor. The resistance of a conductor is therefore described as a measure of how difficult it is for an electric current to flow through the conductor. Resistance is measured in ohms (Ω). The resistance of a conductor can depend on:
- the type of material that the conductor is made of e.g. aluminium and copper are good conductors – they have low resistances;
- the length of the conductor, i.e. long wires have higher resistances than short wires;
- the diameter of the conductor, i.e. wires with small diameters have higher resistances than wires with larger diameters.

The total current flowing around a circuit also depends on the total resistance of that circuit.

Batteries

An electric **cell** is a *store* in which chemical energy can be converted to electrical energy as and when required. A **battery** is a number of cells linked together. There are two forms of cells – primary and secondary.

Primary cells

Primary cells cannot be re-charged. They may be wet or dry.

☐ **Leclanché (wet)** Inconvenient to use because the cell must be kept upright in order to prevent leakage.

☐ **Leclanché (dry)** (Fig. 1) Chemical contents are dry and the cell can therefore be used in any position. The e.m.f. (electromotive force) is 1.5 V.

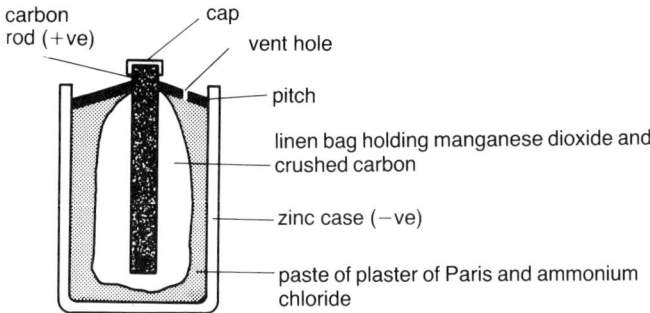

carbon rod (+ve)
cap
vent hole
pitch
linen bag holding manganese dioxide and crushed carbon
zinc case (−ve)
paste of plaster of Paris and ammonium chloride

Figure 1 A dry Leclanché cell

☐ **Leakproof cells** These are enclosed in a second, outer case made of steel.

☐ **Mercury cells** Although these are more expensive and have an e.m.f. slightly lower than

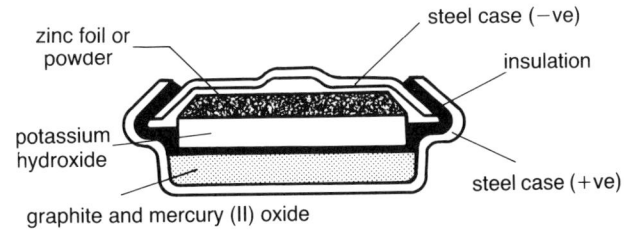

zinc foil or powder
steel case (−ve)
insulation
potassium hydroxide
graphite and mercury (II) oxide
steel case (+ve)

Figure 2 A mercury cell

the above cells (1.3V), they have the following advantages:

1 long life;
2 shock-resistant and leakproof;
3 produced in very small sizes.

Mercury cells are used in watches, calculators and cameras.

☐ **Silver oxide cells** Have similar advantages to mercury cells.

Secondary cells

Secondary cells can be re-charged, when they are discharged (or 'flat') with the aid of a battery charger.

The capacity of a secondary cell is calculated in ampere hours = current × time.

Some common types of secondary cells are:

☐ **Lead–acid cells or accumulators** These are widely used, most commonly in motor vehicles. In a 12 volt car battery six cells are joined together in series.

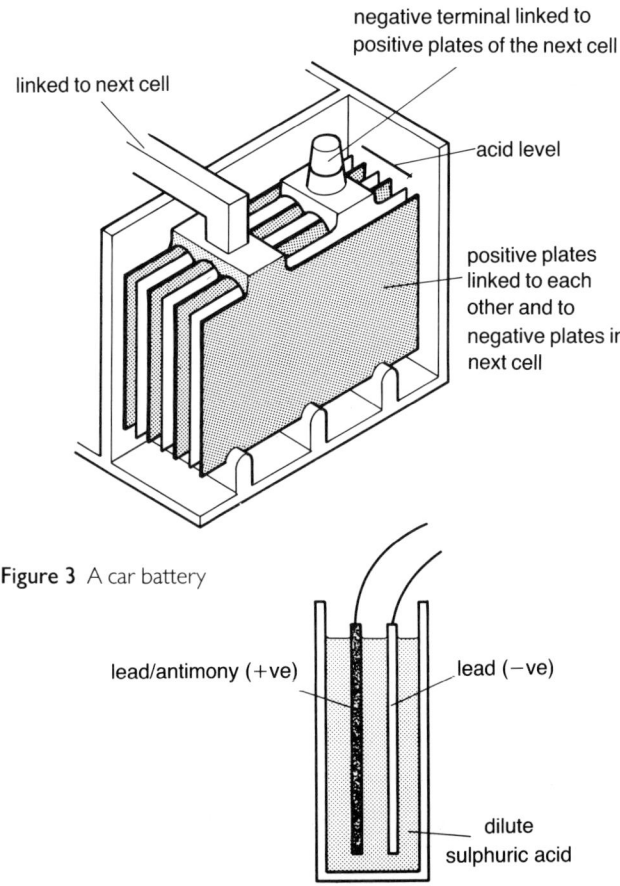

negative terminal linked to positive plates of the next cell

linked to next cell

acid level

positive plates linked to each other and to negative plates in next cell

Figure 3 A car battery

lead/antimony (+ve)

lead (−ve)

dilute sulphuric acid

Figure 4 A single lead-acid cell

☐ **Ni–Cad cells** Nickel–cadmium cells have very low internal resistance. Their excellent resistance to both electrical and mechanical abuse and their high current supply makes them suitable for radio-controlled models.

Projects

1 Use a voltmeter to check the e.m.f. of a single cell battery.
2 Connect a second battery in parallel with the first, and measure the e.m.f. How does it compare with that of the single cell? Will the useful life of these cells in parallel be the same, shorter, or longer than that of a single cell?
3 Connect the two batteries in series and measure the e.m.f. How does it compare with the single cell? What can you say about the life span of cells in series?

Switches

A switch is a way of controlling the electricity supply in a circuit, by breaking or making electrical contact with the rest of the circuit. In a mechanical switch a force *makes* (closes) or *breaks* (opens) metal contacts.

Many switches are marked with two sets of figures, e.g. 250V AC 3A; 30V DC 5A. These are the maximum permissible voltages and currents that should be used. If you exceed them then the working life of the switch will be reduced.

☐ **Push-button switches** May be:
1 push-to-make;
2 push-to-break;
3 push-push, i.e. push for on and push again for off.

☐ **Slide switch** May be of the on/off variety, or it may be off in the centre position only.

☐ **Toggle switch** Worked by a 'toggle' lever operation.

☐ **Rocker switch** A flush fitting switch.

☐ **Key-operated switch** The switch can be locked.

☐ **Microswitch** This can be operated by a small force.

☐ **Rotary wafer switch** The switch is rotated to cause switching.

☐ **Reed switch** Reed switches consist of two strips of easily magnetised metal held inside a glass tube.

Figure 5 Some electrical circuit switches

Fuses

The electricity supply entering a building is connected to a fuse box. A fuse acts as a safety device which will 'blow' in the event of an overload. A fuse is usually a thin wire that connects the electrical supply entering the house with circuits in the building. The thin wire will break if too much current passes through it, so the circuit is broken and the current cut off.

Circuit breakers can be used instead of fuse wire. These work in one of two ways:

1 Thermal tripping: When too large a current passes through the circuit breaker, a bi-metal strip is heated by the current. The strip rapidly expands, causing the switch in the circuit breaker to trip and cut off the current.

2 Magnetic tripping: When too much current passes through the circuit breaker, part of the device is magnetised tripping the circuit breaker contacts.

Should a fuse blow or a switch trip, investigate the cause and remedy any fault before repairing the fuse and switching on again.

To connect household electrical items to a power supply a three-pin plug is often used. The plug must contain a fuse of a rating suitable for the equipment being used.

When connecting a 13A plug to an appliance make sure that:

1 the cable is held firmly in the plug casing;

2 the colours of the wires to the live, neutral and earth terminals are correct and the wires are fixed firmly into the connectors.

Cables

Electrical cables are made of materials that conduct electricity – often copper, which is a good conductor of electricity. Cables are usually coated or covered with an insulating material, such as glass fibre, paper, neoprene, rubber or PVC.

British Standards symbols

The British Standard BS 3939 *Graphical symbols for electrical power, telecommunications and electronics diagrams* shows all the symbols for components which can be used in electrical and electronics circuits. Some symbols from BS 3939 are given in Fig. 6.

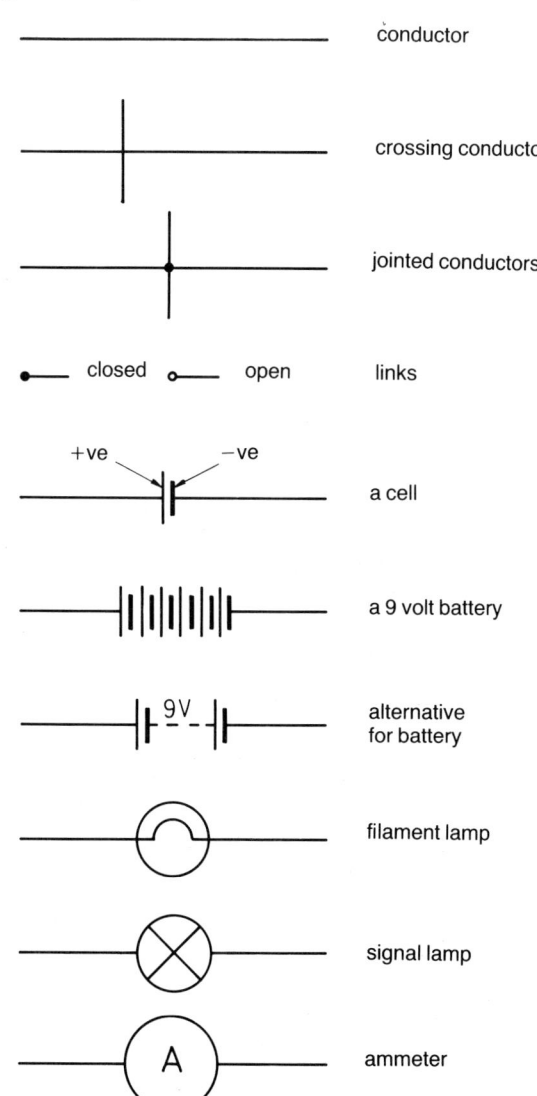

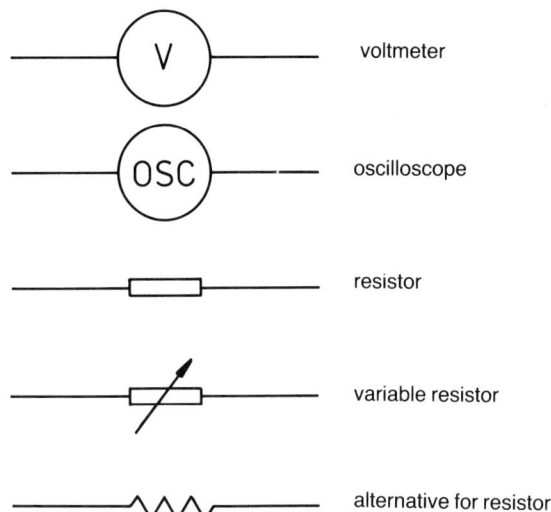

voltmeter

oscilloscope

resistor

variable resistor

alternative for resistor

Figure 6 Some BS 3939 symbols for electrical components

Electrical circuits

Note: The circuits shown in this chapter can be constructed from circuit units such as those in the Danum-Trent system (other systems are also available).

BS symbols for a cell and for batteries are shown in Fig. 6. A cell has two terminals: positive +ve (+) and negative −ve (−).

A power source other than a battery can be used such as a power pack.

By convention, current flows in a circuit from positive to negative.

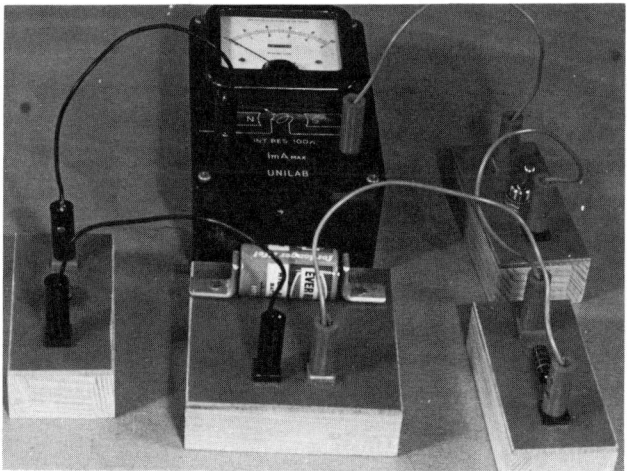

Figure 7 A motor switching circuit made up from Alpha system modules

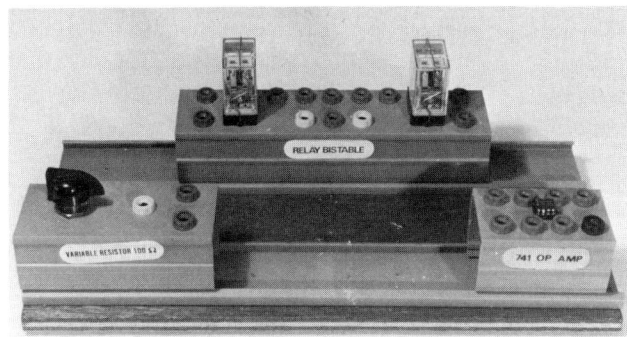

Figure 8 A sample of Danum-Trent circuit modules

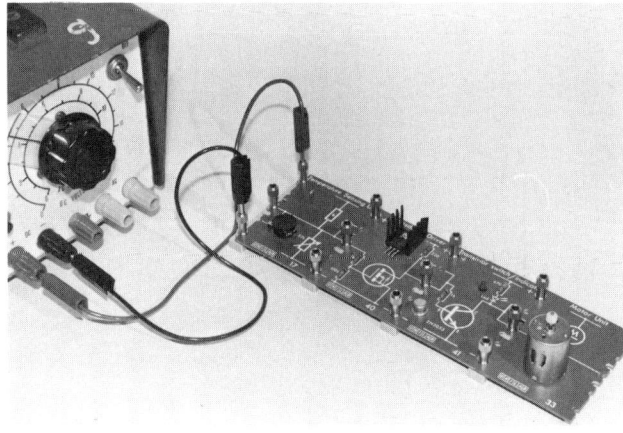

Figure 9 An electrical circuit made up from school-made modules

Circuit 1 – Power source and lamp circuit

Construct Circuit 1 (Fig. 10). If an electrical circuit is to work, it must be connected to both positive and negative sources of the power supply.

The filament of the bulb is made from tungsten. The filament is 'excited' by the current flowing through it so it glows and gives off heat.

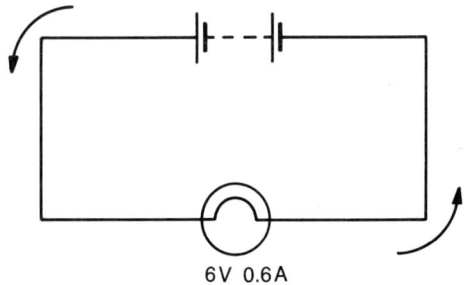

6V 0.6A

Figure 10 Circuit 1 – Power source and lamp circuit

Measuring instruments

Two types of measuring instruments are used to measure current flow and voltage in an electric circuit.

Circuit 2 – Ammeter circuit

An ammeter measures the current (I) in a circuit in amperes (A or amps).

An ammeter must be connected within a circuit so that current passes through it (i.e. in *series* with the circuit). The positive side of an ammeter *must* be connected to the positive side of the power source (Fig. 11).

Construct Circuit 2 (Fig. 11).

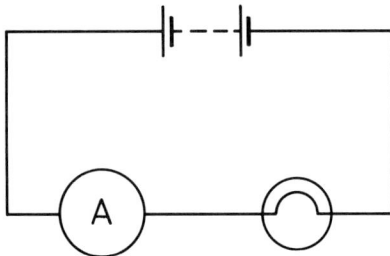

Figure 11 Circuit 2 – Ammeter circuit

Circuit 3 – Voltmeter circuit

Electric current can only flow between two points of a circuit if there is a potential difference (p.d.) between them. A voltmeter measures the potential difference in volts, V.

The higher the p.d. or voltage, the greater the e.m.f. driving current round the circuit.

A voltmeter must be connected across the part of the circuit being measured (i.e. in *parallel* with it), so that

it measures the potential difference between two points in the circuit. The positive side of the meter *must* be connected to the positive side of the power source (Fig. 12).

Construct Circuit 3 (Fig. 12).

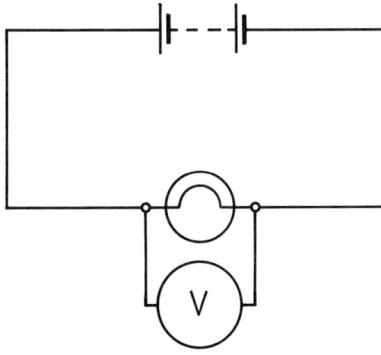

Figure 12 Circuit 3 – Voltmeter circuit

Circuit 4 – Ammeter and voltmeter circuit

Construct Circuit 4 (Fig. 13). The voltmeter is in parallel with the power source. The ammeter is in series with the lamp.

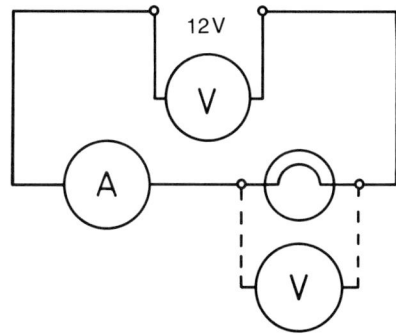

Figure 13 Circuit 4 – Ammeter and voltmeter circuit

1 What is the reading on the voltmeter?
2 What is the reading on the ammeter?
Now place the voltmeter in parallel with the lamp (as shown by dotted lines in Fig. 13).
3 What is the reading on the voltmeter now?

The reading on both meters should be unchanged.

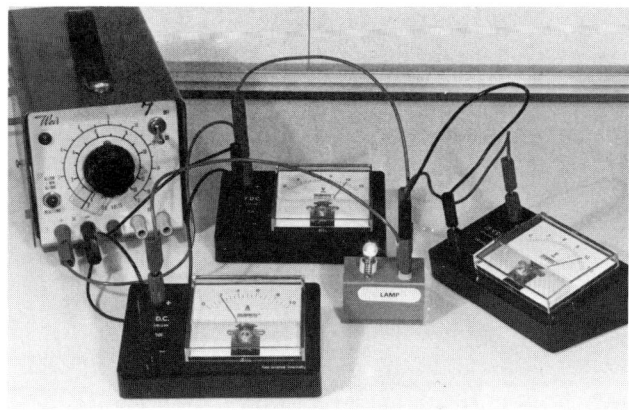

Figure 14 Circuit 4

Circuit 5 – Lamps in parallel

Construct Circuit 5 (Fig. 15) by placing a second lamp in parallel with the first in Circuit 4.

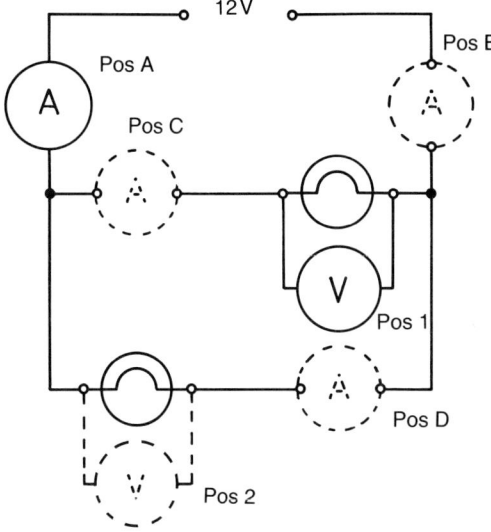

Figure 15 Circuit 5 – Lamps in parallel

1 What is the reading on the voltmeter in position 1?
2 What is the reading on the voltmeter in position 2?
3 What conclusions can you draw from this?
4 What is the reading on the ammeter in positions A and B?
5 What is the reading on the ammeter in positions C and D?
6 What conclusions can you draw from this?

Carefully unscrew one lamp from its holder.

7 Does the second lamp remain lit?
8 What is likely to be the reading on the ammeter if:
■ one extra lamp is added in parallel to the circuit?
■ two extra lamps are added in parallel to the circuit?
Will the voltmeter readings alter?
If you are uncertain, then make and test the circuit. What are your conclusions?

Circuit 6 – Lamps in series

Construct Circuit 6 (Fig. 16).

1 Are the two lamps in series brighter or dimmer than when they were in parallel?

Look at the circuit in detail.

2 What is the reading on the voltmeter in position 1?

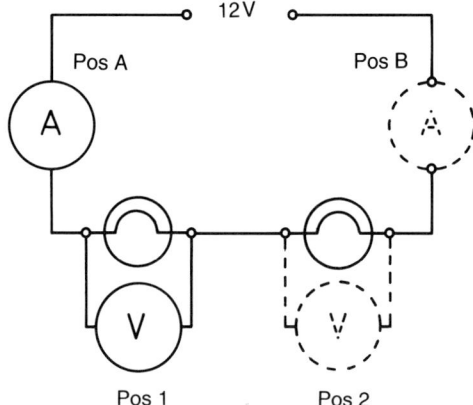

Figure 16 Circuit 6 – Lamps in series

3 What is the reading on the voltmeter in position 2?
4 What conclusions can you draw?
5 What is the reading on the ammeter in positions A and B?
6 What conclusions can you draw?

Carefully unscrew one lamp from its holder.

7 Does the second lamp remain lit?
8 What is likely to be the reading on the ammeter if:
■ one extra lamp is added in series with the circuit?
■ two extra lamps are added in series with the circuit?
Will the voltmeter readings go *up* or *down*?
If you are uncertain then make and test the circuit. What are your conclusions?

A **resistor** is a component which is specially designed to have a certain resistance to current flowing through it. When a current flows through a resistor, a potential difference exists across it.

Circuit 7 – Speed control of an electric motor (1)

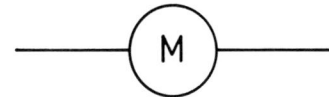

Figure 17 BS 3939 symbol for an electric motor

An electric motor converts electrical energy into mechanical energy. Fix a large pulley wheel to the shaft of an electric motor, and then construct Circuit 7 as shown in Fig. 18, with the resistor having a value of 25 Ω. Switch on and count the number of rotations the pulley

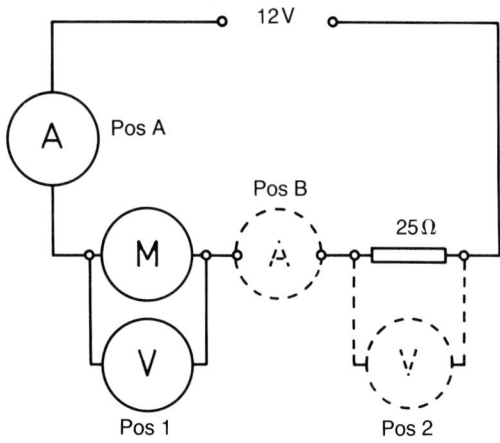

Figure 18 Speed control of an electric motor (1)

makes in 15 seconds. Then calculate the rpm (revolutions per minute). Record the voltage and current in the circuit by taking readings at the places indicated in Fig. 18.

Place other values of resistor in the circuit and then copy and complete the table below:

Resistor value (Ω)	rpm	Current at A (A)	Current at B (A)	Voltage at 1 (V)	Voltage at 2 (V)
25					
50					
75					
100					
125					

Conclusions

The resistance of the motor does not change. The value of resistor placed in series with the motor does change. As the resistance added to the circuit gets higher, how do the readings at positions A and B and 1 and 2 change?

Resistors are components which are specially designed to have a certain value. They allow you to control the current in a circuit. Therefore in this circuit, by using resistors of different values, it is possible to control the speed of the motor. Different speeds are obtained only by changing resistors. However, there are resistors which can have a range of different resistances. These are called *variable resistors*.

Circuit 8 – Speed control of an electric motor (2)

Construct Circuit 8 (Fig. 19). This involves a variable resistor connected in parallel with the motor. Test the circuit.

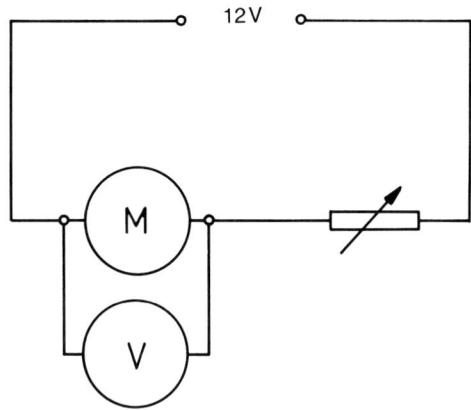

Figure 19 Speed control of an electric motor (2)

Ohm's law

Ohm's law says that at constant temperature, the current flowing in a conductor is proportional to the potential difference between its ends.

$$\text{p.d.} (V) = \text{current} (I) \times \text{constant}$$

The 'constant' is the resistance, R, of the conductor. So $V = IR$. If a circuit is connected to a 1 volt source of supply and 1 ampere of current is flowing through the circuit, then the resistance in the circuit is 1 ohm. Ohm's law shows that resistance, voltage and current are closely linked to each other. Formulae for Ohm's law are shown in Fig. 20. By placing a finger over the unknown quantity in Fig. 20, the formula for achieving the answer is given.

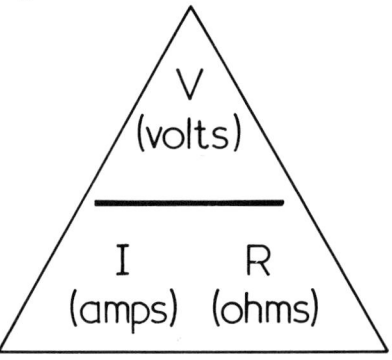

Figure 20 Formulae for Ohm's law – place a finger tip over the value to be found and then read off the formula

Example – a 2 kW electric fire runs from 240 volts.

- How much current (I) flows in the circuit?
- What is the resistance (R)?

power (watts) = p.d. (volts) × current (amps)

Therefore $I = \dfrac{P}{V} = \dfrac{2000}{240} = 8.3\,A.$

Cover R in the triangle (Fig. 20);
the formula reads

$$\dfrac{V}{I}$$

The three formulae for Ohm's Law are:

$$I = \dfrac{V}{R} \qquad V = I \times R \qquad R = \dfrac{V}{I}$$

Formula for resistors in series

In a resistors-in-series circuit it is easy to work out the overall resistance. Just add all the resistances together. Total resistance in the circuit is $R = R_1 + R_2 + R_3 + \ldots$ etc.

Circuit 9 – Resistors in Parallel

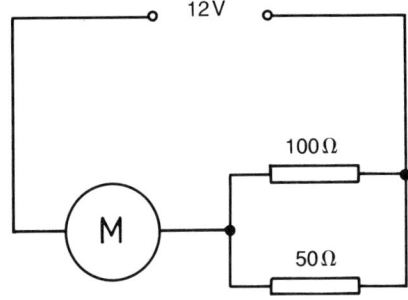

Figure 21 Resistors in parallel

Construct Circuit 9 (Fig. 21). Study the circuit. What do you think the total resistance in the circuit will be?

Measure the total resistance with an ohmmeter. Check your answer using the formula for resistors in parallel as given below:

$$\dfrac{1}{R} = \dfrac{1}{R_1} + \dfrac{1}{R_2}$$

If $R_1 = 100$ and $R_2 = 50$;

Then $\qquad \dfrac{1}{R} = \dfrac{1}{100} + \dfrac{1}{50} = \dfrac{1 + 2}{100} = \dfrac{3}{100}$

$$R = \dfrac{100}{3} = 33.33$$

Now check this against your measured answer.

Choosing resistors

Certain values of resistor (in ohms) are available but their power ratings vary. If a resistor is exposed to more power than it can handle then it will overheat and break. To choose the correct 270 ohm resistor if a p.d. of 12 volts is to be applied to it, use the power formula given above: $P = V \times I$. By using two aspects of Ohm's law: $V = I \times R$ or $I = \dfrac{V}{R}$ to modify the power formula, we get:

$$P = I \times R \times I$$

therefore $P = I^2 R$
and:

$$P = \dfrac{V \times V}{R}$$

therefore $P = \dfrac{V^2}{R}$

To find the power passing through the above resistor, use:

$$P = \dfrac{V^2}{R}$$

$$= \dfrac{12 \times 12}{270} = 0.53 \text{ watt}$$

A 0.5 watt resistor would burn out, so the next highest value must be used.

Exercises

1 What power will a resistor consume if at 6 volts it carries 600 mA?
2 What is the power in a 570 Ω resistor passing a current of 250 mA?

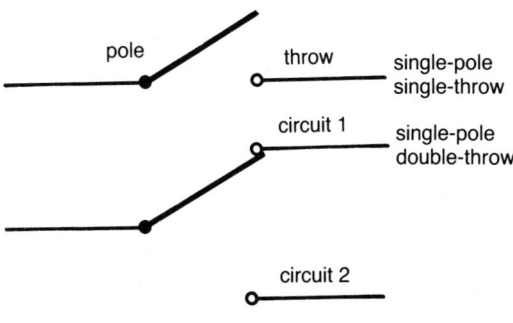

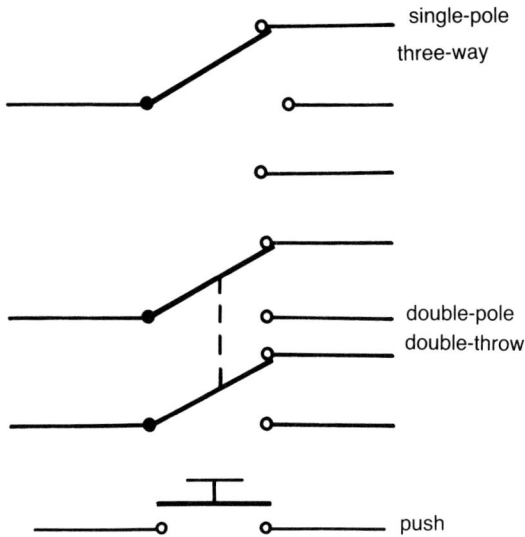

Figure 22 BS 3939 symbols for switches

Circuit 10 and 11 – Switch circuits

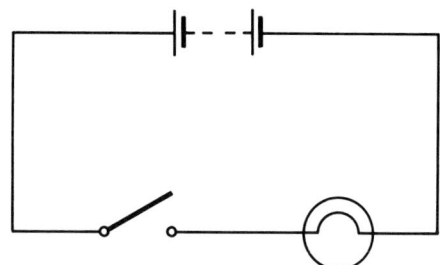

Figure 23 Circuit 10 – Switch circuit

Construct Circuit 10 (Fig. 23).

Note the position of the switch when the lamp is on. Note that the safest method of isolating a power supply is to use a double-pole double-throw switch as shown in Fig. 24.

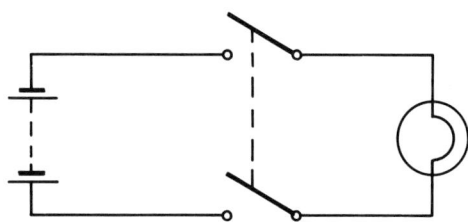

Figure 24 Circuit 11 – Switch circuit

Circuits 12 and 13 – Directional control of an electric motor

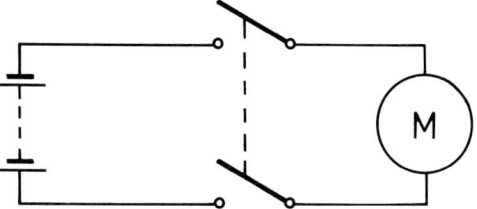

Figure 25 Circuit 12 – Switch circuit for electric motor

Construct Circuit 12 (Fig. 25), using a double-pole double-throw switch.

Circuit 12, Fig. 25, would allow a motorised buggy to stop and start, but not to change direction. Circuit 13 (Fig. 26) allows the buggy to go forwards and backwards as the switch is changed over.

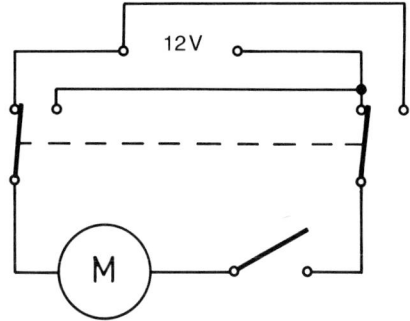

Figure 26 Circuit 13 – Directional control of an electric motor

Exercises

1 Many houses have cupboards under the stairs. These cupboards often lack light. Investigate two types of switch that would automatically switch on a light as the cupboard door is opened. Design and construct a circuit to demonstrate your solution.

2 In a house it is often possible to switch the same light bulb on and off from either of two switches placed at the bottom and at the top of the stairs. Design and construct a circuit to demonstrate this.

Capacitors

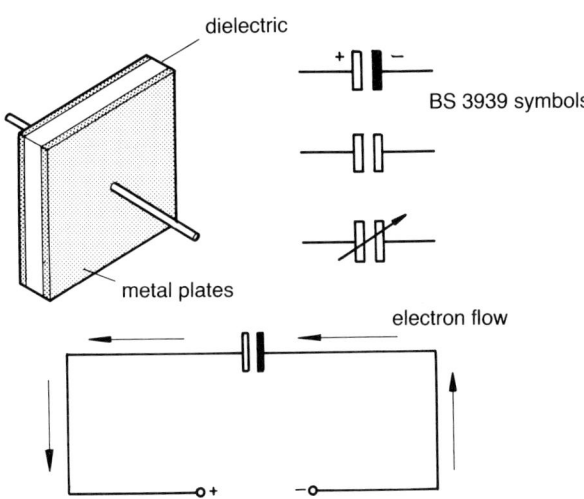

Figure 27 Capacitors

A capacitor can store an electrical charge. In its simplest form, it consists of two flat, parallel metal plates separated by an insulating material. When connected to a power source, one of the plates attracts electrons (negative charge), and the other loses electrons (positive charge). This means that the capacitor is charged. A capacitor stops charging once the p.d. across the plates is equal to that across the power source. When a capacitor is disconnected from a power supply, it does not hold its charge indefinitely, because there is a small leakage of current from the capacitor. If a charged capacitor is connected in a circuit, electric charge will flow from it until the p.d. across its plates is zero.
Note: Electrons cannot flow through the insulating material, therefore current cannot flow through a capacitor. The maximum amount of charge stored by a capacitor is called the **capacitance** of the capacitor. So long as the capacitance is not exceeded, doubling the p.d. across a capacitor will double the charge that it carries.

The capacitance of a capacitor (C) is measured in farads (F).

$$1 \text{ microfarad } (1\,\mu F) = \frac{1}{1\,000\,000} \text{ th of a farad;}$$

$$1 \text{ nanofarad } (1\,nF) = \frac{1}{1000} \text{ th of a microfarad OR}$$

1 thousand millionth of a farad;

$$1 \text{ picofarad } (1\,pF) = \frac{1}{1000} \text{ th of a nanofarad OR}$$

1 million millionth of a farad.

Circuit 14 –Capacitors in series

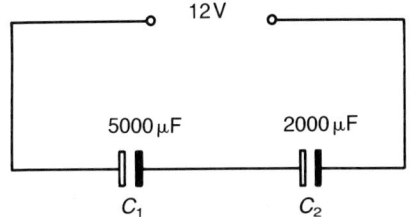

Figure 28 Circuit 14 – Capacitors in series

Circuit 14 (Fig. 28) is a circuit with capacitors in series. To calculate the total capacitance of this circuit:

$$\frac{1}{C} = \frac{1}{C_1} + \frac{1}{C_2} = \frac{1}{5000} + \frac{1}{2000} = \frac{7}{10\,000}$$

Therefore total $C = \dfrac{10\,000}{7} = 1429\,\mu F.$

Circuit 15 – Capacitors in parallel

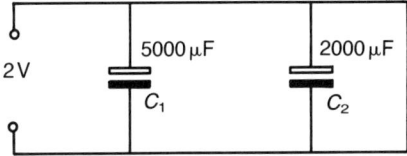

Figure 29 Circuit 15 – Capacitors in parallel

Circuit 15 (Fig. 29) is a circuit with capacitors in parallel. To calculate the total capacitance of this circuit:

$$C = C_1 + C_2 = 5000 + 2000 = 7000\,\mu F.$$

Types of capacitor

☐ **Electrolytic Capacitor** Usually tubular. They are used in low frequency circuits (up to

10 kHz). Look for + on the capacitor. This must be connected to the positive terminal of the power source.

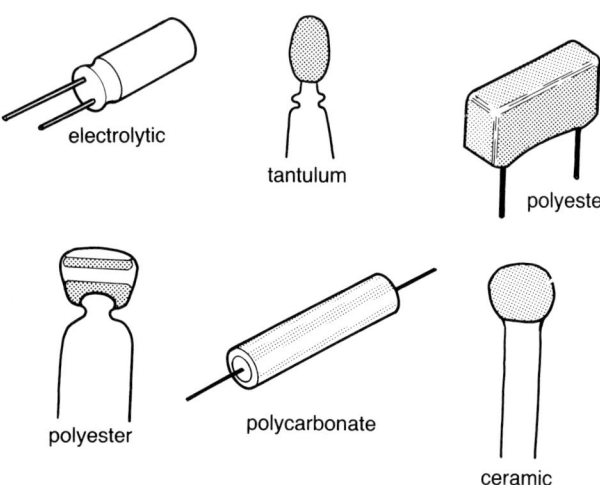

electrolytic

tantulum

polyester

polyester

polycarbonate

ceramic

Figure 30 Some capacitors

☐ **Tantulum capacitor** Usually a squashed spherical shape. Small in size and with capacitance of under $100\,\mu$F. They are used for low voltage circuits, where leakage current must be kept to a minimum. Viewing the capacitor case so

that the dot is visible centrally, the right-hand lead is the +ve connector.

☐ **Polyester capacitor** Tubular plastic or rounded-off square in shape. Used in general purpose circuits, with values up to $0.01\,\mu$F. The case is often colour-coded, using the same code as for resistors but with values in picofarads:

Last colour = *working voltage* e.g. red 240 V; yellow 400 V

Next to last colour = *tolerance* e.g. green 5%; white 10%

Polyester capacitors are not polarised, so they can be connected in a circuit either way round.

☐ **Polycarbonate capacitor** As polyester types, but with values up to $10\,\mu$F. More stable, with smaller leakage currents.

Other types include mica, ceramic and variable capacitors. Variable capacitors are used for tuning radios.

Note: Exercises for electrics are included with those for electronics at the end of the next chapter.

4 Electronics

Introduction

Electricity can flow easily through conductors, but not through insulators. Semi-conductors are materials that behave like both conductors and insulators depending on how they are used. Silicon and germanium are semi-conductors. There are others. Semi-conductor devices are widely used in electronics. Semi-conductors do not obey Ohm's law because the resistance of a semi-conductor varies with the p.d. or voltage across it.

Exercise:

Make a list of other semi-conductors.

Constructing Electronic Circuits

Components

The British standard symbols for some common electronics components are shown in Fig. 1.

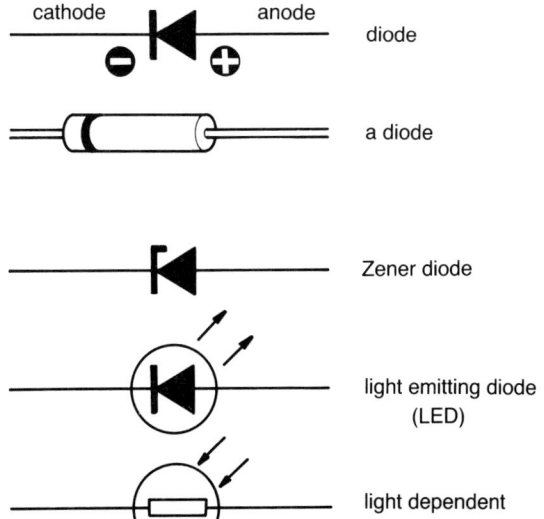

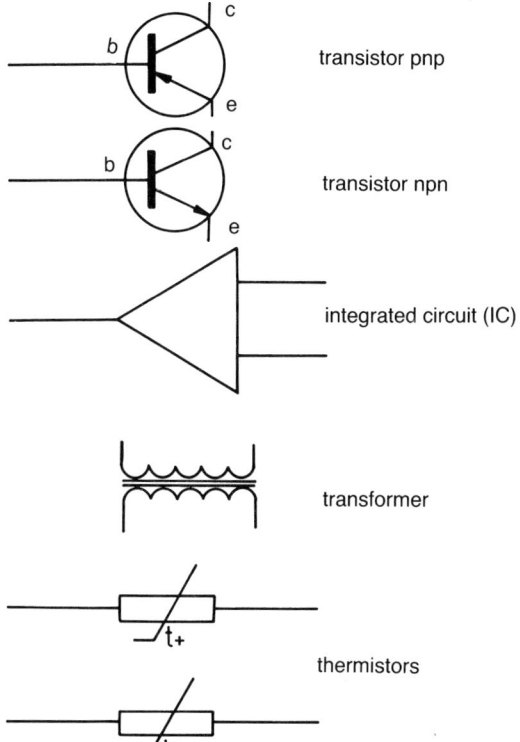

Figure 1   Some BS 3939 symbols for electronic components

Never expect any electronic circuit which you have designed and built to work satisfactorily when it is first tested.

Do not solder up a circuit before it has been tested and found to be working satisfactorily *in the dry*.

The photographs, Figs. 2 to 6, show units suitable for designing, constructing and testing electronic circuits.

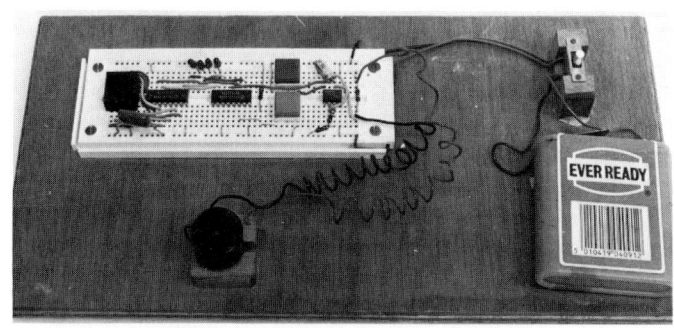

Figure 2 An electronics circuit mounted on 'breadboard'

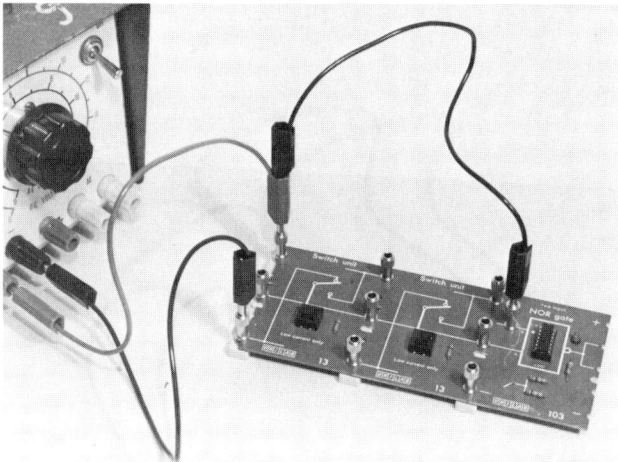

Figure 3 Units from the Alpha units system

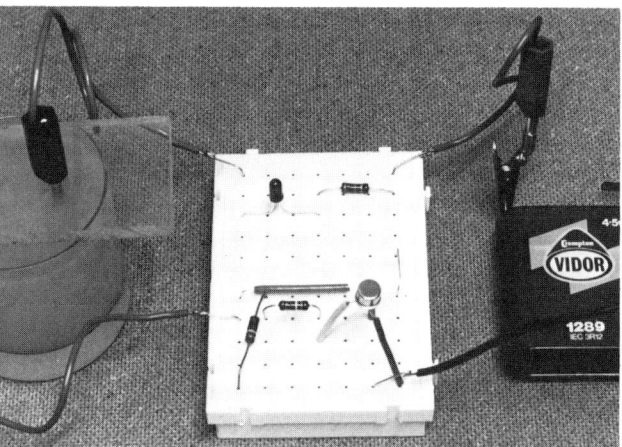

Figure 4 An S-Dec unit with an electronics circuit

Figure 5 Veroboard with a timer circuit

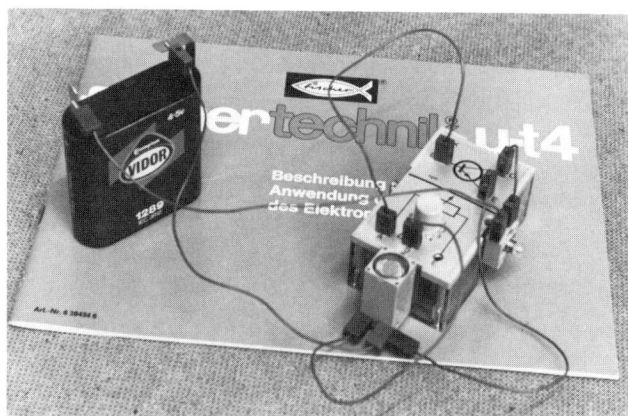

Figure 6 Units from the Fischer-Technik range

Breadboards

Breadboards (Fig. 2) are pieces of plastic sheet in which lines of holes are bored. Metal strips on the underside of the sheet connect the holes in lines. The two outer lines of holes are connected lengthways down the strip. Other holes are connected across the width of the sheet except at a central recess.

If one outer line of holes is used for connecting to the +ve side of a circuit and the other as −ve or 0 volts, circuits can be built between them. The wires of the components are pushed into the holes and those sharing the same lines will be connected to each other.

With the aid of breadboards, circuits can be designed, constructed, tested and modified, before being soldered up. If the circuit is a complicated one, each part of it can be tested on breadboards and then combined to make up the whole circuit.

Soldering

When soldering electronics circuits, use a low-voltage soldering iron, such as 24V 18W, with a fine tip. Use multi-core solder which adds flux to the join, to prevent oxides forming. The oxides can prevent the two parts being properly joined, causing 'dry' joints, with possible failure of a circuit.

The lack of a third hand is often a problem when soldering. Crocodile clips or paperclips are useful for holding parts together while they are being soldered.

Soldering technique

To join two components in a circuit:
1 Use pliers or tweezers to act as a **heat sink**, to conduct away (sink) excess heat which could damage the component.

2 Melt a small amount of solder onto one of the wires of one component with a soldering iron.
3 Melt a small amount of solder onto the wire of the second component.
4 Holding both solder-covered wires together, re-heat them with the soldering iron until the solder melts.
5 Hold the two components together until the solder hardens.

Soldering circuits onto stripboard

1 Insert the wires of the components through the appropriate holes from the top (non-coppered side) of the board.
2 Use a heat sink if necessary, e.g. hold the wires of a transistor in pliers so that heat 'sinks' into the pliers and not into the transistor.
3 Apply the tip of the soldering iron to the point where the copper strip and the component wire meet.
4 Touch cored solder to the coppered side of the stripboard where the wire comes through.
5 Allow the solder time to set before examining the joint.
　　When including integrated circuits (ICs) in a circuit, always use an IC dual-in-line (dil) socket with the correct number of pin holes in which to mount the chip. Then:
1 the chip does not have to be soldered into the circuit, thus avoiding it becoming overheated (this can damage it);
2 the chip can easily be replaced in the event of its failing.

Investigation

In industry, the following techniques are used for joining components in a circuit:
1 flow or wave soldering;
2 surface mounting of components;
3 crimping.
Find out about these methods and make sketches of each one, with explanatory notes.

Voltage divider circuits

Circuit 1

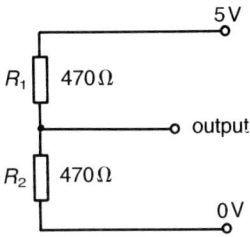

Figure 7 Circuit 1

Construct Circuit 1 (Fig. 7). Measure the p.d. across the 0 V to output sockets. Swap over the two resistors. Measure the p.d. again.

Circuit 2

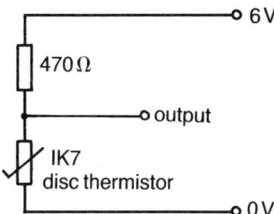

Figure 8 Circuit 2

Construct Circuit 2 (Fig. 8). Connect a voltmeter across the 0 V and output sockets. Note changes in voltage readings as the thermistor is heated. (A thermistor is a resistor where resistance depends on its temperature.)

Diodes

Diodes act as electronic one-way valves. Current can only flow in one direction through a diode. The arrow in the symbol for a diode points in the direction in which it will allow current to pass.

Circuit 3 – Diode circuit

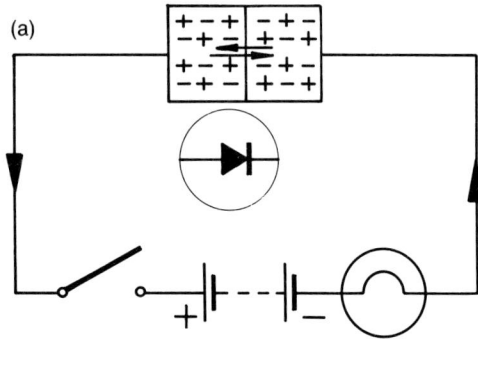

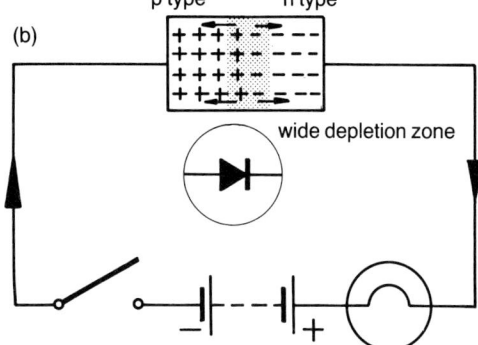

Figure 9 Circuit 3

Construct Circuit 3 (Fig. 9(a)). The lamp will light when the circuit is switched on.

This circuit demonstrates **forward bias**. The resistance of the diode is very low.

Reverse the connections to the battery (Fig. 9(b)). The lamp will not light. The diode now has a very high resistance. The circuit demonstrates **reverse bias**.

The cathode ray oscilloscope

An oscilloscope is an instrument which will show the outline (or *trace*) of a waveform (such as an alternating current) on its screen. The screen is the face of a cathode ray tube – the same device which forms the screen of a computer or of a TV set – hence its full name **Cathode Ray Oscilloscope** or **CRO**. The brightness of the screen, the height and width of the waveform, and the 'freezing' of the waveform are controlled by the various controls on the face of the CRO. Although some CROs appear to be very complex machines, they are really fairly easy to use.

Circuit 4 – Using a diode as a rectifier (1)

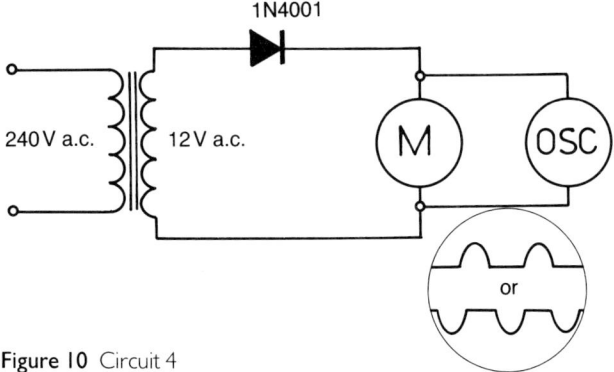

Figure 10 Circuit 4

Before constructing Circuit 4 (Fig. 10) feed 12 V a.c. current to an oscilloscope (a CRO) and note what its trace looks like. Then construct the circuit.

An alternating current can be changed to direct current by the use of diodes. This change is known as **rectification** and may be either **half-wave** or **full-wave** rectification.

Figure 10 shows half-wave rectification, with the traces it produces on an oscilloscope.

Circuit 5 – Using a diode as a rectifier (2)

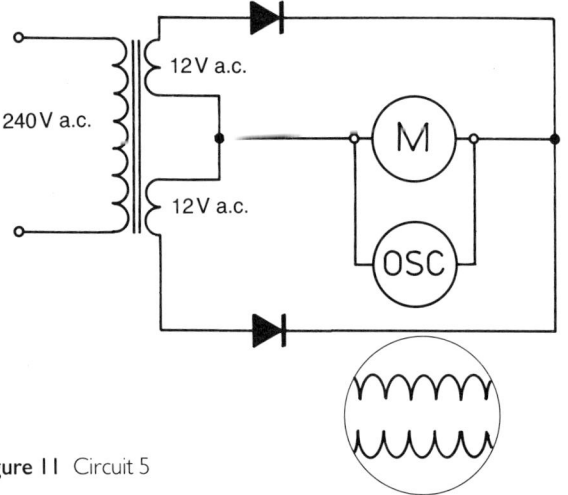

Figure 11 Circuit 5

Now alter the circuit as shown in Fig. 11. The 'noise' from the motor brushes will produce a CRO trace with a very 'whiskery' outline, so change the motor for a resistor equal in resistance to the motor and of about

5 W rating. This will produce a much crisper trace. The diode is alternately on forward and reverse bias. The flat part of the trace shows when the diode is at its highest resistance: because no current is flowing, the p.d. reading is nil.

diodes correctly and note the trace on the oscilloscope. Note that in the photograph Fig. 13, a resistor has replaced the motor in the circuit. What is the value of this resistor? How does its value compare with that of the motor used in Circuit 6?

Circuit 6 – Full-wave rectification

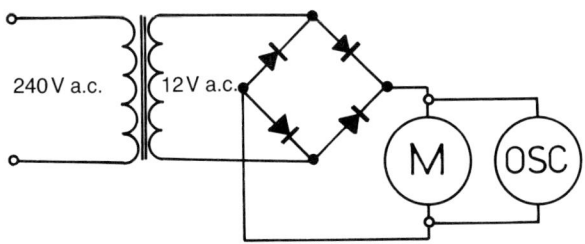

Figure 12 Circuit 6 Full–wave rectification circuit

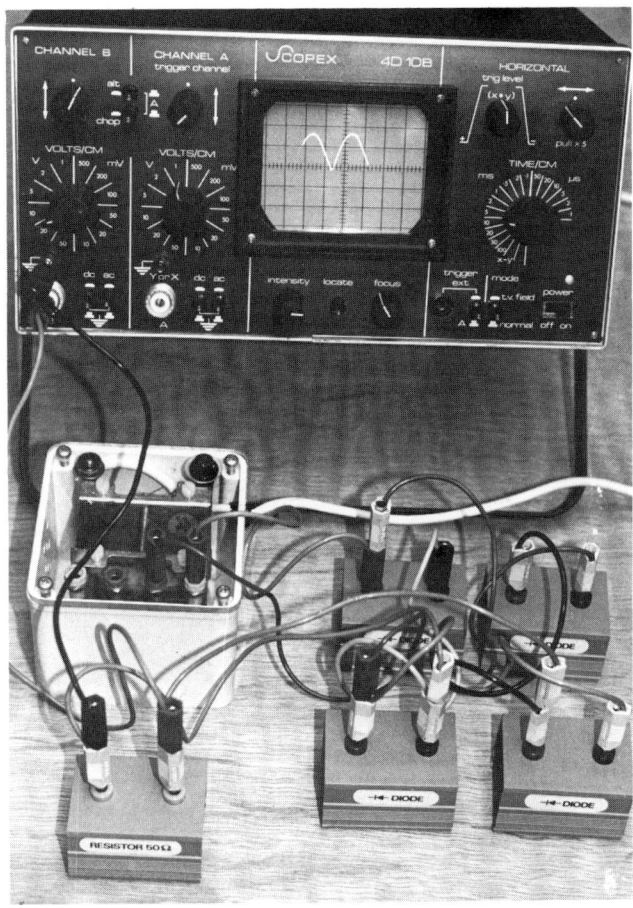

Figure 13 Circuit 6

A full-wave rectifier bridge circuit is shown in Fig. 12. Construct this circuit, making sure you connect the

Project

Fix a disc to the shaft of the motor in each of the circuits 5 and 6 in turn, so that the rpm of the motor may be easily counted.

Make a note of the results. Explain why the rpm is different for the half-wave and full-wave rectified circuits. The waveforms shown on the CRO should help you to come to correct conclusions.

Light emitting diodes

Light emitting diodes (LEDs) are made from gallium arsenide phosphide (red or yellow) or gallium phosphide (green). They are junction diodes which emit light when a current passes through them. A common use for them is in radio alarm clocks, where the numbers are displayed in red, yellow or green.

When using small LEDs, current must be controlled with a resistor. The voltage drop across a LED is between 2 and 3 volts.

In forward bias circuits LEDs require a current no greater than 20 mA. In reverse bias circuits a LED will produce no light output.

Circuit 7 – Light emitting diode

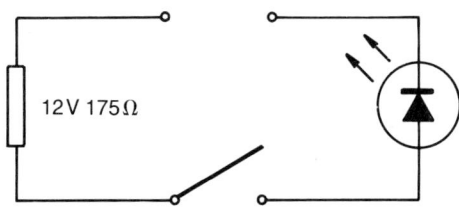

Figure 14 Circuit 7

Construct Circuit 7 (Fig. 14). To calculate the resistor value required:

Potential difference (p.d.) across LED = 2.5 V

The resistor causes a voltage drop $(6 - 2.5) = 3.5\,V$
Current through LED must not exceed $20\,mA$,

therefore resistance must $= \dfrac{3.5}{0.02} = 175\,\Omega$

Circuit 8

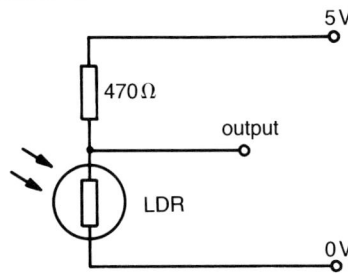

Figure 15 Circuit 8

Construct Circuit 8 (Fig. 15). Connect a voltmeter between the 0 V and output sockets. Shine a light onto the LDR (light dependent resistor – its resistance depends on the intensity of light shining on it). Note the changes in the voltmeter readings as the intensity of the light is varied.

Circuit 9

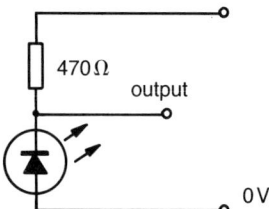

Figure 16 Circuit 9 – Light emitting diode circuit

Construct Circuit 9 (Fig. 16). Measure the voltage 'drop' across the LED.

Exercises

1 Calculate the value of resistor required in a 9V circuit in which an LED is to be placed.
2 Calculate the value of resistor required in a 12V circuit in which an LED is to be placed.
Note: Do not pass too much current through an LED, or place an LED in a reverse bias circuit with more than 5V passing through the LED – if you do, the LED will be destroyed.

Zener diodes

If a range of different input voltages are passed through a Zener diode, it will produce only a fixed output voltage. Zener diodes are always reverse biassed and must be protected with suitable resistors when used in electronic circuits.

Zener diodes are rated to produce a specified output voltage. If an input voltage lower than the rating of a Zener diode is applied to it no current will pass through. Once the input voltage is equal to, or larger than, the rating, the output voltage remains constant. Care must be taken to ensure that the power rating of a Zener diode is not exceeded.

Circuit 10 – Zener Diode Circuit

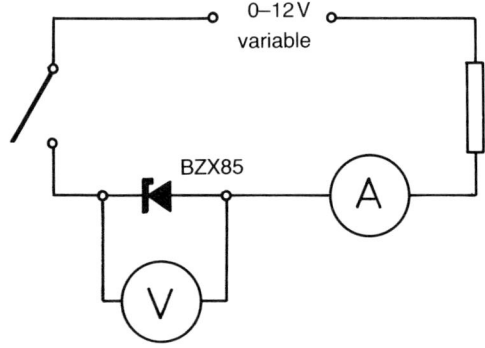

Figure 17 Circuit 10 – Zener diode circuit

Construct Circuit 10 (Fig. 17).
A 400 mW (0.4 W) Zener diode, rated at 5.1 V, can pass:

$$\text{Maximum current} = \frac{\text{Zener rated power}}{\text{Zener voltage}}$$

$$= \frac{0.4}{5.1} = 0.08\,A$$

The size of resistor to be placed in series with the Zener diode can be calculated as follows. If a constant voltage of 5.1 V is needed from a 12 V d.c. supply:

$$R = \frac{\text{input voltage} - \text{Zener voltage}}{\text{maximum current}}$$

$$= \frac{12 - 5.1}{0.08} \qquad = \frac{6.9}{0.08} = 86\,\Omega$$

Light dependent resistors (LDRs)

LDRs, also known as photocells or photoresistors, are made from cadmium sulphide (CdS), which has a high resistance ($10\,M\Omega$) in the dark, but a low resistance ($150\,\Omega$) in bright sunlight. One of the most common forms of LDR is the ORP12, which has a power rating of $200\,mW$.

Circuit 11 – Circuit with light dependent resistor (LDR)

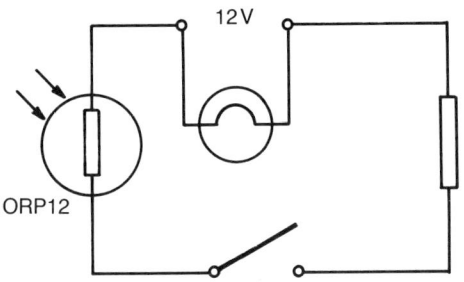

Figure 18 Circuit 11 – Circuit with light dependent resistor

Construct Circuit 11 (Fig. 18). Shine a light on the LDR – from a torch or by taking the circuit into sunlight.

What happens to the light from the lamp bulb in the circuit, when light falls on the LDR?

Circuit 12 – LDR circuit with ammeter

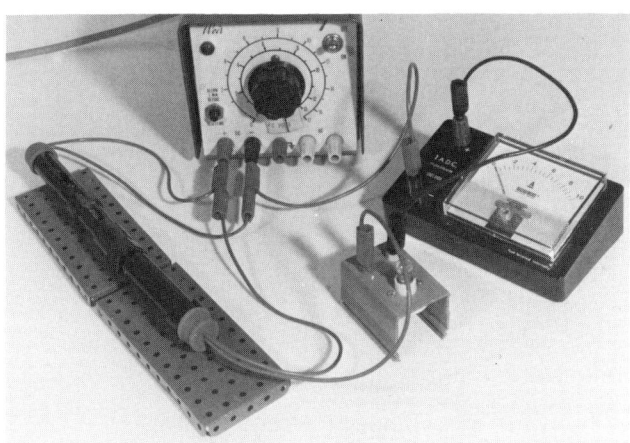

Figure 19 Circuit 12 – Circuit with LDR and ammeter

Change the light bulb in Circuit 11 for an ammeter (see Fig. 19). Note the readings on the ammeter when the light falling on the LDR is varied.

Circuit 13 – Circuit with LDR and two lamps

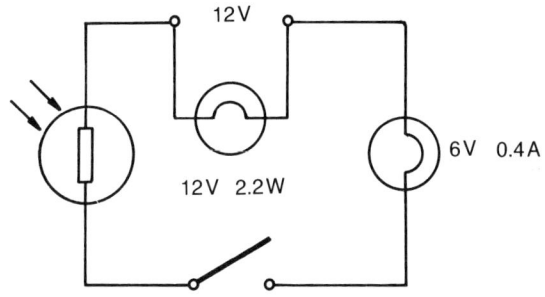

Figure 20 Circuit 13 – Circuit with LDR and two lamps

Construct Circuit 13 (Fig. 20). Shine a light on the LDR by placing in sunlight or shining a torch on it. Note what happens to the brightness of the lamps in the circuit.

Note: Most semi-conductors react to infrared light. This reaction is used in many applications.

Thermistors

The resistance of thermistors changes with changes of temperature. Thermistors are therefore used as heat sensors. There are two types:
NTC (negative temperature coefficient) thermistors – resistance falls as the temperature rises;
PTC (positive temperature coefficient) thermistors – resistance rises as the temperature rises.

When thermistors are used in circuits, do *not* exceed their maximum power dissipation or try to measure temperature beyond their maximum range.

Circuit 14 – Circuit with thermistor

Construct Circuit 14 (Fig. 21). Place a heated soldering iron near the thermistor in the circuit. Note what happens to the brightness of the lamp.

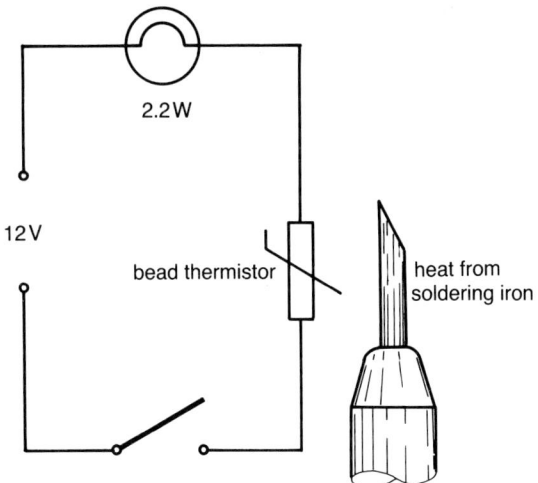

Figure 21 Circuit 14 – Circuit with thermistor

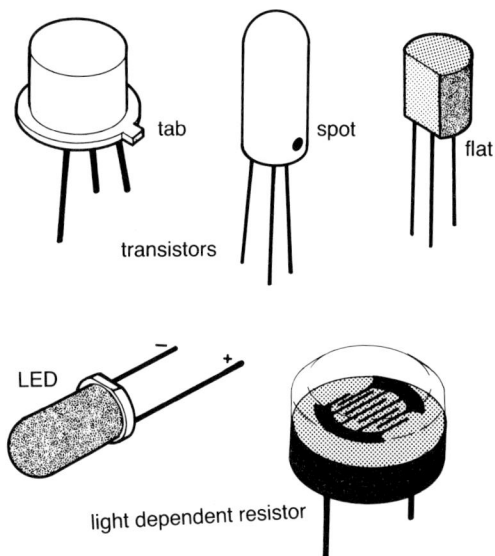

Figure 23 Some transistors, showing identification marks; a LED and a light dependent resistor

Transistors

Transistors are the basic elements in most electronic circuits. They are used in circuits to:
1 act as switches;
2 amplify (increase) electrical currents.

Transistors may be thought of as two junction diodes connected back-to-back.

The order in which the materials in a transistor are joined, determines its type – **pnp** or **npn**.

The symbols for each type of transistor are shown in Fig. 22. The arrows of the symbols show conventional current flow.

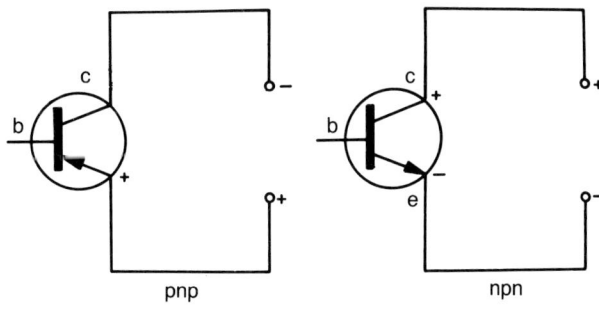

Figure 22 Biasing of a transistor

The shapes of the cases (**encapsulation**) of transistors vary considerably, but all have marks on them which make it easy to identify the connectors.

Transistors usually have three connectors – **base** (b), **emitter** (e), and **collector** (c). Some have more than three.

Connecting transistors in circuits

Transistors must be fitted with their base, emitter and collector leads correctly connected in circuits. If wrongly connected, the transistor will be damaged.

The collector and base of an npn transistor are positive and the emitter is negative. With pnp transistors, the emitter must be positive and the base and collector must be negative. See Fig. 22.

I_c refers to the current flowing through the collector – choose a transistor with an I_c that is about twice the rating you expect to need. P_{tot} refers to the total *power* the transistor can handle.

For transistor switching:

$$P_{tot} = \frac{(\text{supply voltage})^2}{\text{collector resistance}}$$

For transistor amplification:
h_{FE} is the direct current gain of the transistor, where

$$h_{FE} = \frac{\text{collector current}}{\text{base current}} = \frac{I_c}{I_b}$$

By using this formula you can calculate either the collector or base current:

collector current $= h_{FE} \times$ base current

$$\text{base current} = \frac{\text{collector current}}{h_{FE}}$$

Circuits 15 and 16

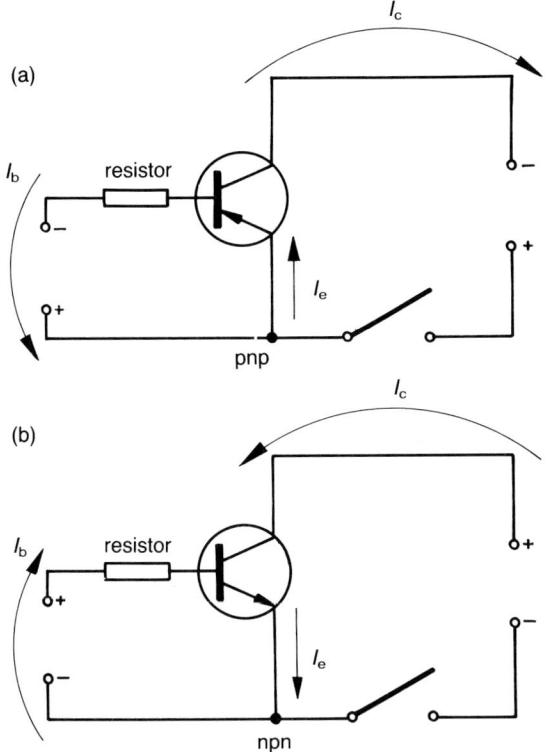

Figure 24 Bias circuits for transistors

The two circuits in Fig. 24 show forward bias circuits – Fig. 24(a) using a pnp transistor and Fig. 24(b) with an npn transistor.

A transistor has two paths along which current can flow:

base to emitter   OR   collector to emitter.

If a small current is applied to the base–emitter circuit (giving it a *forward* bias), current can then flow in the collector–emitter circuit. This is the **common–emitter** mode.

Because the transistor acts as an amplifier, there is a typical current gain of between 100 and 500. The total current flowing in a transistor is given by:

$$I_e = I_b + I_c$$

and the d.c. gain (h_{FE}) is:

$$h_{FE} = \frac{I_c}{I_b}$$

I_b acts as an input current switching on the transistor, so that the output current I_c can flow from it. Unless the turn-on voltage V_{be} (the base–

emitter forward voltage) is larger than 0.6 V (npn) or 0.1 V (pnp), switching will not take place.

Circuit 17 – Basic switching circuit (1)

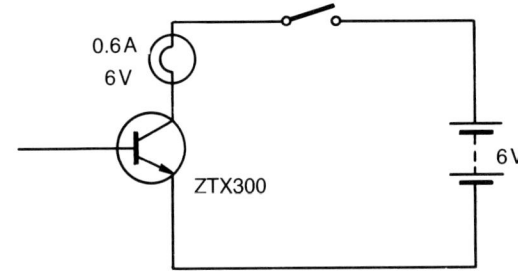

Figure 25 Circuit 17

Construct the circuit shown in Fig. 25. Does the lamp light?

The lamp in Circuit 17 cannot light because the resistance between collector and emitter is very high. Only a small current will flow between the two – too small to allow the bulb to light.

Circuit 18 – Basic switching circuit (2)

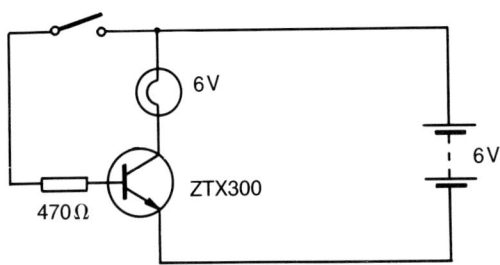

Figure 26 Circuit 18

Modify Circuit 17 as shown in Fig. 26. Switch on.

A small base current reduces the resistance between collector and emitter and so current now flows through the lamp, and it lights.

The resistor will prevent a large base current from flowing. A large base current would reduce the resistance between collector and base to such an extent that all 6 V would flow across the transistor. The heat generated would destroy the transistor.

Circuit 19 – Basic switching circuit (3)

Construct Circuit 19 (Fig. 27). Switch on.

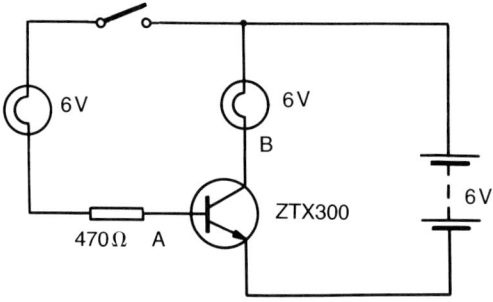

Figure 27 Circuit 19

The tiny current flowing to the base is insufficient to light the lamp in the base circuit. The collector current is sufficient to light the lamp in the collector circuit.

Place an ammeter at A and a micro-ammeter at B. Note the readings on the two ammeters.

From these readings it is possible to calculate the d.c. gain, h_{FE}:

$$h_{FE} = \frac{I_c}{I_b}$$

The current flowing out of a transistor must equal the current flowing in. Thus:

$$I_e = I_b + I_c$$

Circuit 20 – Potential divider control circuit

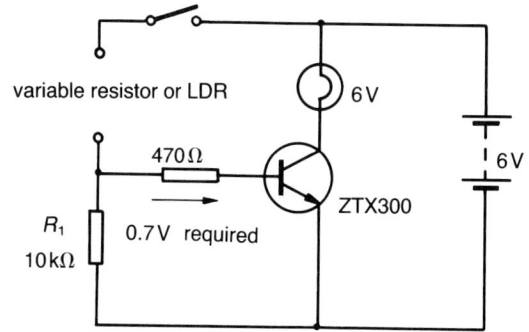

Figure 28 Circuit 20

Construct the potential divider control circuit (Fig. 28). The transistor is switched on when the p.d. across the variable resistor or LDR reaches about 0.7 volts.

If a light dependent resistor is included in the circuit, would the lamp light when the LDR is in darkness? Experiment to find out.

What would be the effect of changing the positions of R_1 and the LDR?

Circuit 21 – A thermistor-switched circuit

Construct Circuit 21 (Fig. 29).

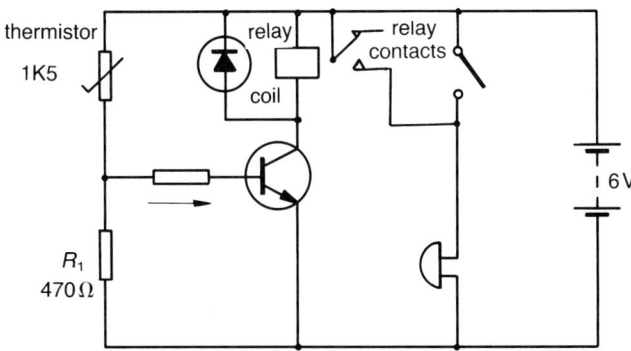

Figure 29 Circuit 21

This circuit will work well as a simple temperature indicator for measuring rapid changes in temperature. But if the rate of change in temperature is slow, or if the rate fluctuates in the critical switch-over range, the relay contacts can burn out.

The relay allows current to flow to the bell to make it ring.

The diode protects the transistor as the collector voltage falls to 0 volts. At close to 0 volts a large e.m.f. is induced in the relay coil. With the diode in the circuit, the induced current is dissipated in the diode–relay circuit.

In the position shown, can the thermistor act as a fire alarm or as an ice alert? What would be the effect of swapping the positions of R_1 and the thermistor?

The Darlington pair

If a second transistor is correctly positioned in a circuit, the switching action of the circuit becomes more sensitive. The second transistor amplifies the emitter current of the first. This two-transistor type of circuit is known as a **Darlington pair**.

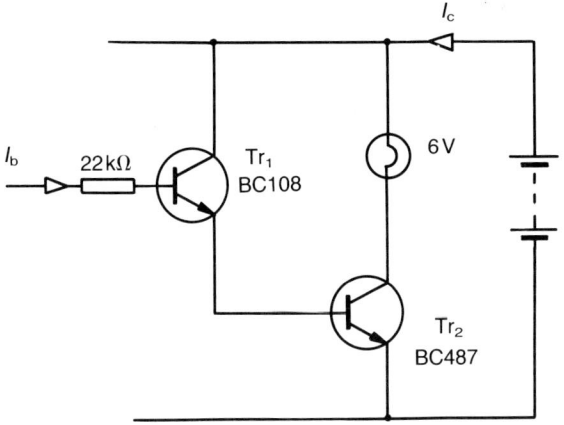

Figure 30 Darlington pair circuit

Figure 30 shows a typical npn transistor Darlington pair circuit. Note the following about this circuit:

1 The emitter of Tr_1 is connected to the base of Tr_2. This means that Tr_2 has the higher collector current of the two transistors (because I_c for Tr_2 is I_e of Tr_1 and I_e is larger than I_c).

2 The switch-on voltage (V_{be}) for a Darlington pair is 1.4V and not 0.7V. The total current gain of the circuit is calculated by:

$$h_{FE} = h_{FE1} \times h_{FE2}$$
$$\text{or } h_{FE} = \frac{I_c}{I_b}$$

The total gain in a Darlington pair circuit typically reaches 10^4.

Circuit 22 – Photoswitch circuit

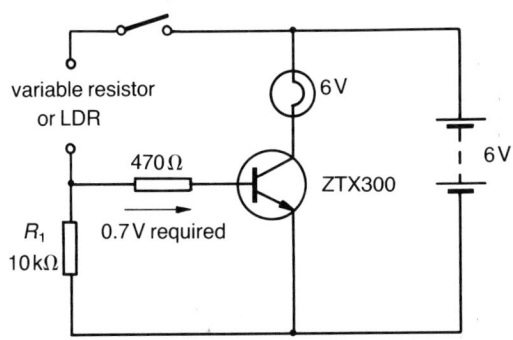

Figure 31 Circuit 22

Construct Circuit 22 – a Darlington pair in a photoswitch circuit (Fig. 31). The effect of having the LDR in the position shown in Fig. 31 is opposite to the effect of having the LDR positioned as in Circuit 20 (Fig. 28), but switching will be more precise.

Test and note the effect of adjusting the 10kΩ variable resistor in this circuit.

Circuit 23 – Automatic switching circuit

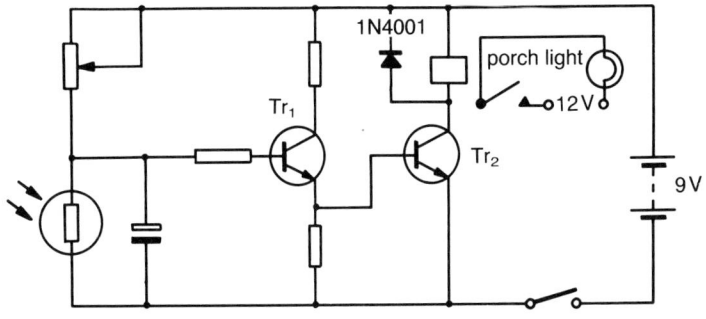

Figure 32 Circuit 23

Construct Circuit 23 (Fig. 32). This circuit could be used to switch on a porch light connected to a house mains supply automatically when dusk falls.

The circuit includes a capacitor. This prevents the LDR from reacting to any temporary improvement in light, say from the headlamps of a passing car.

What other improvements would you need to consider if street lighting some distance away effected the operation of this device?

Circuit 24 – The Schmitt trigger

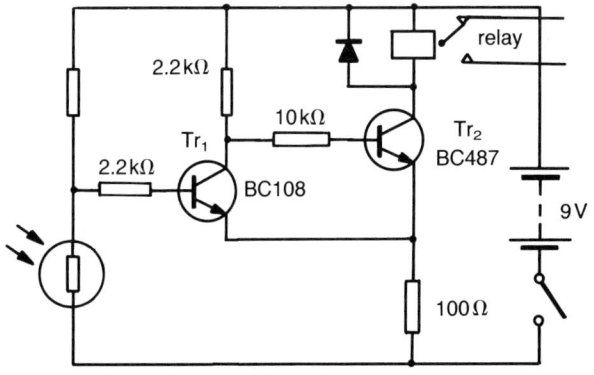

Figure 33 Circuit 24

The simple transistor circuits shown earlier in this chapter may not function well – they may not be fully on or fully off. This can result in bulbs that glow dimly or relays that bounce or chatter. These problems can be solved by incorporating a Schmitt trigger circuit within the system.

A Schmitt trigger circuit may be formed from two transistors as shown in Fig. 33 or it can be purchased as an integrated circuit (see below). The Schmitt trigger is a type of **bistable** circuit with an output that switches rapidly between two states.

If a Schmitt trigger circuit output is connected to an oscilloscope, the transformation of sine waves into square waves can be seen. When the input voltage rises above a set level, the output voltage rises. It will remain in this high state until the input voltage falls to another set level. The difference between the two levels is called **hysteresis** or **backlash**.

In Circuit 24, the collector/emitter current for Tr_2 must pass through the same resistor as the collector/emitter current from Tr_1. Any change in Tr_2 current is fed back to Tr_1 and switching is more rapid.

Integrated circuits

All the circuits included so far in this chapter involve the use of *discrete* components – each component is a separate item, joined to others in circuits.

There are two groups of discrete components: *active* components which alter the nature of the

current flowing, e.g. diodes and transistors, and *passive* components which simply carry the current, e.g. capacitors and resistors.

An **integrated circuit** or **IC** is a complete circuit built onto a **chip** of silicon a few millimetres square and mounted in a plastic case with circuit connections taken to pins set in the case sides. Despite their small size, ICs may each contain many hundreds of components. They are cheap, and very reliable in operation.

Sockets to hold ICs in circuits can be purchased.

555 timer IC

Integrated circuit (IC) timers can produce time delays of from microseconds to hours. The 555 timer IC consists of the equivalent of some 25 transistors, 16 resistors and 2 diodes, surrounded by a case with 8 pin connectors (Fig. 34). This IC timer is very versatile and can be used in both monostable and astable circuits.

The functions of the pins are:
Pin 1 – ground pin: −ve power supply.
Pin 2 – trigger input: when the voltage here falls to less than one-third of the input voltage at pin 8, the output from pin 3 changes from a low to a high voltage. Also pin 7 is disconnected from ground.
Pin 3 – output pin: up to 200 mA and up to the voltage at pin 8 minus 1.7 volts; e.g: input voltage at pin 8 = 9V: max V_s at pin 3 = 9−1.7 = 7.3 V.
Pin 4 – reset pin: only up to 0.4 volts will force pin 3 to go low. Connect to the +ve side of the power supply source when not required, to prevent re-sets.
Pin 5 – usually connected to 0 volts (ground).
Pin 6 – threshold pin: must always have a minimum of 0.1 μA of current supplied. This current affects the maximum value of resistor permissible between pin 6 and the +ve side of the power source. When the threshold voltage input becomes greater than two-thirds of the input voltage at pin 8, the output pin goes low.
Pin 7 – discharge pin: the capacitor used for timing is often connected between this pin and ground (pin 1). As 0 volts is reached at the output pin, pin 7 is connected to ground.
Pin 8 – power supply pin: accepts voltages in the range 3 volts to 15 volts.

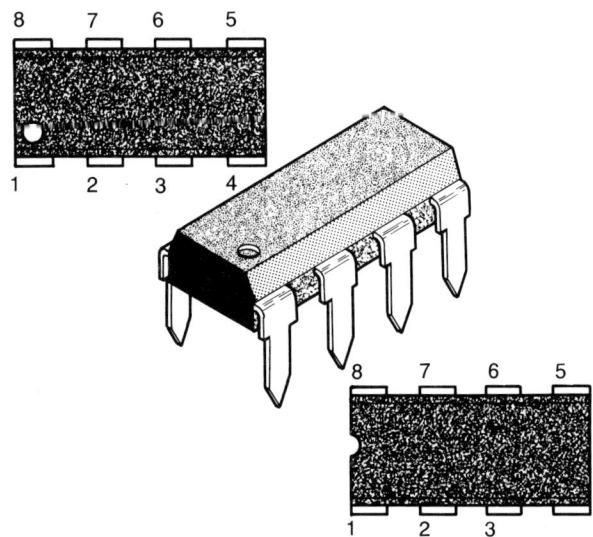

Figure 34 Integrated circuit chips and top markings

IC 555 timer circuits

Two types of circuit are possible using the 555 timer – **monostable** and **astable** multi-vibrators.

Circuit 25 – Monostable circuit with re-set

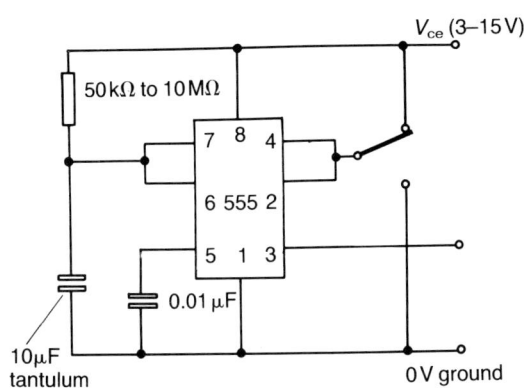

Figure 35 Circuit 25 – Monostable circuit

Construct Circuit 25 (Fig. 35), which is a monostable multi-vibrator circuit. Applying a pulse of current from ground (pin 1) to pins 2 and 4 will stop the cycle. A new trigger pulse will re-start the cycle. The length of pulse is determined by the relationship between resistor R and capacitor C.

Note: The use of capacitors over $1000\,\mu F$ is not to be recommended.

Circuit 26 – To give long and accurate delays

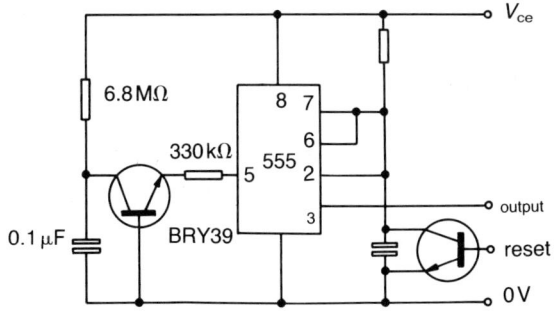

Figure 36 Circuit 26

Construct Circuit 26 (Fig. 36). This will give long and accurate delay timings.

Circuit 27 – Astable circuit

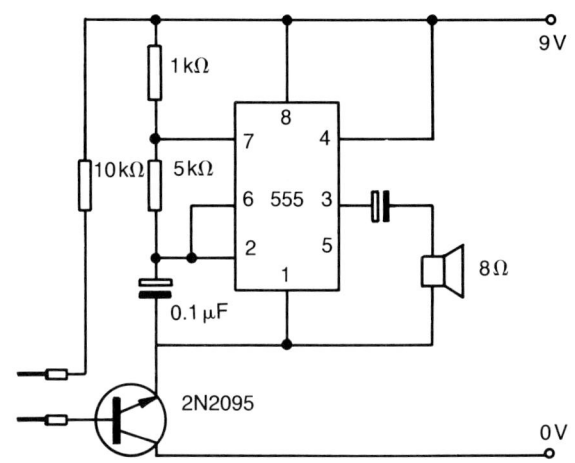

Figure 37 Circuit 27 – Astable circuit

Construct Circuit 27 (Fig. 37). This circuit will allow soil moisture content in a potted plant to be monitored, for example.

If the resistor values are chosen with care, this circuit can be tailor-made for the water requirements of a variety of plants. When the soil becomes too dry, an alarm sounds.

Exercise

Modify Circuit 27 (Fig. 37), so that the detector will be able to monitor the moisture constantly, but can also be used only for occasional monitoring when necessary.

The 741 op. amp.IC

This is an operational amplifier which can amplify small voltage differences by about 100 000 times. It can be used with either a.c. or d.c. voltages.

This IC consists of the equivalent of 20 transistors, 11 resistors and a capacitor, fitted in a casing with 8 connector pins.

The functions of the 8 pins are:

Pin 1 – null pin: not often connected.
Pin 2 – inverting pin: −ve volts here produces +ve output at pin 6. +ve voltage here produces −ve at 6.

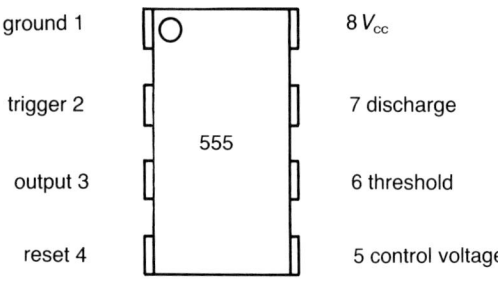

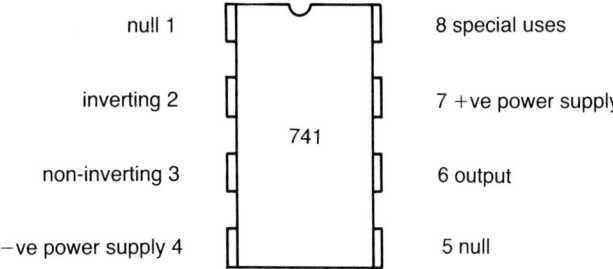

Figure 38 Pin numbering of 555 and 741 IC chips

Pin 3 – non-inverting pin: +ve voltage here produces +ve at pin 6 and vice versa.
Pin 4 – –ve power supply: should not exceed – 15V.
Pin 5 – second null pin.
Pin 6 – output voltage pin: either +ve or –ve.
Pin 7 – +ve power supply
Pin 8 – special use

Circuit 28 – Gain of 741 Op. Amp.

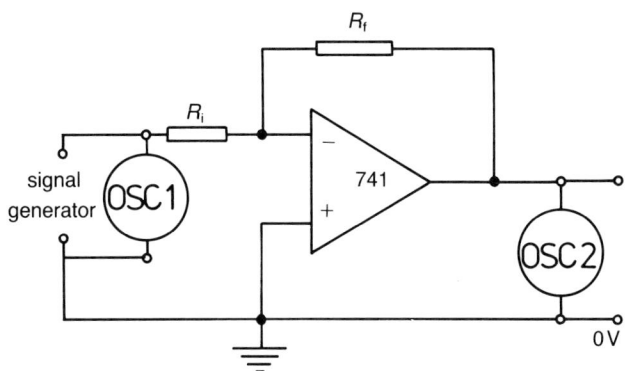

Figure 39 Circuit 28

Construct Circuit 28 (Fig. 39). By using a signal generator and a dual beam oscilloscope, the relationship between input and output voltages (the gain) can be displayed. To calculate the gain:

If R_f (feedback resistor) $= 50\,\text{k}\Omega$;
and R_i (input resistor) $= 5\,\text{k}\Omega$;
Then Gain
$$= \frac{V_{out}}{V_{in}} = \frac{-R_f}{R_c}$$
$$= \frac{-50}{5} = -10$$

Hence the output voltage will always be 10 times the input voltage.

Circuit 29 – Op. Amp. circuit

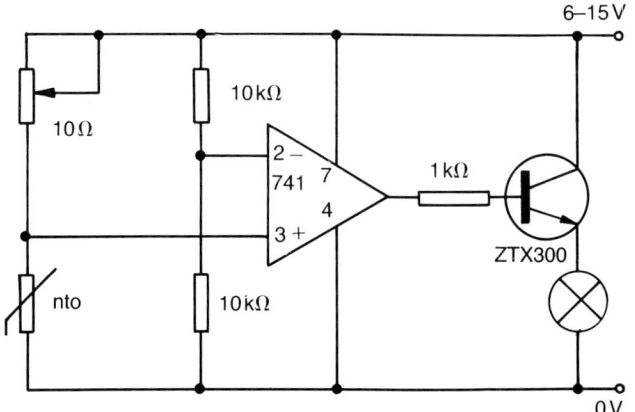

Figure 40 Circuit 29 – Op. Amp. circuit

Construct Circuit 29 (Fig. 40). The circuit will act as an ice-alarm when calibrated. Place the thermistor on an ice cube and adjust the variable resistor so that the lamp is just on. The lamp comes on when the temperature falls below freezing point.

Experiments

1 Modify the design of the ice-alarm circuit so that a buzzer would sound at low temperatures.
2 Modify the ice-alarm circuit by reversing the pins 2 and 3. This makes the circuit suitable as a boiling water indicator. Adjust the variable resistor so that the lamp is on when the thermistor is placed in boiling water.

Circuit 30 – Ice alarm with relay

Construct Circuit 30 (Fig. 41). This circuit is adapted from the ice-alarm circuit (Circuit 29). A relay is energised when ice is encountered.

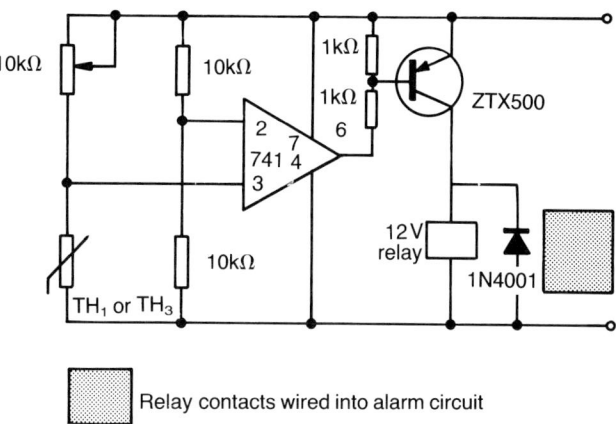

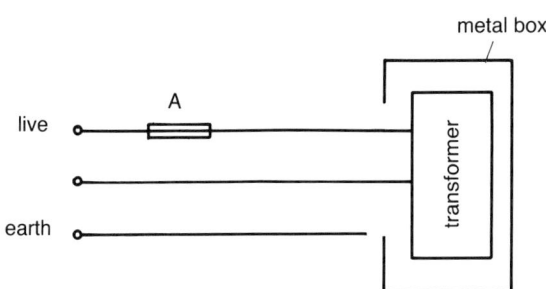

Relay contacts wired into alarm circuit

Figure 41 Circuit 30

Could this circuit now be used to act as an alert in a car when ice is encountered on the road? The relay switches into the second circuit to sound the alarm.

Exercises (Electrics and Electronics)

Figure 42 Exercise 1

1 Figure 42 shows 240 V connections to a transformer.
■ Name and state the purpose of component A.
■ The equipment has just been installed. How would you identify the live and earth wires?
■ Copy and complete the drawing by correctly connecting the earth wire.
■ State the two safety precautions which must be taken:
a as the cable enters the outer box;
b once the cable is inside the outer box.
■ If the transformer primary coil contains 480 turns and the secondary coil contains 24 turns, what is:
a the turns ratio;
b the secondary voltage?

■ Name a component that could be used to give a visual indication that the transformer is plugged into the mains.
2 ■ The circuit diagram in Fig. 43 represents a reversing circuit using a double-pole double-throw switch, a 12 V electric motor and two 6 V batteries. Re-draw the circuit eliminating the errors and ensuring the motor can operate at full capacity.
■ Mark the circuit with an X where you would insert a bulb to act as a reversing light.

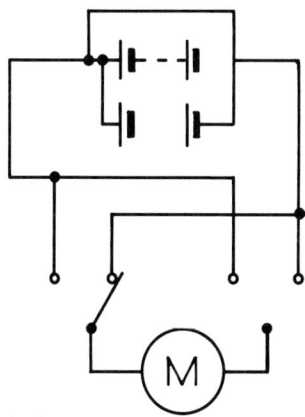

Figure 43 Exercise 2

3 ■ In an electronic component catalogue one cable is listed as 1/0.6 and another as 7/0.2. Explain these figures.
■ How does the cross-sectional area of a conductor affect the current and voltage it can carry?
■ Why is it important to use a stranded conductor for a cable used to wire a door bell fixed to a door?
4 ■ You have four 1.5 V cells. Draw a diagram to show how you would connect them to produce:
a the highest possible voltage;
b the longest running time.
■ Draw circuits showing four bulbs placed in:
a series; **b** parallel.
■ If each bulb draws 0.2 A which circuit would produce an ammeter reading close to 0.8 A? Mark the position of this ammeter with an A.
5 Explain the differences you would expect to see between a cable used to supply power to an electric cooker and cable supplying battery power to a wall-clock.
6 Calculate the total resistance in each resistor circuit in Fig. 44 and the total capacitance in the capacitor circuits.
7 The current to an LED is to be limited to 20 mA. If 12 V is available, what size of resistor is needed?
8 A 15 Ω 0.33 W resistor is connected across a 12 V power supply:

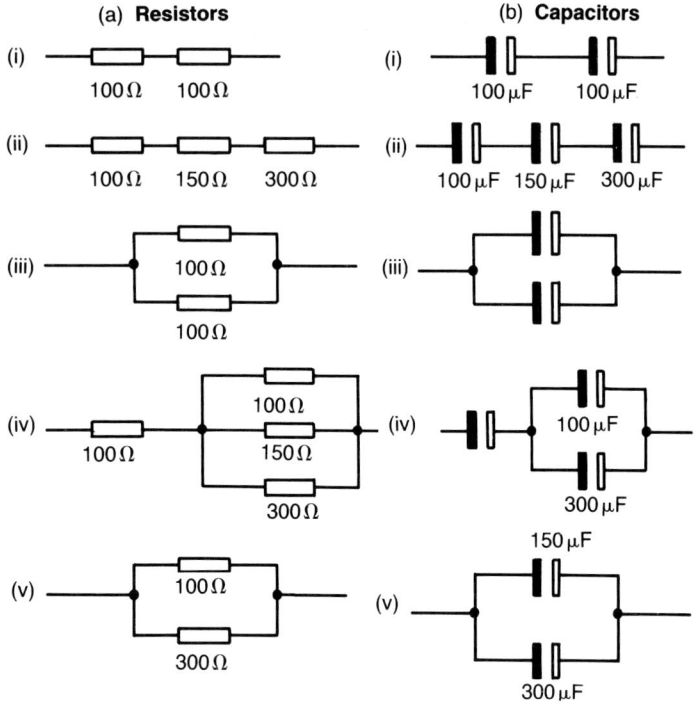

Figure 44 Exercise 6

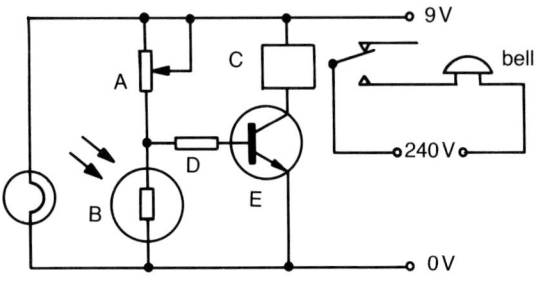

Figure 46 Exercise 12

12 The circuit in Fig. 46 has been designed by a student to act as a burglar alarm. When the light beam is broken the alarm is sounded.
- Name parts A–E and state their function in the circuit.
- It is electrical bad practice to place a relay coil in series with a transistor. Explain why this is so. Re-draw this part of the circuit so as to eliminate the fault.
- When the burglar steps out of the light beam the bell will stop ringing. Draw a circuit using a double-pole double-throw relay to act as a latch, holding the alarm on.
- Two simple improvements could be made to this circuit. Explain how *one* of them could be achieved.

13 ■ Copy Fig. 47 and label the primary and secondary windings.

- Calculate the power dissipated in the resistor.
- Give the nearest value of resistor (2% tolerance) that could be used in this situation.

9 The four bands on a resistor are green, blue, orange and silver.
- What is the value of the resistor?
- What is the tolerance of the resistor?
- What are the highest and lowest values this resistor might have?

10 What is the purpose of including each of the following components in a circuit?
- a fuse;
- a resistor;
- a diode.

11 ■ Name the components shown in Fig. 45.
- Name each leg on component 1.

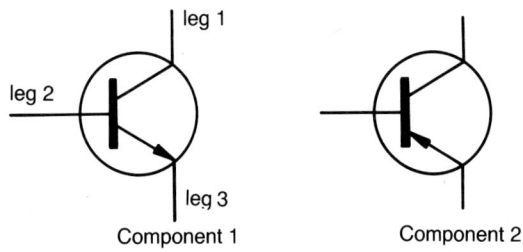

Figure 45 Exercise 11

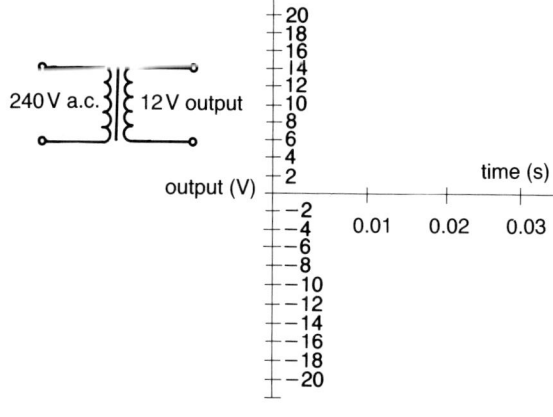

Figure 47 Exercise 13

- If the a.c. output is 12V calculate the r.m.s. potential difference.
- Copy the graph axes in Fig. 47, then draw and name the waveform such an output would produce on an oscilloscope.
a Add the symbol for a diode to the diagram so that a rectified current will be produced as the output.
b Name this form of rectification.
c Use a different colour to plot the rectified waveform on your graph.

14 Make a copy of the graph axes from Fig. 47. You wish to run a model car with a 12V d.c. motor from a 12V output.
- Draw the output circuit needed to give a fully smoothed and rectified current.
- On the graph axes draw the waveform such a circuit would produce.
- Add to the circuit a switch and three values of resistor so that on/off switching and speed control of the model is achieved.

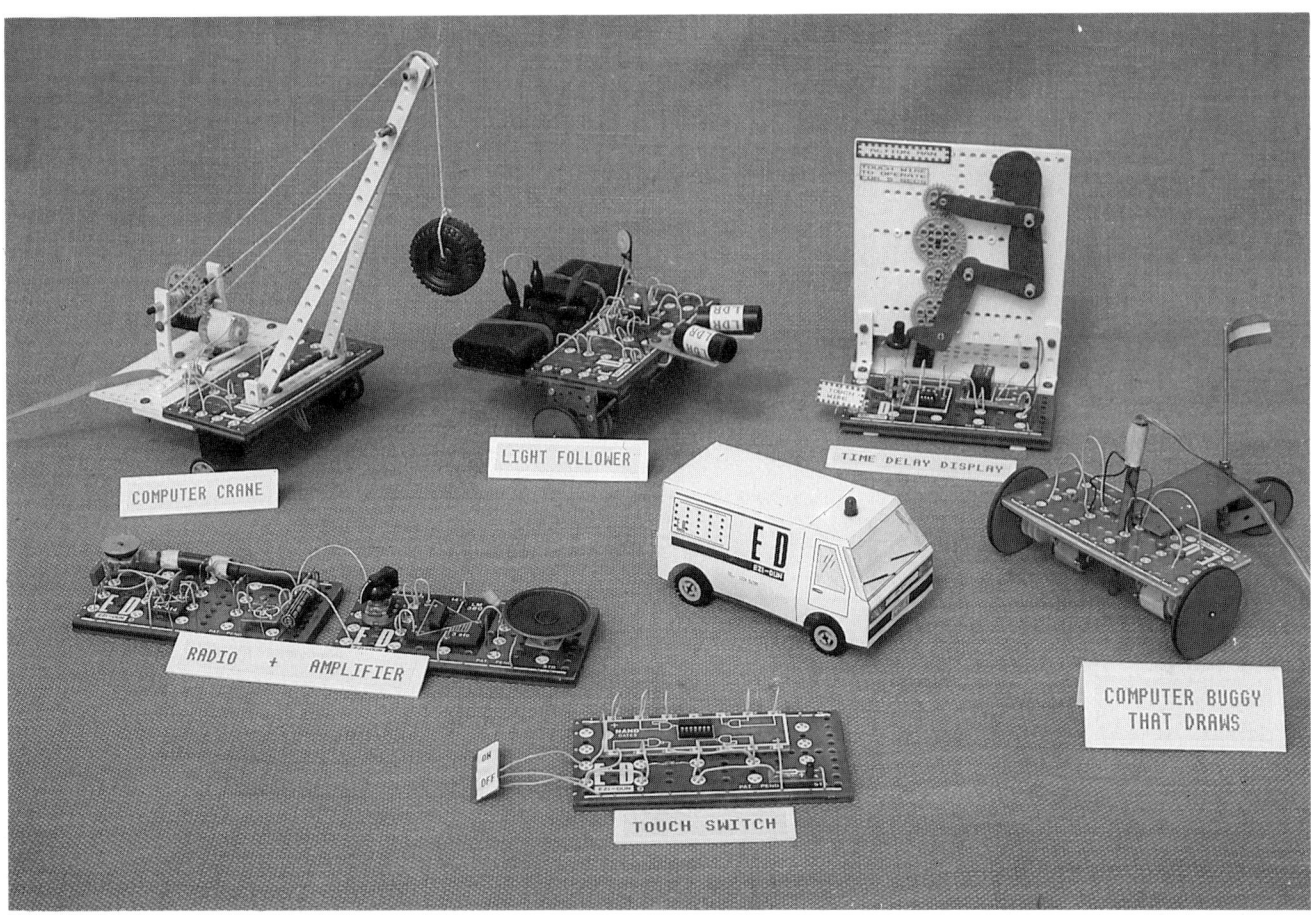

Figure 48 Units from the recently developed EZI-DUN construction kits

5 Pneumatics

Introduction

When connecting up circuits

1. Pipes connecting parts of a circuit must be long enough for pipes not to become creased.
2. Remove pipes from quick-release fittings, by first pushing the pipe IN to depress the inner ring, then OUT by pulling *in line* with the pipe.

The compression of air

Compressed air is supplied from a compressor. The compressed air is passed to storage tanks. Fig. 1 shows a typical school/college system. The air pressure in the storage tank is shown on a pressure gauge. A second pressure gauge shows the pressure in air lines from the storage tank.

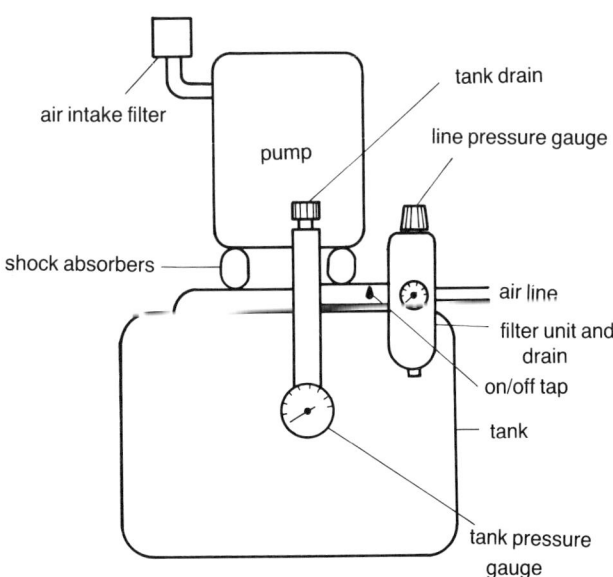

Figure 1 Details of a compressor

Figure 2 A compressor

Pneumatic Components

Cylinders

Pneumatic cylinders in a circuit will produce movement in a straight line (linear). As the piston of a cylinder moves in and out, it is said to *in-stroke* and *out-stroke*. At the end of an in-stroke the piston is said to be *negative* (−ve). At the end of an out-stroke, the piston is said to be *positive* (+ve).

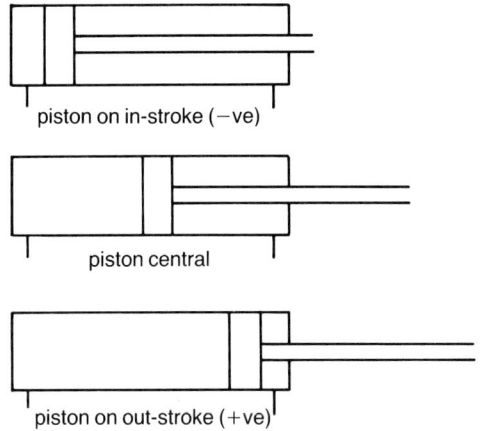

Figure 3 Positions of piston in pneumatic cylinders

Single-acting cylinders

Compressed air is supplied to only one **port** of a single-acting cylinder. When compressed air causes the piston to move along the cylinder to its +ve position, a spring inside the cylinder returns the piston to its −ve position. Air is **exhausted** (released) from the cylinder, through the **exhaust port**, as the spring is compressed.

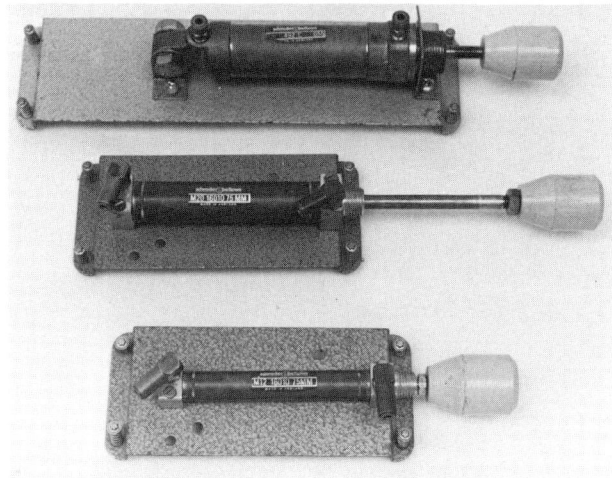

Figure 4 Pneumatic cylinders

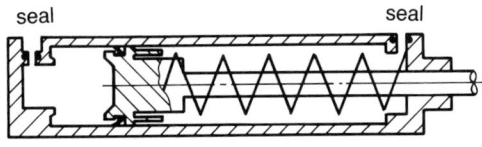

Figure 5 Diagrammatic section through a single-acting cylinder

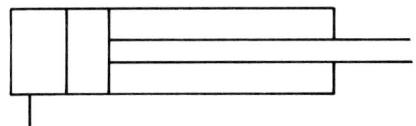

Figure 6 BS symbol for single-acting cylinder

Double-acting cylinders

A double-acting cylinder requires a supply of compressed air to both ports in turn. Air exhausts from one port as compressed air enters the other. Double-acting cylinders provide fast, forceful movement on both the in-stroke and the out-stroke. Double-acting cylinders can provide better speed control of the piston, in both directions, than is possible using single-acting cylinders.

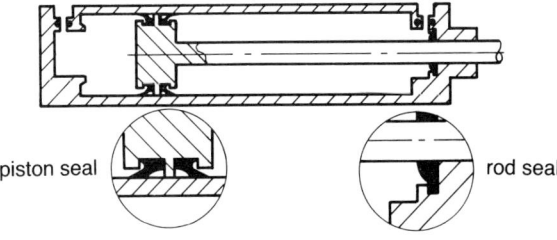

Figure 7 Diagrammatic section through a double-acting cylinder

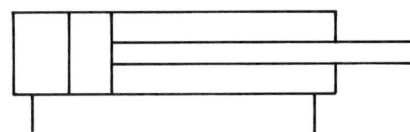

Figure 8 BS symbol for a double-acting cylinder

Seals

Sealing rings seal the joint between the piping and the components and also between parts of the components themselves. Their purpose is to prevent air leaking from circuits. The most common seals are O rings, so called because of their O-like shape (Fig. 7).

Push and pull forces

The piston of a cylinder always moves slightly more slowly on its out-stroke than on its in-stroke. This is because there is a difference between the forces pushing and the forces pulling the piston head.

Force = pressure × area

The *pressures* on each side of the piston are equal, but the *area* on each side is different.

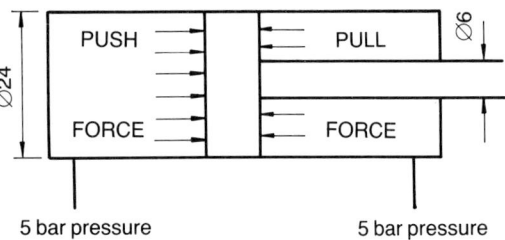

Figure 9 Push and pull forces

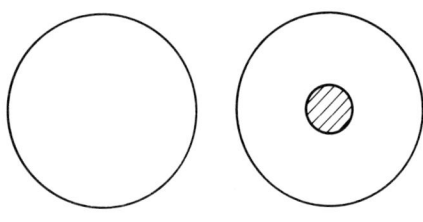

area occupied by section of piston

Figure 10 Area of *push* and *pull* ends of cylinder

Area on PUSH side of piston	Area on PULL side of piston

area $= \pi \times r^2$ hatched part $= \pi \times r^2$

$\quad = \pi \times 12^2 \qquad\qquad\qquad = \pi \times 3^2$

$\quad = \pi \times 144\,mm^2 \qquad\qquad = \pi \times 9\,mm^2$

$\qquad\qquad\qquad\qquad\qquad\qquad = 28.27\,mm^2$

$\quad = 452.38\,mm^2$ Piston face $= 452.38 - 28.27$

$\qquad\qquad\qquad\qquad\qquad\quad = 424.11\,mm^2$

This difference in area on which the force acts on each side of the piston results in a greater force to the PUSH out-stroke than to the PULL in-stroke.

Valves

All valves are designed for controlling the flow of liquids or gases or, in the case of electrical circuits, for controlling the flow of electrons. Pneumatic valves are placed in circuits to control the flow of compressed air through a circuit. They permit, prevent or divert the flow of air.

Compressed air passes in two directions:

1 air under pressure goes into the circuit and passes around it;

2 air is exhausted from the circuit.

Valves are made to **change-over** by:

■ pressing an **actuator;**

■ the action of one valve passing compressed air into another.

Figure 11 BS symbols for main air supply and exhaust

Poppet valve

Poppet valves have three ports, each having a certain function.

Port 1 – **inlet** port: connects the circuit to compressed air.

Port 2 – **cylinder** port: passes compressed air to other components.

Port 3 – **exhaust** port: allows exhaust air to escape.

When compressed air is let into port 1, a spring valve keeps the path between port 2 and port 3 open.

When the actuator button is pressed, the path between ports 2 and 3 is closed and air flows between port 1 and port 2.

Figure 12 A poppet valve

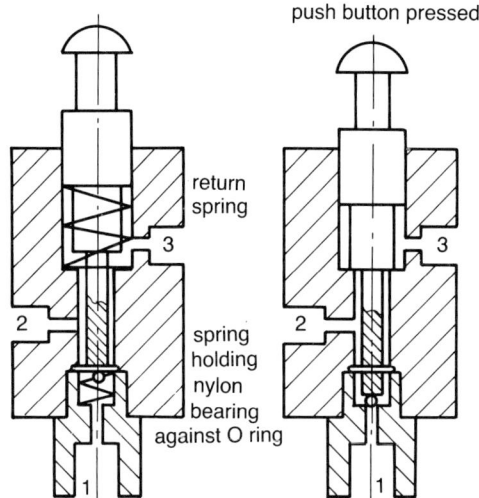

push button pressed

return spring

spring holding nylon bearing against O ring

Figure 13 Sectional views through a poppet valve

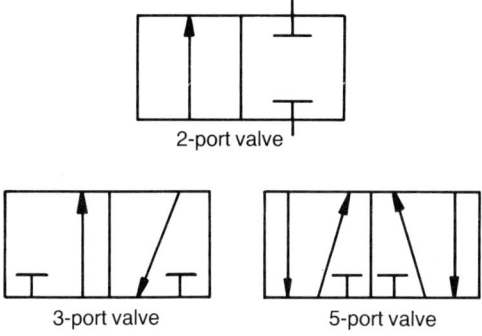

2-port valve

3-port valve

5-port valve

Figure 14 BS symbols for pneumatic valves – may represent either poppet or spool valves

Spool valve

Although there are a number of different types of spool valve, they all have similar effects.

The action of a double-pilot operated 5-port spool valve is as follows. Compressed air is connected to port 1; port 1–2 and port 1–4 are supplied with air from other valves in the circuit. This supply is shown by broken lines in all circuit

Figure 15 A spool valve

diagrams and is known as the **pilot signal line**. When an air signal is received at port 1–2, the spools move sideways and compressed air flows in through port 1 and out of port 2; exhaust air leaves port 4 through port 5. When an air signal is received at port 1–4, the spools move sideways and compressed air flows through port 1 and out through port 4; exhaust air flows through port 2 and out of port 3.

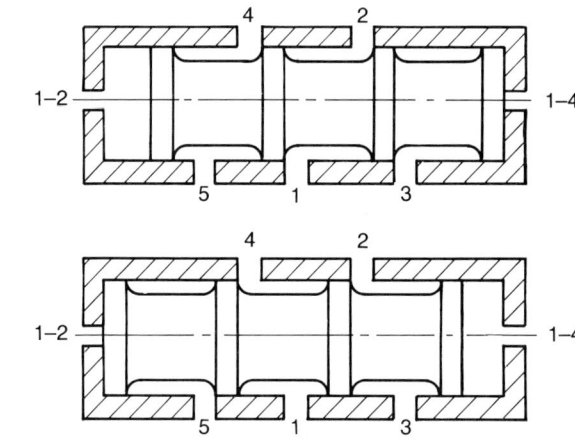

Figure 16 The action of a spool valve

Shuttle valve

Shuttle valves are used in circuits when a single-acting cylinder is to be controlled from two positions. An air signal to port 1a (Fig. 18) moves the shuttle across the valve, stopping the passage of air from 1b and making air leave through port 2. An air signal to port 1b moves the shuttle across the valve, stopping the passage of air from 1a and

Figure 17 A shuttle valve

making air leave port 2. If air is provided to both ports 1a and 1b at the same time, the shuttle remains in its last position.

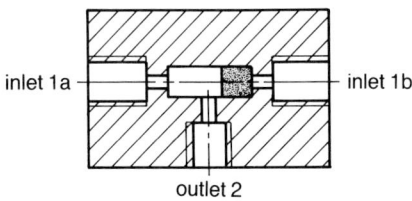

Figure 18 A sectional view through a shuttle valve

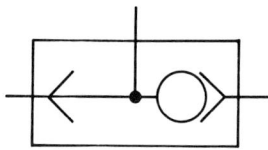

Figure 19 BS symbol for a shuttle valve

Parts of valves

A valve is made up of:
1 the body containing the mechanism;
2 an actuator to cause change-over;
3 a return device, to make the valve return to its original state.

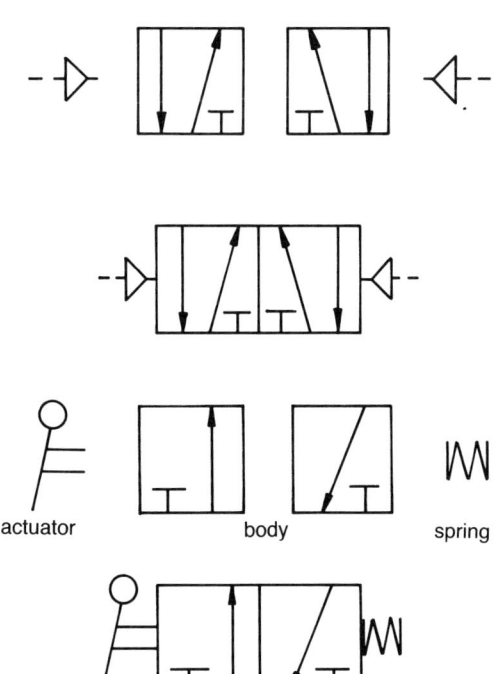

Figure 20 BS symbols for the three parts of a valve

□ **Actuators** Actuators switch valves to change-over allowing:
1 air to pass through the valve to force a change in other components in a circuit;
2 air to exhaust through the valve and return the circuit to its previous state.

Drawing Pneumatic Circuits

Pneumatic circuits are *always* drawn in their **un-operated** state – before the actuator is pressed.

The circuit in Fig. 22 shows a plunger-operated 3–port valve connected to a single-acting cylinder. The bottom half of the symbol for the valve is shown in use, with the piston of the cylinder in a −ve position – the path between cylinder port and exhaust port is complete. There is no compressed air in the circuit, so the spring of the cylinder keeps the piston in the position shown (+ve).

When the actuator is pressed, the top half of the valve symbol operates (Fig. 23). Circuits should *not* be drawn in this **operated** state. The circuit in Fig. 23 shows that the path between the main air port and the cylinder port is connected. Compressed air has passed through the valve causing it to change-over and the piston is now in its +ve position.

Drawing valves

When drawing valves, using the British Standards BS 2917 symbols, start by drawing a pair of touching squares. One of the squares shows the valve in its un-operated state, the other in its operated state. **Remember:** pneumatic circuits are *always* drawn in their *un-operated* state.

The direction of flow of the compressed air through the valve is shown by the arrows in the symbols.

Broken lines show valves passing pilot signals to other components in the circuit.

Construction of Pneumatic Circuits
Manifolds

In the photographs of circuits given here, note that each circuit is connected to the compressed air supply from a compressor through a component known as a **manifold** (Fig. 21). Compressed air to components in the circuits is supplied via any of the eight ports of the manifold. Ports which are not required are blanked-off. Some manifolds have a switch for turning the air on or off, and a control for adjusting the air pressure delivered from its ports. A meter in the manifold measures the air pressure from its ports.

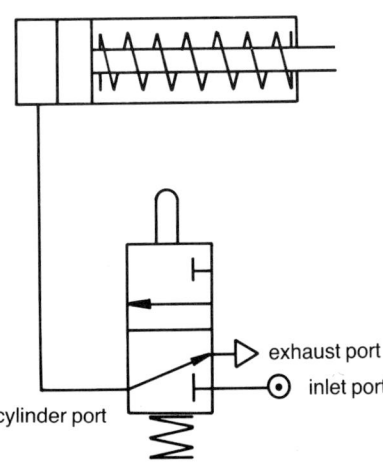

Figure 21 A manifold

Circuit 1 – Control of single-acting cylinder

Construct the circuit shown in Fig. 22.

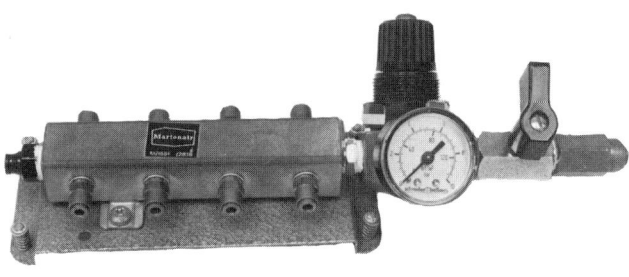

cylinder port

exhaust port

inlet port

Figure 22 Circuit 1 – Control of single-acting cylinder

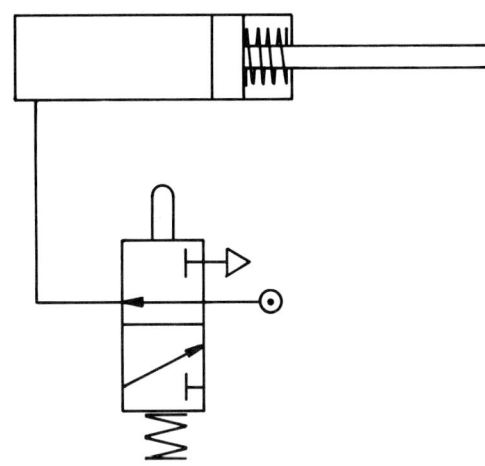

Note: circuits must not be drawn in this operated state

Figure 23 Circuit 1 – when actuator has been pressed

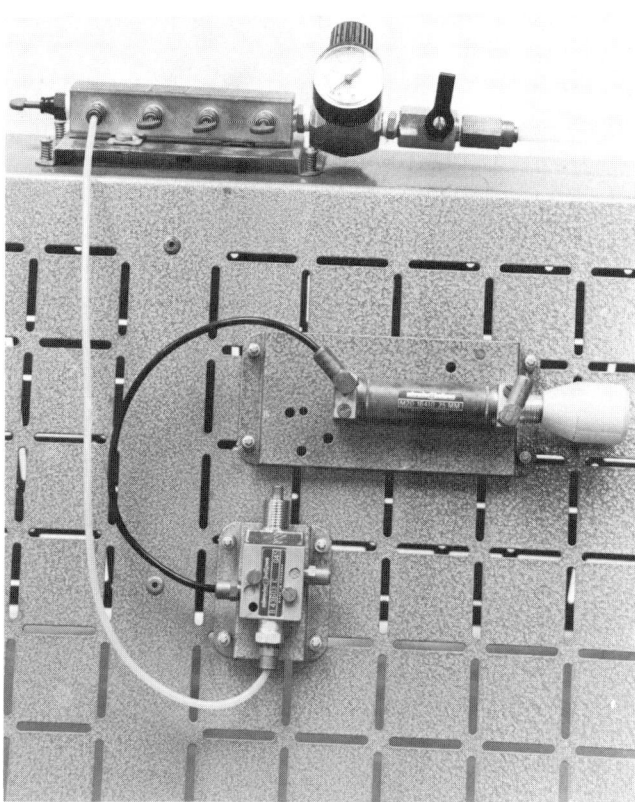

Figure 24 Circuit 1

Note:

1 The piston remains +ve when the plunger is pressed.

2 Air leaves the exhaust port as the piston in-strokes.

Project 1

Design and construct a circuit using a single-acting cylinder, two plunger valves and a shuttle valve to enable the cylinder to be operated from either valve.

Project 2

Now alter your circuit. Remove the shuttle valve and replace it with a tee-junction. How does this affect the operation of the circuit?

Note: In both Project 1 and Project 2 the shuttle valve performs the vital operation of directing air to the cylinder port and therefore preventing the air from following the route of least resistance by passing out through the exhaust port of the opposite valve.

Circuit 2 – Control of double-acting cylinder (1)

Construct the circuit shown in Fig. 26.

Press valve A and release. Does the piston move in when you apply hand pressure to the piston rod? If it does, what are the likely consequences to a machine operator? Now test the remainder of the circuit.

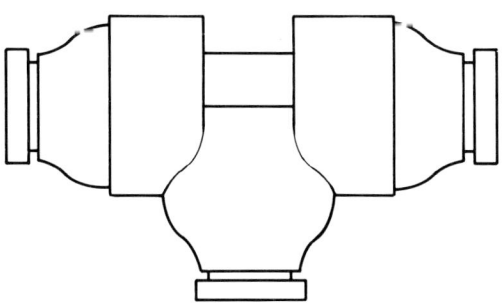

Figure 25 A tee junction

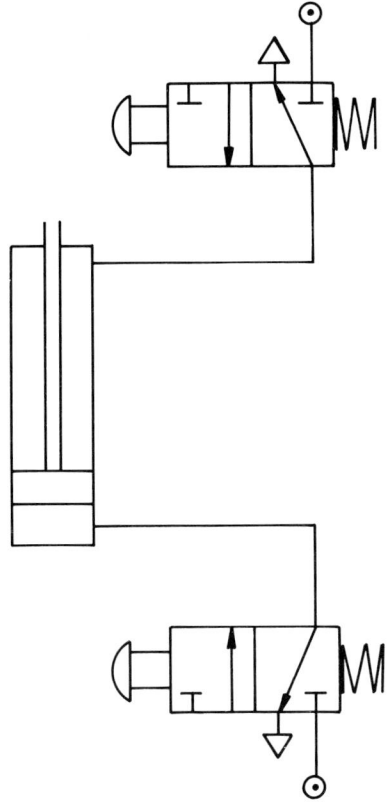

Figure 26 Circuit 2 – Control of double-acting cylinder (1)

Questions

1 Could you use this circuit to apply force to a machine vice?
2 Would this circuit control a sliding door, if a valve was placed each side of the door? Imagine you have opened the door with one valve, stepped through and closed the door using the other valve.

Project 3

Design a circuit using 4 valves, 2 tee connectors (see Fig. 25) and a double-acting cylinder, which will allow a sliding door to be opened and closed from either side. Construct the circuit you have designed. What dangers are there in having a door move at the speed at which the piston rod moves?

Circuit 3 – Control of double-acting cylinder (2)

Construct the circuit shown in Fig. 27.

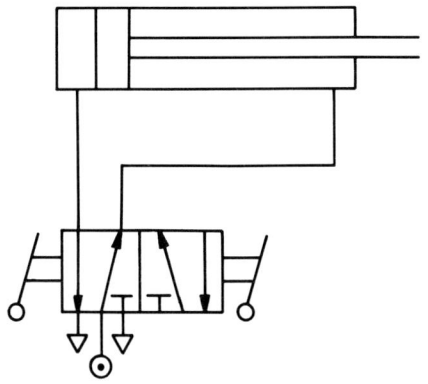

Figure 27 Circuit 3 – Control of double-acting cylinder (2)

Make sure the circuit operates and leave the piston in its +ve position. Do not switch off the air supply to the valve. Grasp the end of the piston rod and, applying a little force, try to pull the piston out-stroke. Release the piston rod.

Now switch the valve so that the piston is in its −ve position and attempt to push it in-stroke. You might move the piston by exerting a great force. The advantage of a lock-down valve (see Appendix 2) is that it applies compressed air in both ON and OFF positions. Therefore the piston is always under pressure in this circuit.

Question

Could this circuit be used to control the movements of a vice? But note that the system is manually operated.

Question

Would the jaws of the vice continue holding the work firmly even when the IN valve was released? Note that this circuit is still manually operated.

Circuit 5 – Semi-automatic circuit

Construct the circuit shown in Fig. 29.

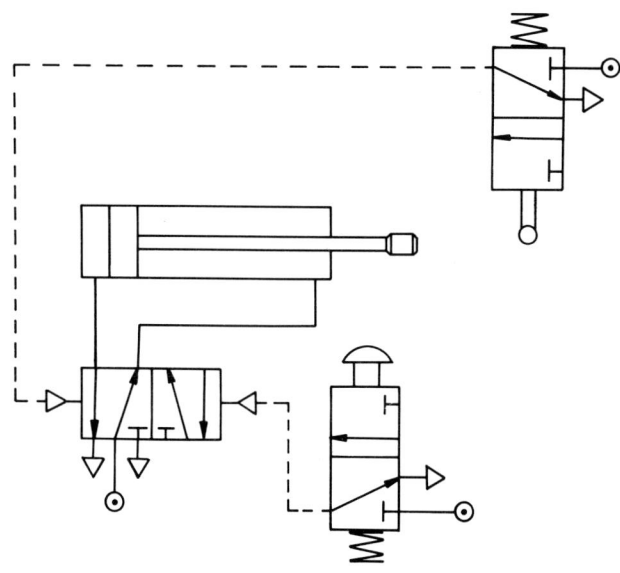

Figure 29 Circuit 5 – Semi-automatic circuit

Circuit 4 – Vice circuit

Construct the circuit shown in fig. 28.

Operate the circuit to check whether it would be suitable for controlling a vice.

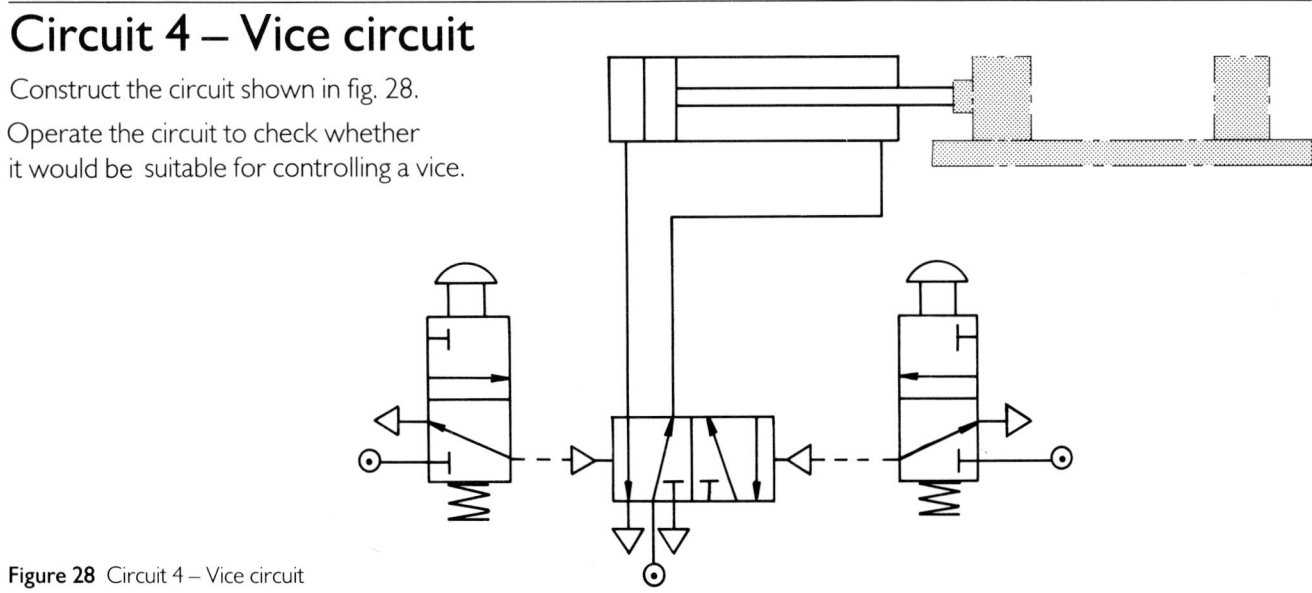

Figure 28 Circuit 4 – Vice circuit

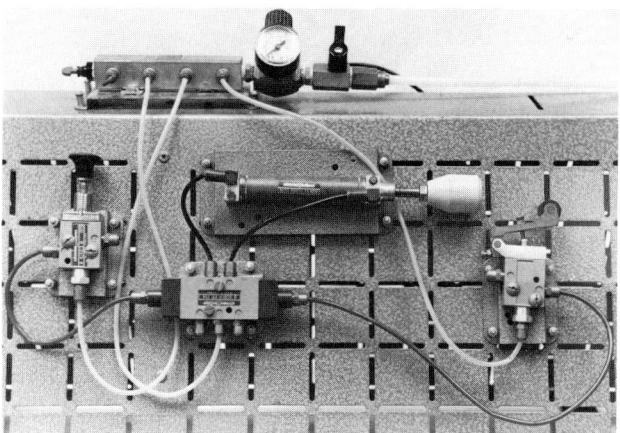

Figure 30 Circuit 5

By carefully positioning the roller lever valve (see Appendix 2), make the piston return in-stroke automatically. One half of this circuit requires manual operation, making it *semi-automatic*.

Project 4

Re-position the components of Circuit 5 to produce a fully automatic circuit.

Flow control valves

When using a flow control valve (see Appendix 2):
1 it is positioned in the exhaust stream and not in the main air supply;
2 always place the flow regulator as close to the cylinder as is possible.
Make sure that you have placed the control flow regulator valve with full flow from cylinder to valve. Keep pressing the push button ON and OFF while slowly adjusting the rate of air flow over the complete range.

Circuit 6 – Speed control using a flow control regulator valve

Construct the circuit shown in Fig. 31.

Question

What is the effect of connecting the flow control regulator valve the other way round?

Disadvantage

The time spent in pressing the button.

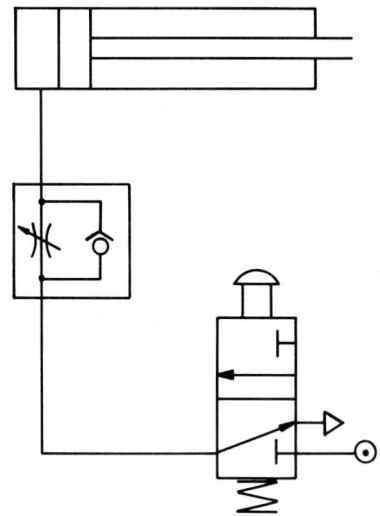

Figure 31 Circuit 6 – Speed control using a flow control valve

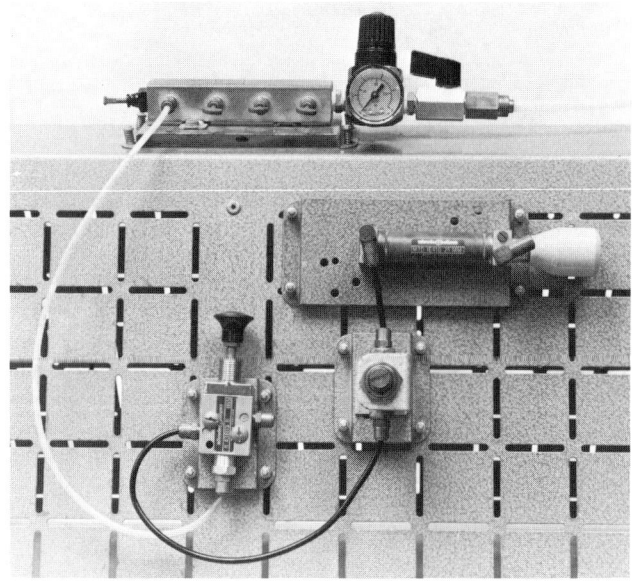

Figure 32 Circuit 6

Project 5

Replace the valve shown in Fig. 31 with a 3-port lock-down lever valve.

Circuit 7 – Drilling machine circuit

Construct the circuit shown in Fig. 33.

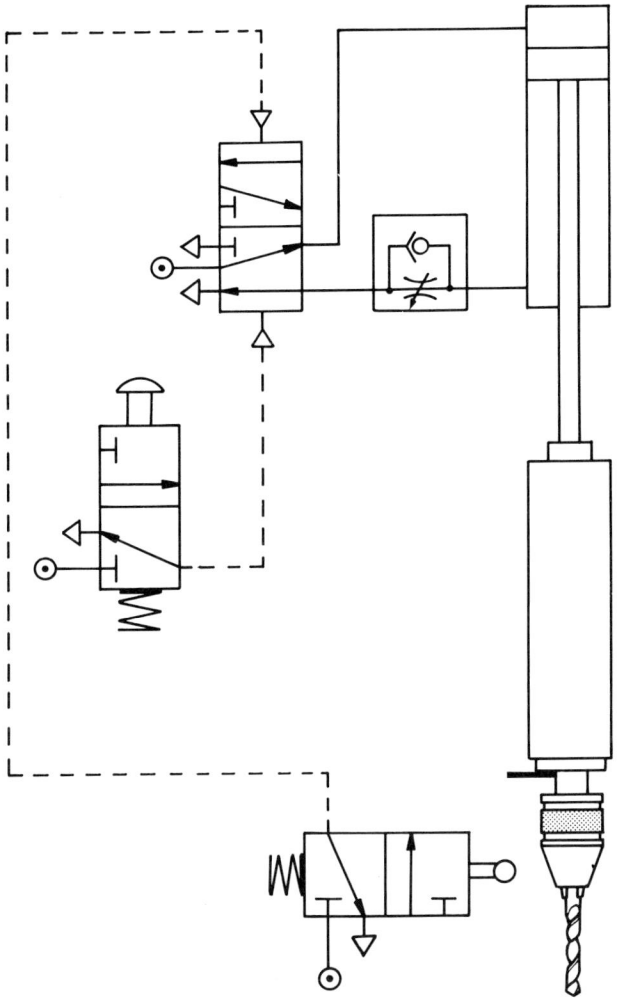

Figure 33 Circuit 7 – Drilling machine circuit

You found when constructing a previous circuit that constant pressure in either +ve or −ve positions of the piston was possible, but the speed of movement of the piston was fast in both directions. When operating a drilling machine, the drill must enter the material being drilled slowly and steadily, but the return stroke taking the drill out from the material can be fast.

Question

Does the circuit you have constructed meet these requirements?

Project 6

Design and construct a pneumatic system to control the speed of movement in both directions of a sliding glass door. It must be possible to control the movement of the door when approaching it from either side.

Circuit 8 – Speed control using an exhaust restrictor

Construct the circuit shown in Figs. 34 and 35. This is a fully automatic circuit which includes an exhaust restrictor to control the speed of movement of the piston.

To achieve speed control, adjust the exhaust restrictor. Find adjustments of movement are possible.

Figure 34 Circuit 8

Question

Is the in-stroke or the out-stroke being slowed down?

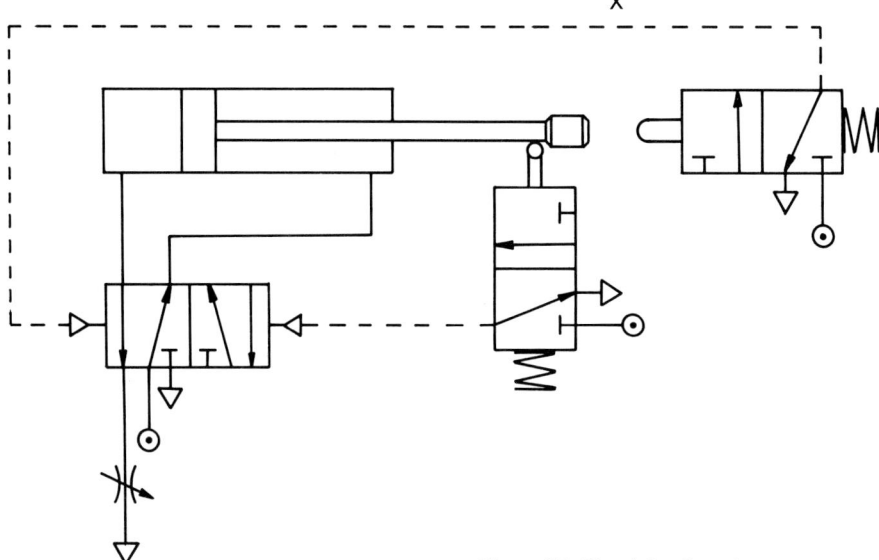

Figure 35 Circuit 8 – Speed control using an exhaust restrictor

Project 7

Insert a 3-port lock down-lever valve at position X in Fig. 35. To switch off this circuit switch over the lock-down lever valve. Note that the piston will always stop in the same position.

Project 8

Place a second lock-down lever valve in the circuit so that you can fully control the position at which the piston stops.

Circuit 9 – Delay circuit

Construct the circuit shown in Fig. 36.

The amount of air being stored in the reservoir is controlled by the flow adjuster.

The purpose of a reservoir in a pneumatics circuit is similar to the purpose of a capacitor in an electric or electronics circuit. It introduces a delay. Measure the times of the maximum and minimum intervals as the flow control regulator is adjusted.

Project 9

Insert a second reservoir in series with the first. How does the maximum time interval compare with that of the original circuit?

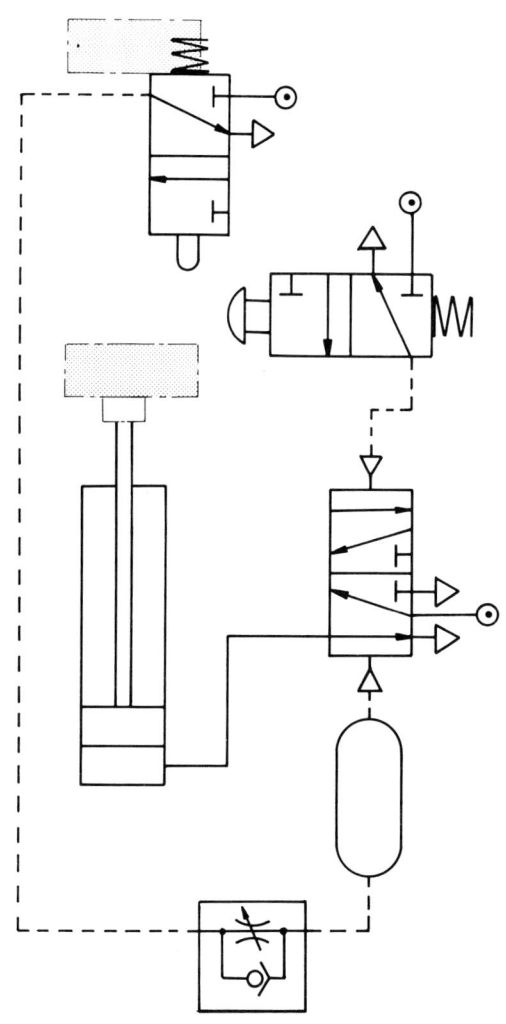

Figure 36 Circuit 9 – Adding a reservoir into the circuit

The circuit in Fig. 36 shows a pneumatically controlled vice or clamping device. The flow regulator would allow an operator to vary the time element in working the vice/clamp. The circuit is semi-automatic, because, after a delay, the piston returns to its negative position.

Circuit 10 – Air bleed occlusion

Construct the circuit shown in Fig. 37.

In the circuits that you have constructed so far, all the pressure-operated valves have required a minimum pressure of 3 bar before change-over could be achieved. This is due to the small areas on which air signals can act in those circuits. Diaphragm valves can react to low air signal pressures.

Figure 38 shows a school-built air bleed which can be attached to a component mounting frame. Partly covering the end of the pipe in the air bleed produces a back pressure sufficient to operate the diaphragm valve (see Appendix 2).

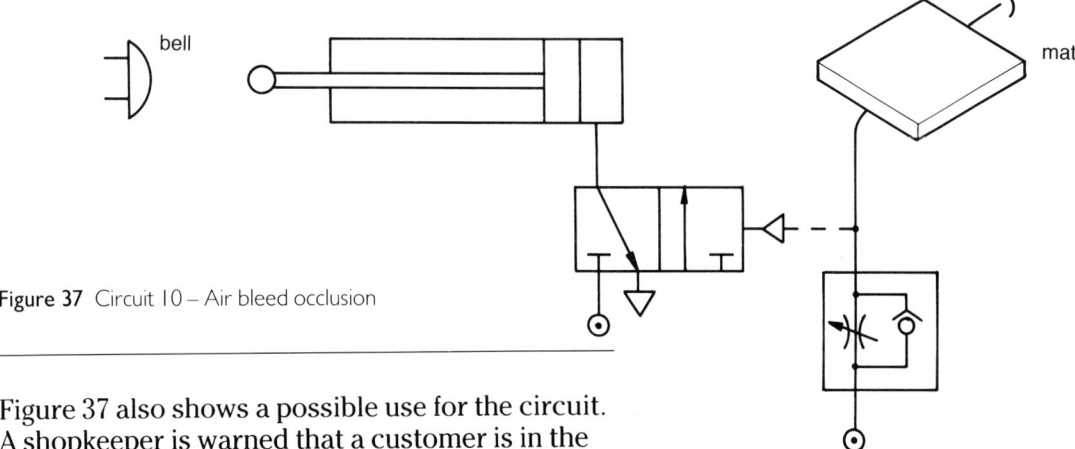

Figure 37 Circuit 10 – Air bleed occlusion

Figure 37 also shows a possible use for the circuit. A shopkeeper is warned that a customer is in the shop when a bell is struck by the end of the piston as the customer steps on a mat, restricting the air bleed.

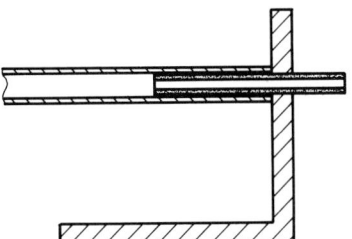

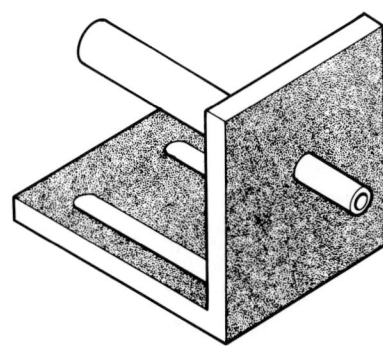

Figure 38 School-built air bleed

Circuit 11 – A circuit with a solenoid valve

Construct the circuit shown in Fig. 39. The circuit illustrates a simple automatic system, which uses a solenoid valve (see Appendix 2) as a control to a 5-port double pressure-operated valve.

Project 9

Modify the circuit to obtain speed control in either direction by introducing a time delay to make the piston stop in either a +ve or a −ve position.

Exercises

Take each of the six situations given in Fig. 40 in turn. For each, draw a circuit diagram of a pneumatic system which could be used as a

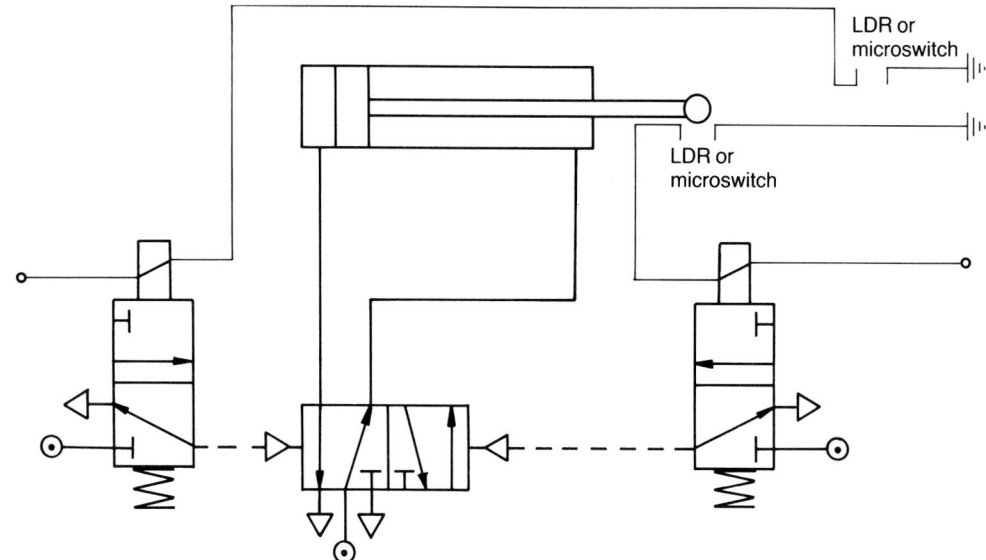

Figure 39 Circuit 11 – Circuit with a solenoid valve

control system. Work freehand or with instruments on square grid paper or on graph paper.

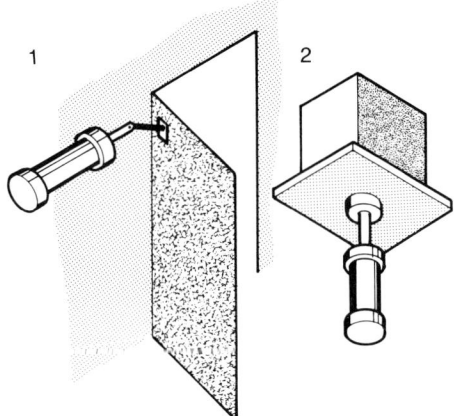

Open and close a door   Lift a heavy object

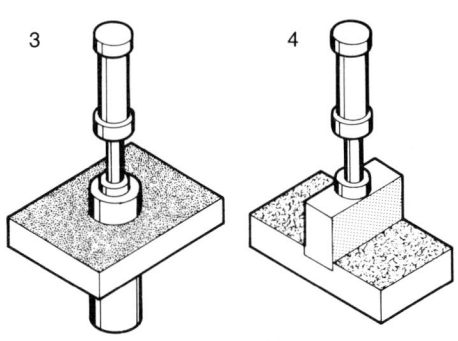

Secure a press-fit   Clamp parts together

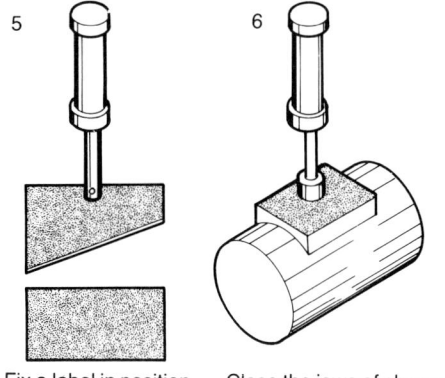

Fix a label in position   Close the jaws of shears

Figure 40 Diagrams showing six industrial applications for pneumatic circuits

Exercises

Note: Exercises 1 to 7 below are Core type exercises. Exercises 8, 9 and 10 are Module type exercises.

1 Draw the symbol for each of the following pneumatic components. Explain briefly its method of operation, and where it could be used in a circuit.
- single-acting cylinder
- shuttle valve
- roller-trip operated valve
- 5-port lock-down lever valve
- diaphragm valve
- 5-port pressure operated valve

2 What is the prime purpose of an exhaust restrictor?

3 Give four examples of the industrial use of compressed air.

4 Calculate the maximum load that can be lifted by a 25 mm diameter × 75 mm stroke cylinder subject to a pressure of $0.4 \, N \, mm^2$.

5 Explain why, when a supply of compressed air is connected directly to both ports of a double-acting cylinder, the cylinder will always out-stroke.

6 Draw two circuits to illustrate the following conditions:
- a double-acting cylinder used where constant positive pressure is not required;
- a double-acting cylinder used where the piston must remain under pressure in the negative and positive positions.

7 Draw and explain a circuit which includes a reservoir and a flow restrictor to produce a time delay.

8 Figure 41 shows a conveyor belt used to carry different sizes of parcel. Using a selection of components from the list below, design a pneumatic circuit which will automatically remove parcels over a certain height from the conveyor belt.

   1 single and 1 double-acting cylinder
   1 flow regulator
   1 air-occlusion device
   2 solenoid valves
   2 light dependent resistors and light source
   1 diaphragm valve
   1 5-port pressure operated valve
   pipes – as needed

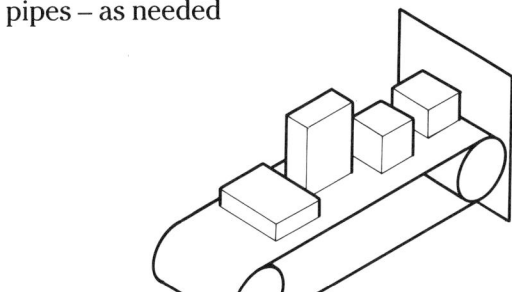

Figure 41 Exercise 8

9 Figure 42 shows a pair of sliding doors of the type found in supermarkets. A sensor 'senses' the approach of a customer and signals for the door to be opened by a pneumatics circuit.

In your answer to this question do *not* draw the electronic circuit.

Design a pneumatic circuit to operate as follows:
- The doors open quickly, remain in the open position for a period of time, then close automatically.

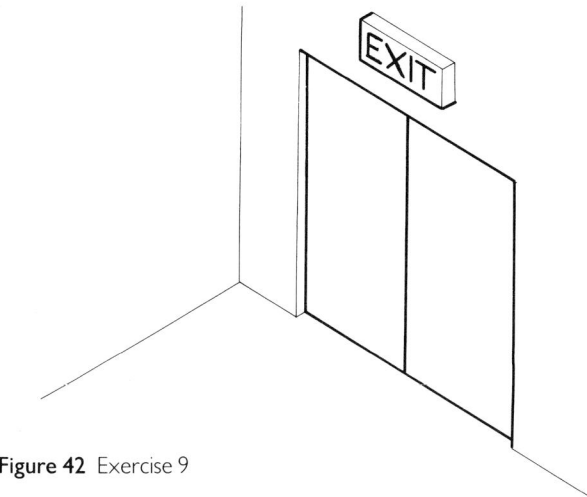

Figure 42 Exercise 9

- A manual over-ride valve is to be fitted. The purpose of this valve is to hold the door open in the event of an emergency.

10 Figure 43 shows the circuit diagram of a pneumatic machine punch. The work-piece is held in place by the flat ends of the rods in cylinders A and B. The end of the piston rod in cylinder C forms the punch.
- Name components A, D, E, F and G.
- State the sequence of events that occurs when valve E is operated.
- State the sequence of events that occurs when valve E is switched back to the un-operated position.

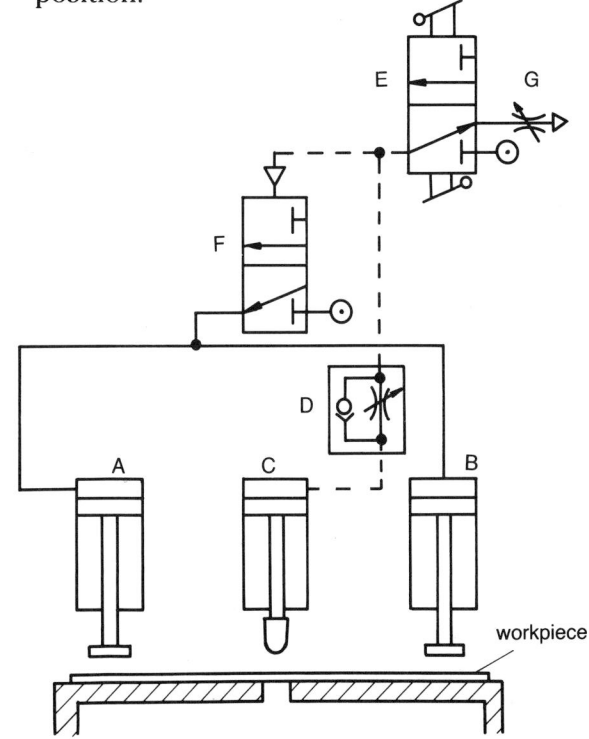

Figure 43 Exercise 10

6 Computers

Introduction

The integrated circuit chips (ICs) in modern microcomputers contain thousands (sometimes millions) of tiny transistors (among other components) arranged in **gates**. Gate circuits either allow or prevent the flow of electric currents. They can be open or shut – ON or OFF. ON or OFF can be represented by **binary** numbers. There are only two digits used in binary, 1 and 0. Binary 1 can stand for ON and binary 0 for OFF.

Binary numbers

Rules for binary numbers

1 They are read from right to left.
2 The right hand digit represents 0 or 1; the second from right represents 0 or 2; the next represents 0 or 4; the next represents 0 or 8, and so on.
This results in the following:

binary   0 = decimal 0
binary   1 = decimal 1
binary   10 = decimal 2 (2 + 0)
binary   11 = decimal 3 (2 + 1)
binary   100 = decimal 4 (4 + 0 + 0)
binary   101 = decimal 5 (4 + 0 + 1)
binary   110 = decimal 6 (4 + 2 + 0)
binary   111 = decimal 7 (4 + 2 + 1)
binary 1000 = decimal 8 (8 + 0 + 0 + 0)
binary 1001 = decimal 9 (8 + 0 + 0 + 1)
binary 1010 = decimal 10 (8 + 0 + 2 + 0)
binary 1011 = decimal 11 (8 + 0 + 2 + 1)
binary 1100 = decimal 12 (8 + 4 + 0 + 0)
binary 1101 = decimal 13 (8 + 4 + 0 + 1)
binary 1110 = decimal 14 (8 + 4 + 2 + 0)
binary 1111 = decimal 15 (8 + 4 + 2 + 1)

Binary numbers may have more than four digits:

Binary 1111 1111 = 128 + 64 + 32 + 16 + 8 + 4 + 2 + 1
$$= \text{decimal } 255.$$

Binary 1001 1100 = 128 + 0 + 0 + 16 + 8 + 4 + 0 + 0
$$= \text{decimal } 156.$$

Binary numbers help to explain how computers work through the ON or OFF nature of their gate circuits.

pulse – binary 1   no pulse – binary 0

Figure 1 Digital binary pulses of current (diagrammatic)

Bits and bytes

Each pulse of electric current ON or OFF is called a **bit** (Binary digit). Eight bits make a **byte** (Fig.2). 1024 bytes make a **kilobyte** (kb). 1024 kb make a **megabyte** (Mb).

When a key on a computer keyboard is pressed, a binary number is sent (in digital electrical pulses) to the **operating system** (OS) of the computer. On a BBC computer, when key **A** is pressed pulses of electrical current representing the number 65 are sent to the OS. The key **a** sends 97. Other keys each send their own personal numbers. On receiving the digitised signal the OS causes the letter Λ (or a) to appear on the computer screen or to be printed by a printer on paper.

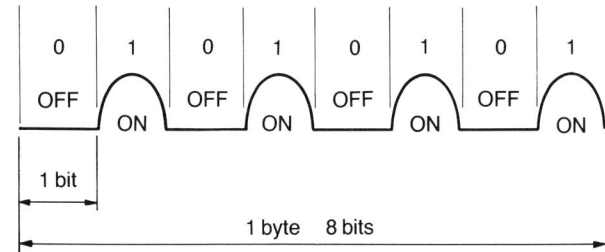

Figure 2 A byte of digital pulses

Figure 3 shows the two signals sent to the OS by the numbers 65 and 97 – letters A and a. The binary number representing each byte is shown above each diagram.

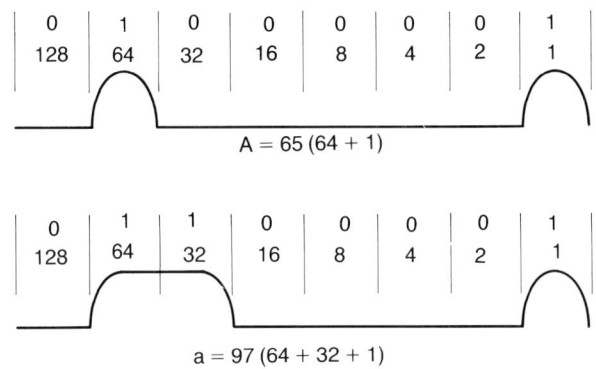

A = 65 (64 + 1)

a = 97 (64 + 32 + 1)

Figure 3 Digital pulses from key **A** and key **a**

Hexadecimal numbers

Hexadecimal numbers are used to show the positions of bytes in the **memory** of computers. Hexadecimal numbers have a base of 16 with numbers 0 to F having the same values as 0 to 15 in decimal. An & in front of a number shows that it is hexadecimal. The first 16 hexadecimal numbers (counting 0 as the first number) have the following decimal equivalents.

Hexadecimal		= decimal	
	0	= decimal	0
&1	=		1
&2	=		2
&3	=		3
&4	=		4
&5	=		5
&6	=		6
&7	=		7
&8	=		8
&9	=		9
&A	=		10
&B	=		11
&C	=		12
&D	=		13
&E	=		14
&F	=		15

If decimal, binary and hexadecimal are compared:

binary 1111   = decimal   15 = hexadecimal F
binary 1111 1111 = decimal 255 = hexadecimal FF

To find the decimal (or *denary*) value of a hexadecimal number, the values of each letter or number must be worked out as follows:

Example 1: &FF

The first F = 15 × 16 = 240
The second F = 15
adding the two together &FF = 240 + 15 = 255

Example 2: &FFF

The first F = 15 × 256 = 3840 (256 = 16^2)
The second F = 15 × 16 = 240
The third F = 15
adding all three, &FFF = 3840 + 240 + 15
= 4095

Example 3: &3E

The first figure 3 = 3 × 16 = 48
The second letter E = 14
Adding, &3E = 48 + 14 = 62

Example 4: &4AF

The first figure = 4 × 256 = 1024
The second letter A = 10 × 16 = 160
The third letter F = 15
Adding, &4AF = 1024 + 160 + 15
= 1199

Exercises

1 What is the decimal equivalent of the following binary numbers:
1101 1000; 1001; 1111 1000 1010; 0001; 1111 0011?
2 Give the numbers: 204; 20; 198; 2045; 17; in binary.
3 Give the decimal numbers for:
&B; &FE; &AO; &B5; &C3; &FFF; &AF4.

Logic gates

A forward-bias transistor circuit (page 41) can be used as a switch. This switch has no moving parts, and as the electrons move through the circuits at a speed of 300 000 kilometres per second the switch operates very quickly. In fact, the switching action of transistor circuits appears to be instantaneous. The tens of thousands of transistor switching circuits in an integrated chip can operate in millionths of a second. The switching action of a transistor can be shown as in Fig. 4.

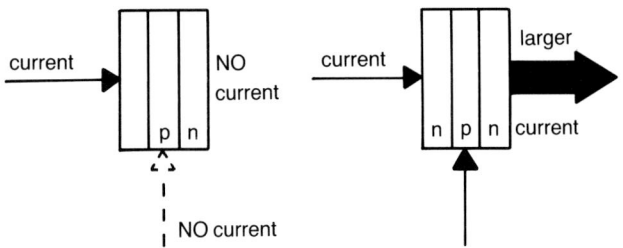

Figure 4 Diagrammatic representation of a transistor as a switch

Logic gates in computers contain large numbers of switching circuits. The switches in the circuits are either ON (binary 1) or OFF (binary 0). Six types of logic gate are found in computer gate circuits – AND, OR, NOT, NAND, NOR and Exclusive OR (EOR) gates. Figure 5 shows, in diagrammatic form, the action of an AND gate and Fig. 6 the action of an OR gate. A NAND gate can be thought of as an AND gate followed by a NOT gate and a NOR gate can be thought of as a NOT gate following an OR gate (Fig. 7).

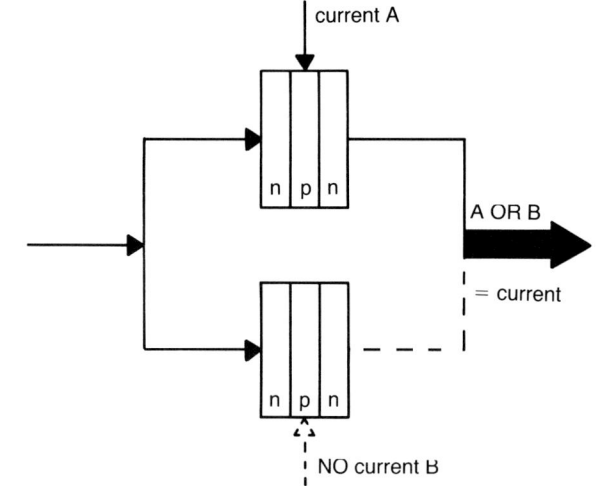

Figure 6 Diagrammatic representation of an OR gate

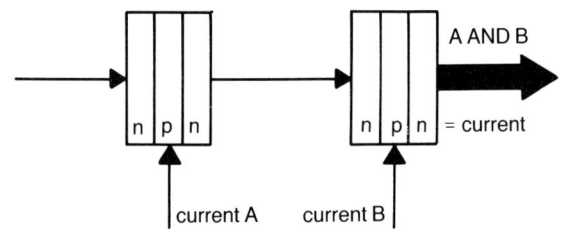

Figure 5 Diagrammatic representation of an AND gate

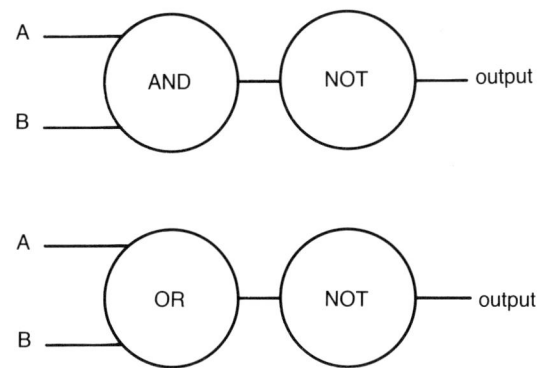

Figure 7 Diagram of action for NAND and NOR gates

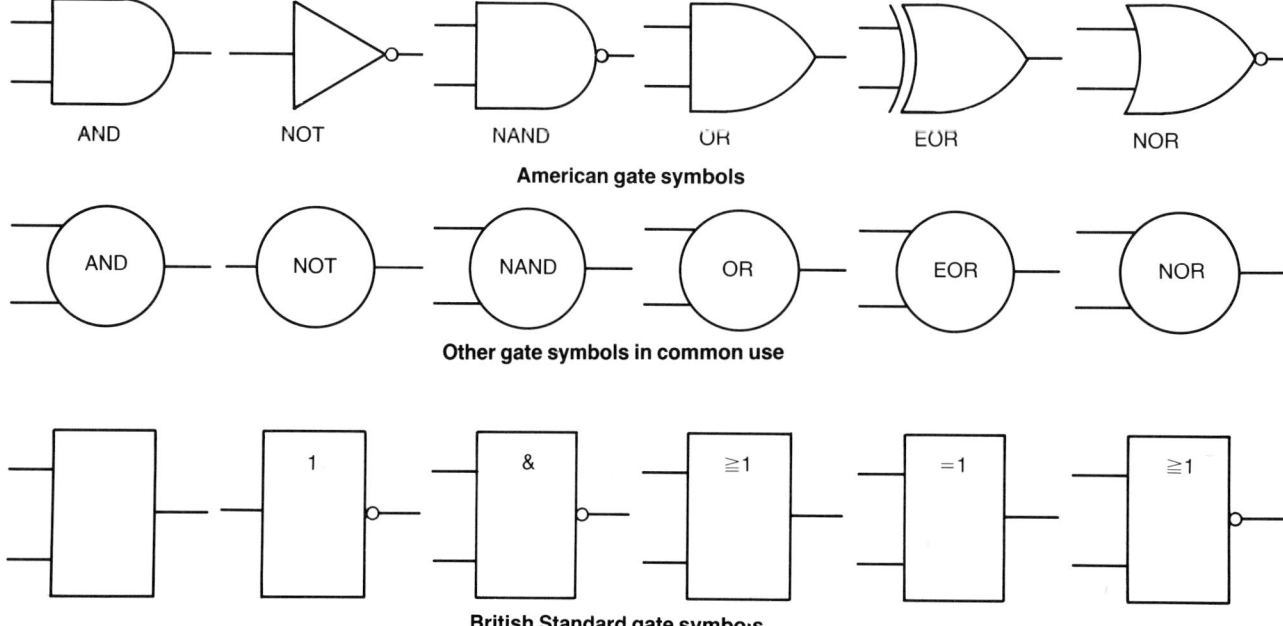

Figure 8 Symbols for diagrams of logic circuits

Circuit diagram symbols for logic gates

A variety of symbols is used for drawing diagrams of logic circuits. See Fig. 8.

Truth tables

Truth tables are used to show what the output from a logic gate will be for any combination of inputs. Each input and the output can be one of two states: ON or OFF, representing binary 1 or 0.

AND gate

In an AND gate, if switch A is ON AND switch B is ON then current passes through the gate. If either A or B is OFF, then no current passes through the gate. The four possible states of an AND gate are given in the truth table:

A	B	Output
0	0	0
0	1	0
1	0	0
1	1	1

OR gate

In an OR gate, if either switch A OR B is ON, then current passes through the gate. If neither A OR B is ON then no current passes through the gate. The truth table for an OR gate will be:

A	B	Output
0	0	0
0	1	1
1	0	1
1	1	1

NOT gate

In a NOT gate, if switch A is ON then the gate is OFF. If switch A is OFF then the gate is ON. A NOT gate therefore has only two states, as shown by its truth table. A NOT gate is sometimes known as an **inverter** (because it 'inverts' the input).

A	Output
0	1
1	0

NAND gate

A NAND gate can be thought of as an AND gate followed by a NOT gate. This gives the following truth table:

A	B	Output
0	0	1
0	1	1
1	0	1
1	1	0

NOR gate

A NOR gate can be thought of as an OR gate followed by a NOT gate:

A	B	Output
0	0	1
0	1	0
1	0	0
1	1	0

EOR gate

In an Exclusive OR gate, if either, but not both, inputs are 1, then the output is 1; i.e. the output is 1 if inputs are *different*. Otherwise the output is 0.

A	B	Output
0	0	0
0	1	1
1	0	1
1	1	0

The operation of gates

When digitised pulses of current are sent through a maze of different types of gates, logical and mathematical operations can be carried out by a computer. But even simple addition problems such as adding two numbers together may need circuits with 12 or 13 gates of different types. The pulses of current switch transistors ON or OFF in the logic gates. The pulses from the output of the gate system represent the sum of the two numbers. Other, more complicated logical and mathematical operations are carried out in a similar manner.

Calculating the effect of a logic circuit
Example

Fig. 9 is a diagram of a simple logic circuit. Calculate the truth table for the output from this circuit.

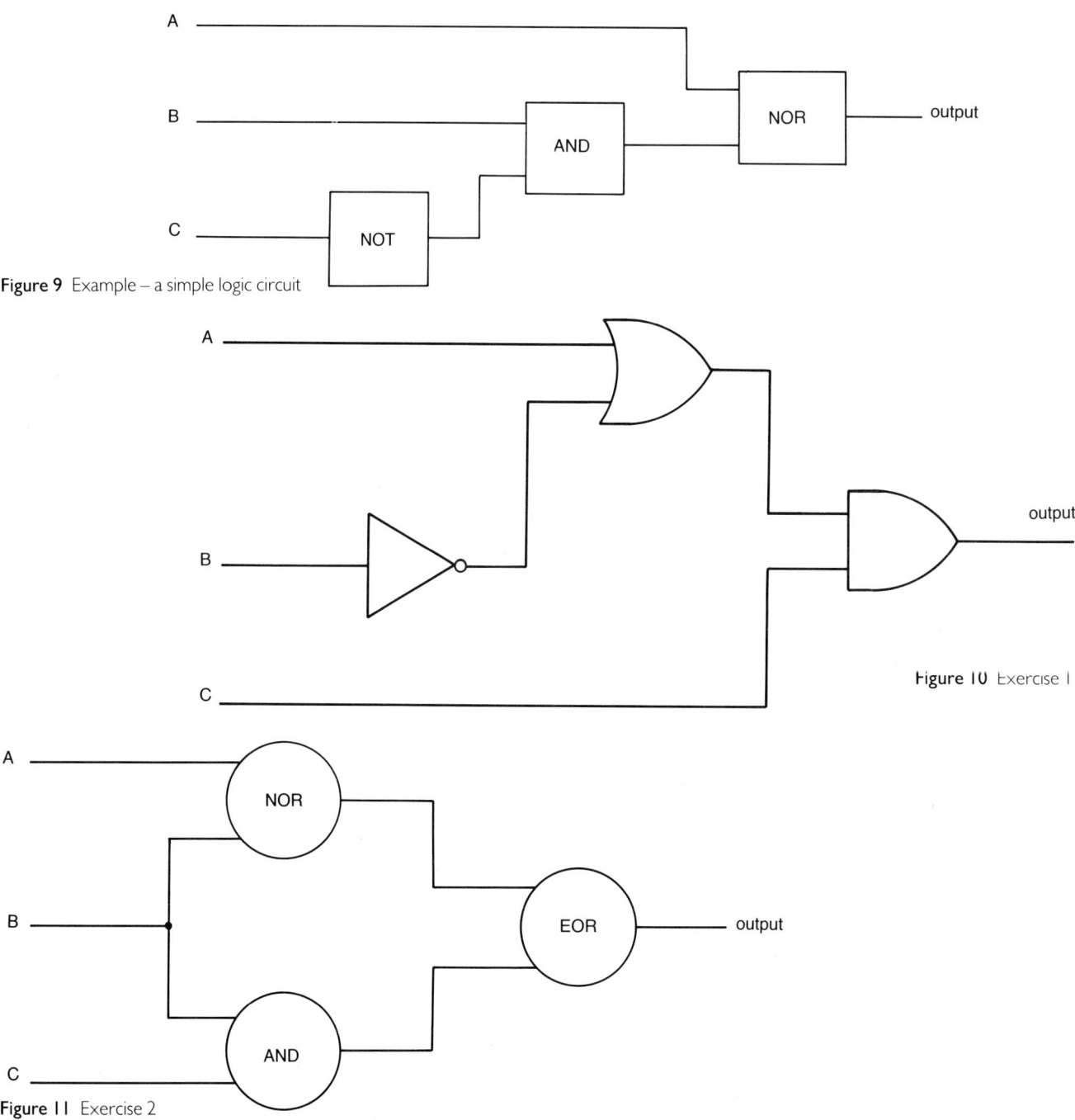

Figure 9 Example – a simple logic circuit

Figure 10 Exercise 1

Figure 11 Exercise 2

Exercises

1 Using symbols from those given in Fig. 8, draw the logic circuit correctly from the details given in Fig. 10. Construct a truth table for the circuit you have drawn.
2 Construct the truth table for the logic circuit given in Fig. 11.

Computer memory

A computer **memory** consists of electronic circuits in integrated circuits with huge numbers of logic gates. These circuits carry out all the calculations and operations of which modern microcomputers are capable. There are two types of computer memory – **Read Only Memory (ROM)** and **Random Access Memory (RAM)**.

ROM – Read Only Memory

ROM can only be *read* by a computer. You cannot write information into a ROM. Memory in ROM chips is permanent. When a computer is switched on, the ROM is ready to be read from. ROM usually contains the operating system (OS) programs by which a computer operates. The programs held in ROM are made available for use when the appropriate commands are typed in on the computer's keyboard. The programs are read by the ROM and operated on when the necessary commands are given.

RAM – Random Access Memory

RAM can be *written* to and *read* from. Information can be typed in from the keyboard or read from a disc system into RAM. When the computer is switched off, all RAM memory is lost. It can, however, be **saved** on a disc or tape, or on some other form of storage system, before switching off.

CPU – Central Processing Unit

The heart of any microcomputer is its processing unit. All signals from the many gates in other circuits are processed by the CPU. Figure 12 shows the relationships between CPU, ROM and RAM.

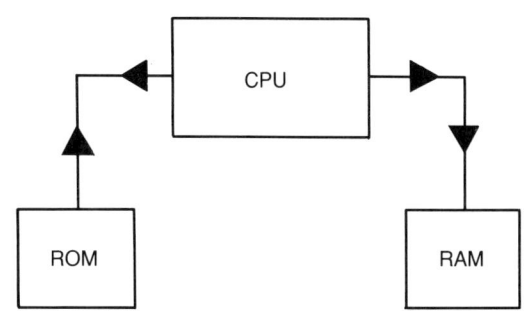

Figure 12 Diagrammatic representation of the relationship between the CPU, ROM and RAM

Timing systems within computers

The unit of timing is the hertz, or cycles per second – abbreviation Hz. A megahertz (MHz) is one million cycles per second.

Without a timing device, or **clock**, a computer could not function. The rate at which the timer operates is in the order of millions of times per second. Each of the numerous operations that must be carried out by the gate circuits of the machine is completed in one of the clock's time intervals. When the first operation is completed, the second starts, then the third and so on. It must be remembered that some seemingly simple computer instructions may require many such operations to work. The human brain cannot detect timing intervals much above 30 Hz. With timing rates of millions per second, the results appear to be instantaneous to the human eye.

A computer is switched on. In the first interval of the clock timing, the operating system (OS) sends a stream of electrical pulses to the first key on the keyboard to find if it has been pressed. If it has not, then in the second clock interval, the second key receives the request. This process goes on at each clock interval until a key is pressed. The process continues over and over again every millionth, two millionths, or four millionths of a second, according to the type of computer being used. When a key is pressed, its personal number is sent back by the OS to the CPU as a byte of information. The OS places this byte into RAM memory in the next clock interval. Then in the next interval the OS program causes the letter of the key to appear on the VDU. Taking the letter A as an example, the computer follows a procedure such as:

Time interval 1 – sense that A has been keyed;
Time interval 2 – OS writes A into RAM memory;
Time interval 3 – OS processes the A in RAM;
Time interval 4 – A appears on the VDU screen.

Microcomputer systems

A microcomputer cannot carry out any processes on its own. It requires **input** and **output** devices to allow information to be passed in and out. Figure 13 shows some input and output devices.

Programs

Programs are *loaded* into a computer either from **discs**, run on disc drives, or from ROMs. Programs on disc are known as **software**. Software in ROMs is sometimes known as **firmware**. Computers and their operating systems (VDU screens, disc drives and other such equipment) are called **hardware**.

Input devices

Keyboard

A keyboard allows information to be *written* to the computer by a person typing in commands, text or other information.

Tape

Plastic tape coated with a magnetisable material allows information to be *written* to and *read* from a computer. Audio tapes are suitable. At present it is more common to use discs (disks).

Discs

Floppy discs are made from thin sheets of plastic (mylar), coated with magnetic oxide. Information is stored as molecular-sized magnetised particles. This information is *read* from a disc by a magnetic sensor in a head which moves across the disc. Discs operate much faster than tapes. They can transfer information between computer and disc at over 20 000 bytes a second.

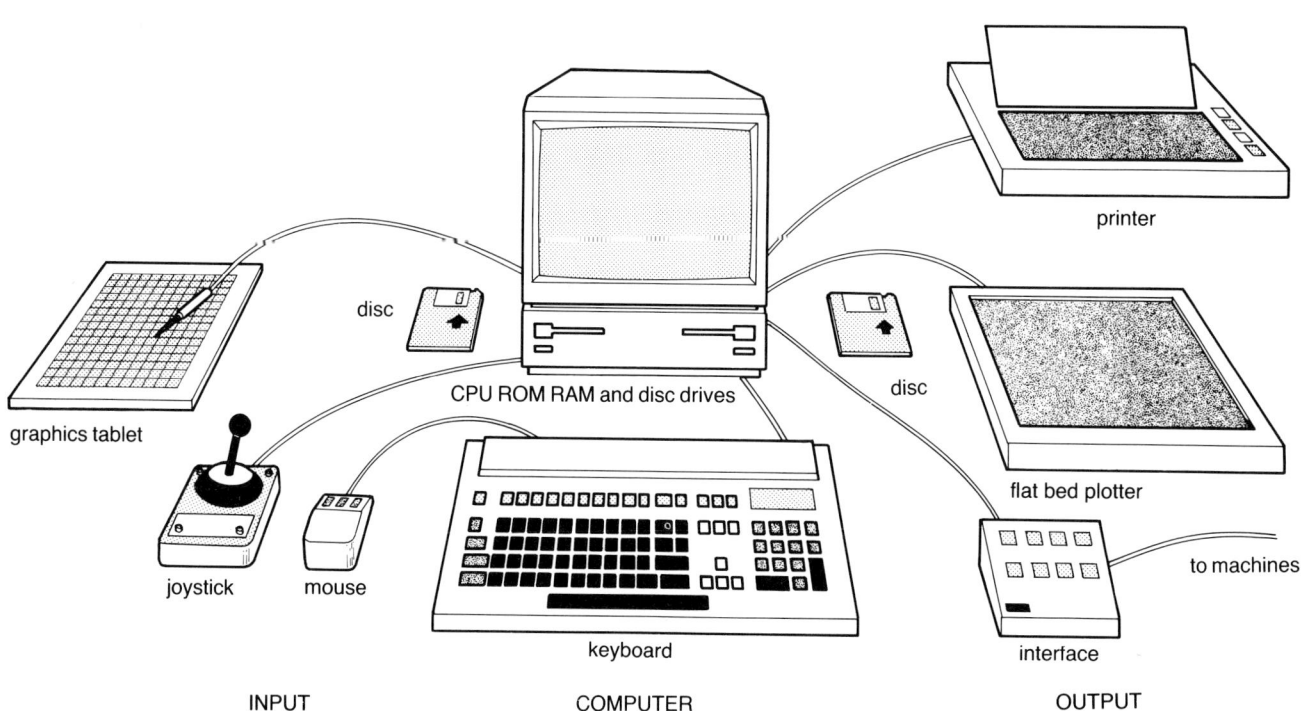

Figure 13 Computer input and output devices

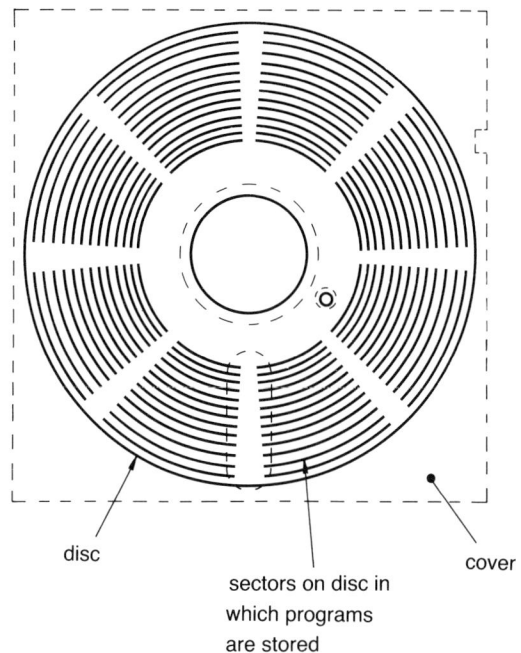

disc

sectors on disc in
which programs
are stored

cover

Figure 14 A floppy disc

Discs are made in three diameters: 5.25 inch
(133 mm), 3.5 inch (89 mm) and 3 inch (76 mm).
5.25 inch discs will hold about 0.5 Mb of
information, 3.5 and 3 inch discs up to 0.8 Mb.
Another type of disc, the **hard disc**, is coming into
more common use. These work at higher speeds
than floppy discs and will hold up to 20 Mb or
more of information.
Note: Tapes and discs allow information written
to RAM to be **saved**. When a computer is switched
off, all information within its RAM is lost. At
switch-off all switches in the gates within RAM go
to 0. When needed, information on tape or disc
can be loaded back into RAM and so used again.

Kimbal tags

These are cardboard tags with holes punched in
them. They are attached to goods in shops for
stock control purposes. A machine reads the holes
as binary code and transfers the information to a
magnetic tape for input to the computer.

Magnetic character readers

A computer can read specially printed characters
such as the numbers on cheques, or bars on a
price code. The characters and bars are printed
with magnetic ink so that the computer can read
them.

Mouse

Some computer programs can be controlled by a
mouse – a small, hand-held block, connected to
the computer by a cable. A ball under the mouse is
connected to two variable resistors. As the ball
rolls over a surface, a cursor on the screen follows
the position of the mouse. Two to four buttons
control the action of the program at the point to
which the cursor is aimed.

Joystick

Some programs can be run by a joystick which
controls a cursor on the VDU screen. Its action is
similar to that of a mouse, except that the cursor
is controlled by movements of a handle on the
joystick.

Figure 15 A joystick

Trackerball

An input device in which the position of a cursor is
controlled by a ball rolled by hand. It works as if it
were on an upside down mouse.

Graphics tablet

A board, under the surface of which is a network of wires, connected to the computer by a cable. A pointing device – a puck with cross-hairs, or a stylus that touches the surface of the board – is used with the board to determine the position of a cursor.

Figure 16 A graphics tablet

Light pen

The pen touches the VDU screen and pulls a cursor over it, so drawing a line that follows the cursor position.

Output devices

VDU screen

A VDU (Visual Display Screen) can be either a monochrome or a colour screen.

Hard copy

This is the name given to copies made on paper of what is shown on a computer's VDU screen. Hard copy is printed by a printer or a plotter attached by a lead to the computer.

Printers

There are two main types. One is the **dot matrix** printer, in which a head containing a number of pins prints patterns of dots through a coated ribbon (Fig. 17). The second is the **daisy wheel** printer in which typefaces are mounted on a wheel-like platter which is revolved to the correct position for printing. A hammer pushes the letters against a ribbon to mark the paper. A third type coming into more common use is the **laser** printer, in which the printing is carried out by a laser beam. Laser printers produce very high quality printing at high speeds.

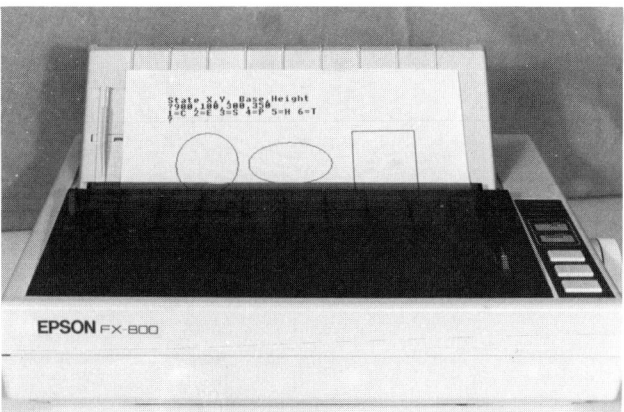

Figure 17 A printer

Plotters

Plotters produce good quality linework. They are available for printing on paper from A4 size upwards. Pens are held in a frame which allows them to move along x and y axes, the x movement drawing vertical lines and the y movement drawing horizontal lines.

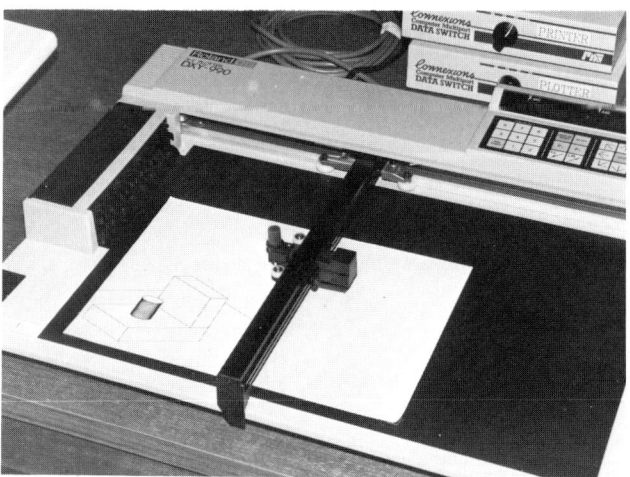

Figure 18 A plotter

Interfaces

Computers can be used to control mechanisms such as robots, vehicles, cutting machinery and

other mechanical devices (see CAM, below).
However, before a computer is programmed for
this form of control, note the following:
1 The current produced by a computer at its
output ports is digital in nature (i.e. the current
flows in pulses).
2 Machinery is controlled by analogue current –
one that can be shown as a continuous, although
perhaps varying, line on a graph.
3 The power output of a computer is too small to
control machinery effectively.
4 There is a danger of feedback power from the
machinery damaging the computer.
5 Computers work by electrons travelling at the
speed of light. Machines are run by mechanical
methods, and work at a much slower rate.

Interfaces make machines and computers more
compatible. They do this by the following means.

Interfaces are designed to amplify and change
the digital form of the output into an analogue
form suitable for controlling machines. They also
balance the different rates at which the machines
and computers work, to block any possibility of
feedback surges. With properly designed
interfaces, a computer becomes a powerful
control tool.
Note: All computers have interface circuits on
their own circuit boards. These are needed to
make sure that the output to printers, disc drives,
loudspeakers and other such parts is in an
analogue form suitable for their correct
functioning.

CAD/CAM

CAD Computer Aided Designing
CAM Computer Aided Machining
CAD is sometimes taken as meaning Computer
Aided Draughting.

Graphic with computers

Types of computing graphics

There are three main types of computer graphics
suitable for use in school technology project work:
1 graphics from programs that you have written;
2 graphics for circuit diagrams, also self-
designed;
3 graphics drawn with the aid of CAD programs.

Example 1:

A simple BBC BASIC II graphics program run on a
BBC B computer (Fig. 19). The value of such
graphics in school technology work is very
limited, but practice in developing this type of
simple program give useful computer graphics
experience.

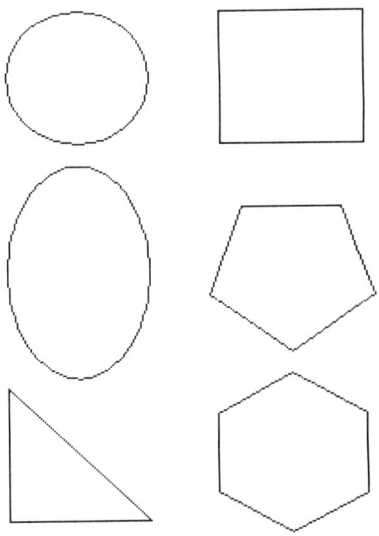

Figure 19 Printer 'printout' of the result of RUNning a BASIC II program for a BBC B computer

Example 2:

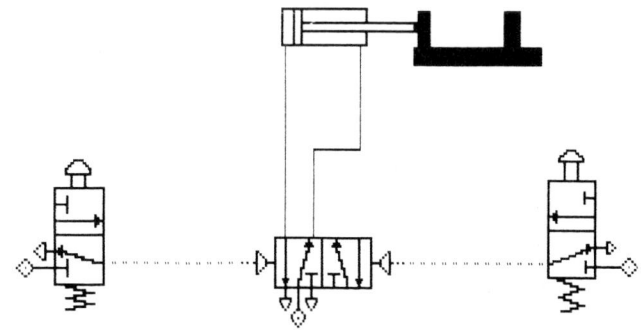

Figure 20 Printer 'printout' of the result of RUNning a BASIC V program on an Archimedes computer.

```
>L.
   10 MODE 12
   15 *SLOAD S4
   20 *SCHOOSE S4
   30 PLOT &ED,500,800
   40 *SLOAD S11
   50 *SCHOOSE S11
   60 PLOT &ED,100,400
   70 *SLOAD S10
   80 *SCHOOSE S10
   90 PLOT &ED,440,393
  100 *SLOAD S9
  110 *SCHOOSE S9
  120 PLOT &ED,900,417
  130 *SLOAD S12
  140 *SCHOOSE S12
  150 PLOT &ED,700,772
  160 LINE 507,800,507,523
  170 LINE 620,800,620,650
  180 LINE 620,650,550,650
  190 LINE 550,650,550,523
  195 VDU23,6,5,5,5,5,5,5,5,5
  196 MOVE 237,486
  198 PLOT &11,197,0
  200 VDU23,6,5,5,5,5,5,5,5,5
  205 MOVE 686,486
  210 PLOT &11,214,0
  220 LINE 645,834,700,834
  230 LINE 645,820,700,820
   >
```

Figure 21 The program which produced the 'printout' of Fig. 20

A circuit diagram from a program written in BBC BASIC V for an Archimedes computer (Fig. 20). Its program is given in Fig. 21. This type of simple design drawing can be programmed on some computers without the need for CAD programs. Note that use has been made of the SPRITE program which is part of this computer's 'firmware'. The two commands *SLOAD and *SCHOOSE, load and run previously designed SPRITES into the program.

Example 3:

A photograph of the screen result of an electronic circuit produced from another BASIC V program on an Archimedes computer is shown in Fig. 22.

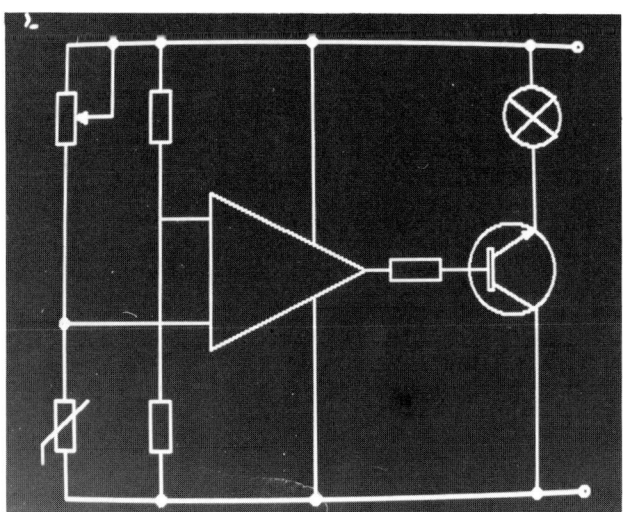

Figure 22 A photograph of the screen results of an electrical circuit program written for an Archimedes computer

Again the SPRITE facility of the computer was used to obtain the symbols for the circuit.

Example 4:

A number of CAD programs can be purchased, examples being Autocad, Bitstik, Novocad, TechnoCAD, Techsoft Designer. There is insufficient space in this book to go into details about such programs. An example of a drawing produced with the aid of a CAD program is given in Fig. 24. The value of using computers in design work is the ease and speed with which changes can be made to drawings produced by computer, and the fact that you need not possess considerable drawing skills to be able to produce clear designs. A procedure such as the following would be suitable for revising drawings on a computer.

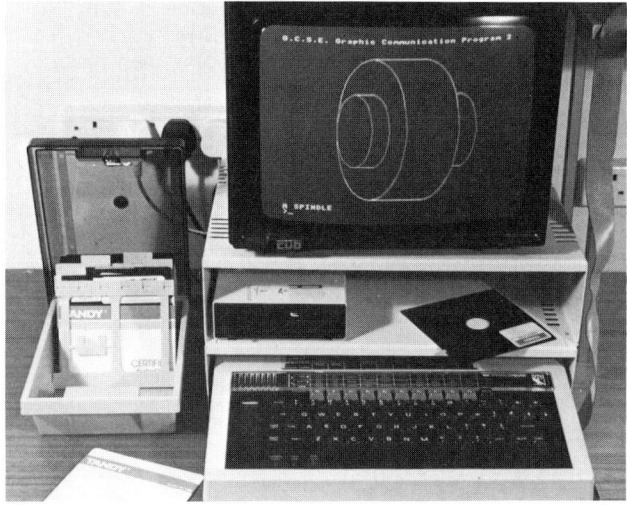

Figure 23 Screen photograph of the results of running another BASIC II program

1 Produce the drawing on the VDU of the computer, either with or without the aid of a CAD program.
2 SAVE the drawing to disc.
3 Print a hard copy of the drawing on a printer or plotter.
4 Use the drawing to assist you to make a model or prototype or the actual finished design.
5 Upon evaluating the resulting design, you may find that modifications are required.
6 LOAD the drawing from disc into the computer.
7 Make any necessary changes to the drawing.
8 SAVE the amended drawing to disc.

You will now have a record of the original design drawing and of the modified drawing. This process

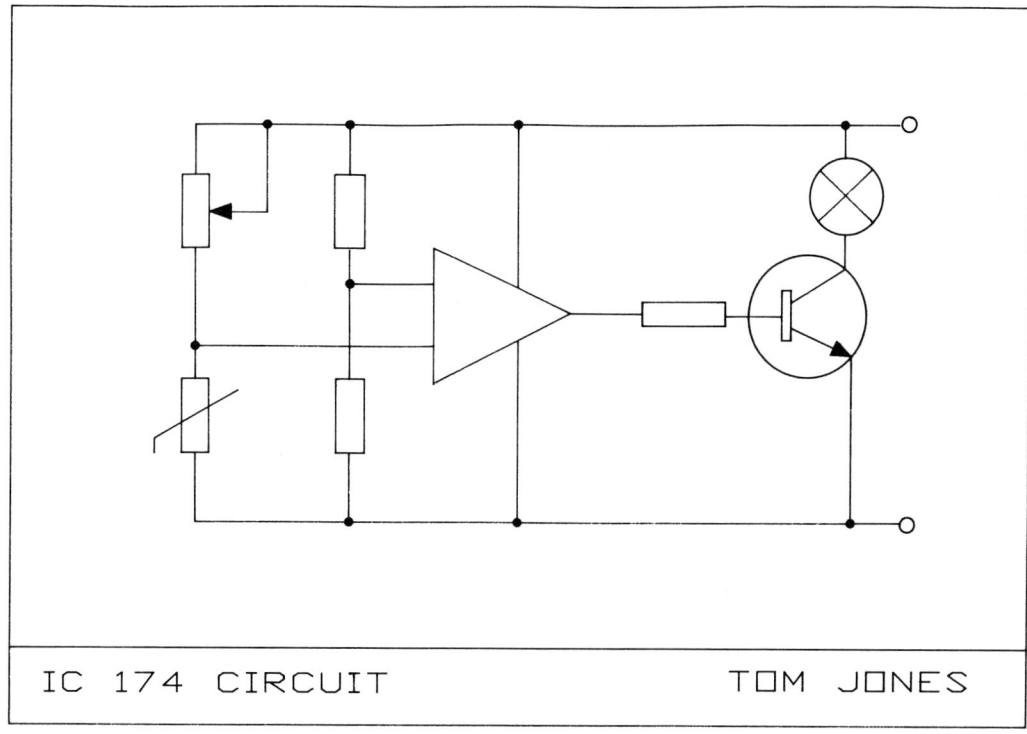

IC 174 CIRCUIT   TOM JONES

Figure 24 A drawing produced by a plotter from a CAD program

can be repeated as many times as you wish. At no stage is it necessary to re-draw the whole drawing – only to modify it. Yet you will have a record of all drawings from the original to the final modified and satisfactory result.

CNC Machining

The initials CNC stand for Computer Numerical Control. CNC programs can be used to operate machine tools, such as lathes and milling machines, from outlines designed at a computer on its VDU screen. CNC machining can be

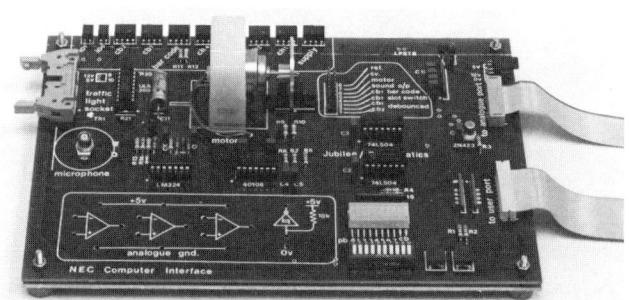

Figure 25 The Applications Board from the National Extension College's Interface Adaptor kit

regarded as a CAM process, although the term CAM covers the whole area of machining under the control of computer programs.

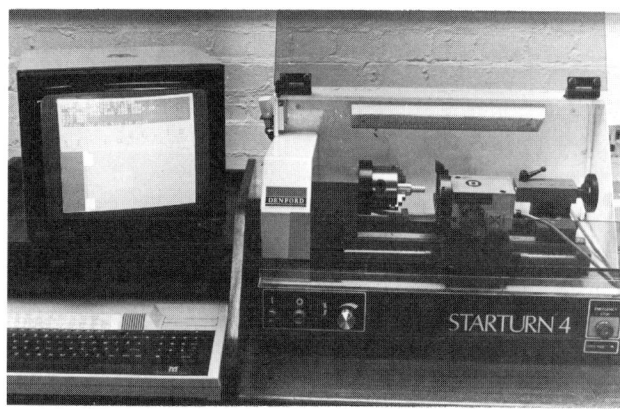

Figure 26 Denford Machine Tool's Starturn 4 with a BBC Master running their CNC Starturn program

Word processing

Some of the notes for the project shown on pages 150 to 155 were typed on a computer with the aid of a word processor. Word processing can be carried out on most microcomputers. Students at school and college should find this form of

Figure 27 Robot arm model from Lego® set 1092, Lego® Interface No: 1093 connected to a BBC computer

'typewriting' fairly easy. What is being typed can be seen on the screen of the computer and can be easily checked, altered, erased or amended. The appearance of notes printed on a printer from a computer provides a clear, easily read record.

Exercises

1 Microcomputers contain a large number of silicon *chips*. What is meant by 'chip'?
2 There are many *logic gates* in a microcomputer. What is meant by the term 'gate'?
3 What are the full names for the following:
■ CPU
■ ROM
■ RAM
■ CAD
■ CAM
4 What do the following terms mean:
■ a bit
■ a byte
■ a kilobyte
■ a megabyte?
5 ■ What number base do the following numerical systems use?
a denary
b binary
c hexadecimal
■ What is the number &A6 in decimal?
■ What is the binary number 1100 0111 in hexadecimal?
■ What is the denary number 1988 in hexadecimal?
6 Many *peripherals* can be added to microcomputers to enhance their performance.
■ Name three *input* devices;
■ Name three *output* devices.
7 If you wish to use a computer to control the action of a mechanical robot, you will need to fit an interface between the computer and the robot.
   Why is it necessary to fit the interface?
8 Why is it necessary to be able to understand binary and hexadecimal numbers if you are designing an interface?
9 Briefly explain the meanings of the two computer terms – *Central Processing Unit* and *Operating System*.

7 Mechanisms

Introduction

Mechanisms such as levers, pulleys, wheels and gears are **machines**. People use machines to take the effort out of work. Machines can be used to change the magnitude and/or direction of forces.

Levers

A lever is a rigid beam pivoted about a point called the **fulcrum** (Fig. 1). It is usually arranged so that an **effort** will move a much larger **load**.

$$\text{mechanical advantage (M.A.)} = \frac{\text{load}}{\text{effort}}$$

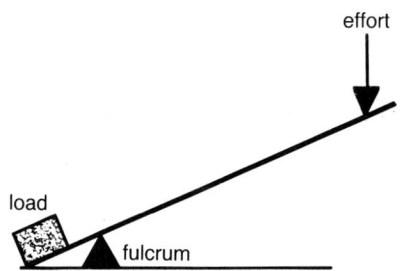

Figure 1 A lever

Example 1: (Fig. 1)

A load of 50N can be moved by an effort of 10N. Calculate the M.A.

$$\text{M.A.} = \frac{\text{load}}{\text{effort}} = \frac{50}{10} = 5$$

$$\text{velocity ratio (V.R.)} = \frac{\text{distance moved by effort}}{\text{distance moved by load}}$$

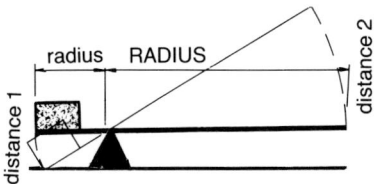

Figure 2 Calculating velocity ratio (V.R.) of a lever

Example 2: (Fig. 2)

The load is moved through 95 mm when the effort moves through 760 mm. Calculate the velocity ratio.

$$\text{V.R.} = \frac{760}{95} = 8$$

Note: The V.R. of a lever can also be calculated by:

$$\text{V.R.} = \frac{\text{RADIUS}}{\text{Radius}}$$

The **efficiency** of a machine is measured by:

$$\text{efficiency} = \frac{\text{M.A.} \times 100\%}{\text{V.R.}}$$

Thus, combining Examples 1 and 2,

$$\text{efficiency} = \frac{5 \times 100}{8} = 62.5\%$$

Note: No machine can be 100% efficient. From the three examples given above:

Work done *by* the effort = force × distance moved
= 10N × 0.76 m
= 7.6 joules

Work done *on* the load = force × distance moved
= 50N × 0.095 m
= 4.75 joules

$$\text{Efficiency} = \frac{\text{work done on the load}}{\text{work done by the effort}} \times 100\%$$

$$= \frac{4.75}{7.6} \times 100\%$$

$$= 62.5\%$$

$$\text{or:} \frac{\text{energy got out of the machine}}{\text{energy put into the machine}}$$

Classes of lever

There are three classes or kinds of lever, depending on the relative positions of the load, the effort and the fulcrum.

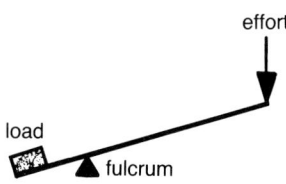

Figure 3 First Class Lever

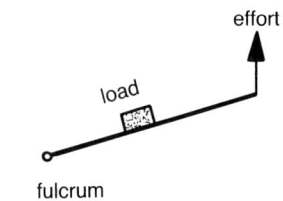

Figure 4 Second Class Lever

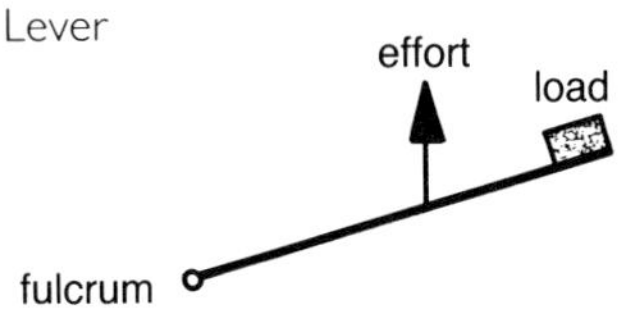

Figure 5 Third Class Lever

1 *First class lever* (Fig. 3) Examples – claw hammer, parcel trolley, tin opener.
2 *Second class lever* (Fig. 4) Examples – wheel barrow, microswitch, jar/bottle opener.
3 *Third class lever* (Fig. 5) Examples – tweezers, a shovel when used for lifting earth.

Project

Design and make a testing device which shows the differences between the three types of lever. Figure 6 shows a suggestion for a first class lever.

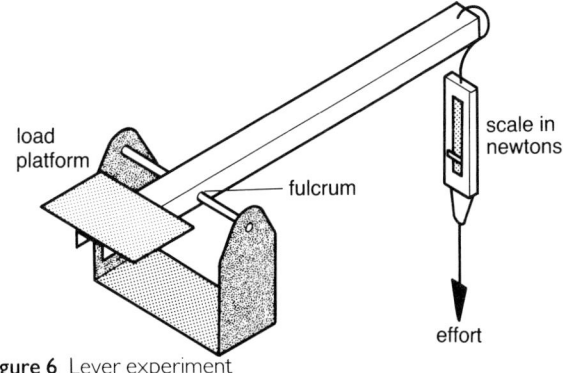

Figure 6 Lever experiment

Bell-crank lever

A car brake pedal works on a lever principle. The bell-crank lever (Fig. 7) gives an increase in leverage *and* a change in direction. Its velocity ratio is given by

$$\frac{\text{distance 2}}{\text{distance 1}}$$

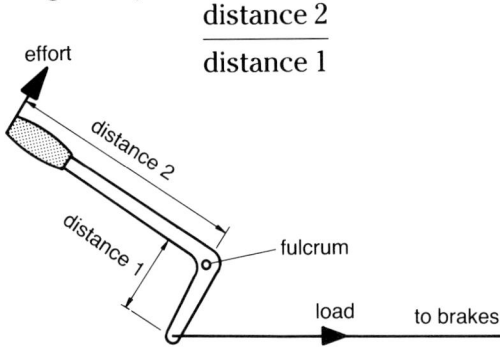

Figure 7 Example of a bell-crank lever – the brake pedal of a car

Moments of force

A **moment** is a turning force. In the case of the see-saw (Fig. 8):

moment = force (load) × distance from the fulcrum

When the see-saw is in **equilibrium** (i.e. balanced) clockwise moments must equal anti-clockwise moments.

Figure 8 Moments – a see-saw

Exercise:

How far from the fulcrum must the 450N load in Fig. 9 be placed to achieve equilibrium?

clockwise moment = anti-clockwise moment

$$200 \times 6 = 450\text{N} \times \text{distance}$$

$$\therefore \quad \text{distance} = \frac{1200}{450} = 2.66\,\text{m}$$

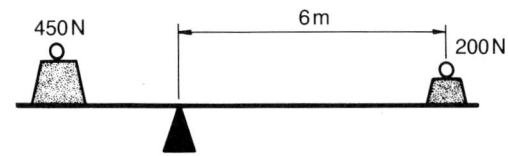

Figure 9 Exercise – moments

Pulleys

The central support of a pulley acts like the fulcrum in a first class lever. In Fig. 10:
1 the load equals the effort;
2 the distance moved by both load and effort will be the same.
    Thus, M.A. and V.R. = 1.

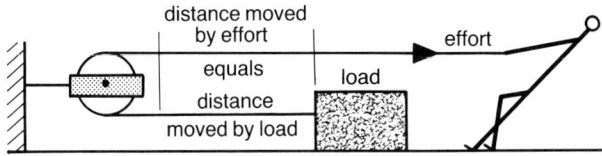

Figure 10 Pulley system

    By changing the system so that the pulley block moves (Fig. 11) both M.A. and V.R. become equal to 2. The person applying the effort in Fig. 11 will have to move twice as far as the person in Fig. 10, because the sides of the pulley share the total movement equally between them.

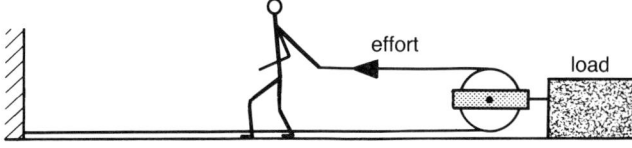

Figure 11 Pulley system

Projects

1 Design and construct a single movable pulley system which shows that both M.A. and V.R. = 2, no matter what the load.

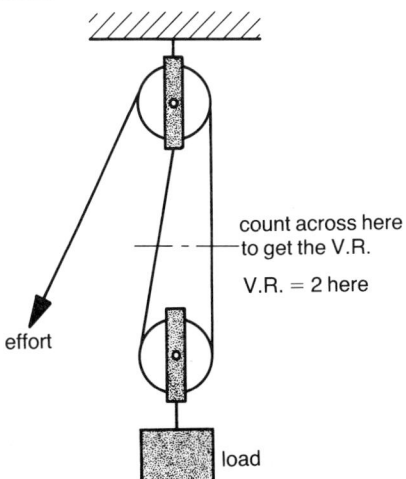

Figure 12 Project – pulleys

2 Construct the double -pulley system shown in Fig. 12. Estimate its V.R. Now experiment to confirm your estimate.

Pulleys and belts

Pulley belts may be made from a variety of materials; examples are leather, woven fabric, leather/nylon, rubber, sintered metal/rubber.

Flat belts

Flat belts (Figs 13, 14) can be used for crossing or reversing drives. Figure 15 shows how flat belts can be kept tight.

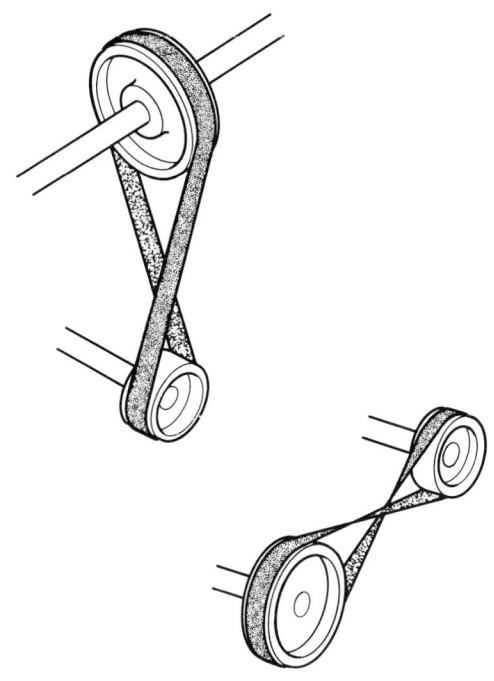

Figure 13 Flat belts on pulleys

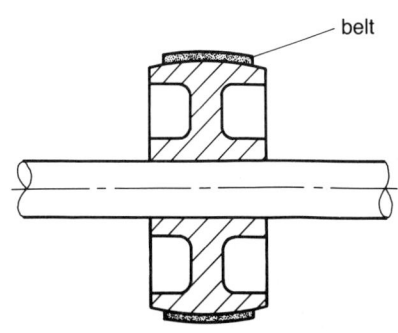

Figure 14 Position of a flat belt on a pulley

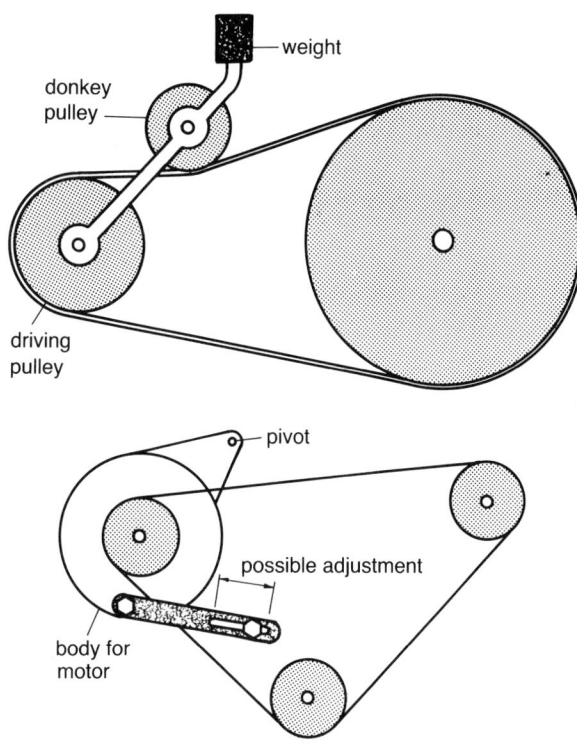

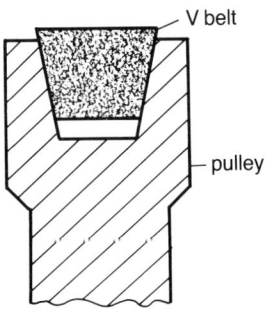

Figure 15 Methods of keeping pulley belts tight

Vee belts

V belts grip the side walls of the grooves in their pulleys, giving a better grip than flat belts (Fig. 16).

Figure 16 V belt in its pulley groove

Pulley cones

Sets of pulley **cones** are fitted to wood and metal lathes and drilling machines.

Project

Make a sketch of a set of pulley cones. Show:
 1 which pulley cone is the driver (attached to the motor);

2 which pair of pulleys produces maximum speed, and which produces minimum speed. Explain your choices.

Chain and sprocket

Chain and sprocket, toothed belts or gears are used where speed control is needed.

Belt-driven machinery provides an in-built safety factor. The belt will slip if a machine is over-loaded. However, a chain drive cannot slip.

Wheel and axle

A small effort applied to the wheel will raise a much heavier load attached to the axle (Fig. 17).

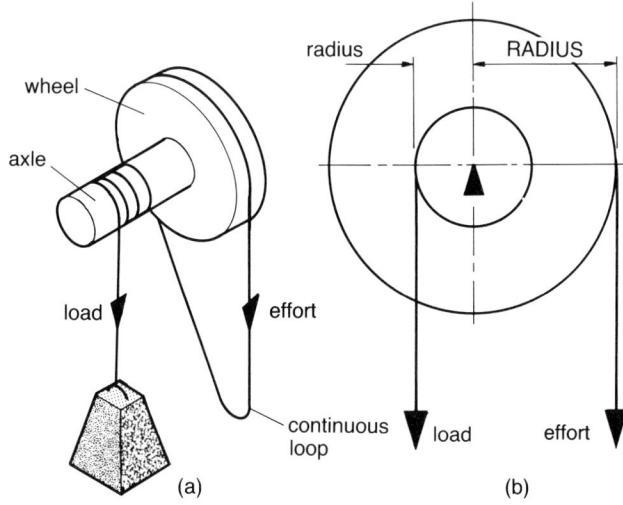

Figure 17(a) Wheel and axle system **(b)** To achieve equilibrium in a wheel and axle system

When a wheel and axle system is in equilibrium:

clockwise forces = anti-clockwise forces

So,   RADIUS × effort = radius × load.

This may also be expressed as:

$$M.A. = \frac{load}{effort} = \frac{RADIUS}{radius}$$

Example: If the wheel has a diameter of 280 mm and the axle a diameter of 80 mm, calculate the V.R., M.A. and efficiency of the system.

$$M.A. = \frac{RADIUS}{radius} = \frac{140}{40} = 3.5$$

$$V.R. = \frac{\text{distance moved by effort}}{\text{distance moved by load}}$$

$$= \frac{\text{circumference of wheel}}{\text{circumference of axle}}$$

$$= \frac{2 \times \pi \times RADIUS}{2 \times \pi \times radius}$$

$$= \frac{RADIUS}{radius}$$

$$= \frac{140}{40} = 3.5$$

$$\text{Efficiency} = \frac{M.A.}{V.R.} \times 100 = \frac{350}{3.5} = 100$$

No system can be 100% efficient. These results are for a system in equilibrium.

Pulleys and sprockets

In a pulley and belt system:

$$V.R = \frac{\text{distance moved by effort}}{\text{distance moved by load}}$$

$$= \frac{\text{no. of revs of driving pulley}}{\text{no. of revs of driven pulley}}$$

$$OR = \frac{\text{diameter of driven pulley}}{\text{diameter of driving pulley}}$$

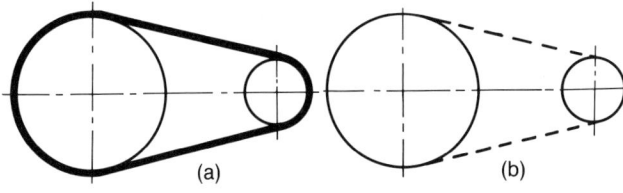

Figure 18(a) Pulley and belt system (b) Sprocket and chain system

For a chain and sprocket system:

$$V.R. = \frac{\text{no. of teeth on driven sprocket}}{\text{no. of teeth on driving sprocket}}$$

Example: In Fig. 18(a), if the driven pulley has a diameter of 60 mm and the driving pulley has a diameter of 30 mm, what is the V.R?

$$V.R. = \frac{\text{diameter of driven pulley}}{\text{diameter of driving pulley}}$$

$$= \frac{60}{30}$$

$$= 2$$

Project

Construct the pulley and belt and sprocket and chain systems shown in Fig. 18. Rotate the smaller diameter sprocket or pulley and check the answer given above.

Revolutions per minute (rpm)

For pulley and belt systems:
rpm of driven pulley × diameter of driven pulley = rpm of driving pulley × diameter of driving pulley.
For chain and sprocket systems:
rpm of driven sprocket × no. of teeth on driven sprocket = RPM of driving sprocket × no. of teeth on driving sprocket.

Example: Imagine the chain and sprocket system shown in Fig. 18(b) is on a bicycle. If the chain wheel sprocket has 54 teeth and is rotating at 45 rpm, and the rear wheel sprocket has 18 teeth, calculate the rpm of the rear wheel.

$$\text{rpm of driven sprocket} = \frac{45 \times 54}{18} = 135$$

Exercise: If the diameter of the rear wheel of the bicycle is 0.7 m, calculate the speed of the bicycle in kilometres per hour.

Gears
Spur gears

Spur gears are used in light duty applications. Their drive shafts run in parallel. Figure 19 shows part of two meshing spur gears.

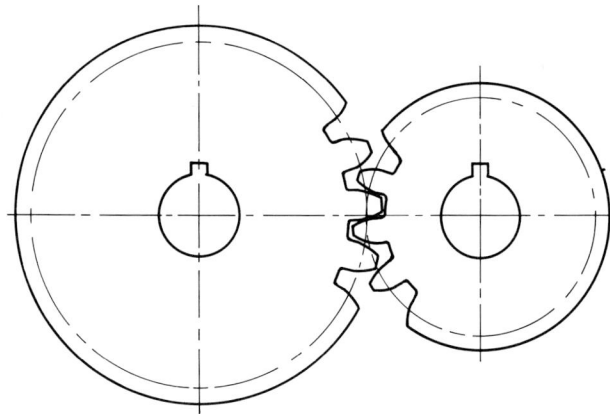

Figure 19 Meshing spur gears

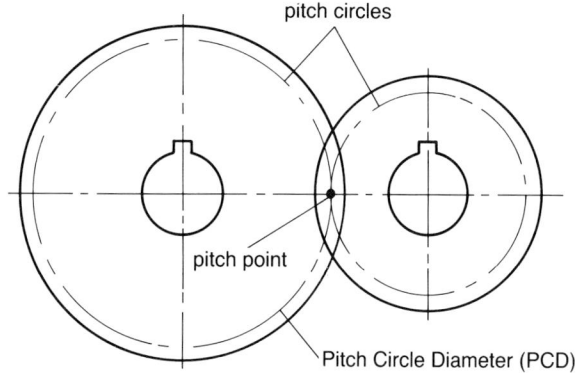

Figure 20 British Standard conventional drawing of meshing spur gears

Helical gears

Helical gears are used for high-speed applications. Some of the load is transferred sideways to the drive shafts and suitable bearings must be used. Their drive shafts run in parallel.

Figure 21 A Meccano helical gear system with shafts at right angles

Figure 22 A contrate gear – the Meccano equivalent to a bevel gear

Spiral gears

The drive shafts of spiral gears are not parallel and these gears are often used to transfer rotation through 90°.

Worm gears

Worm gears allow large reductions in the speeds of drive to driven shafts to be made. The shafts are usually at right angles. The drive shaft turns the worm gear. If the worm is placed in the driven shaft, the system will lock up. Because of this, worm gears can be used as fail-safe gears in lifting devices.

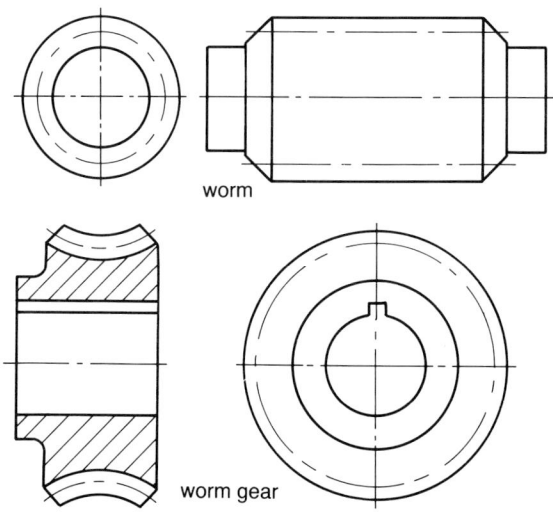

Figure 23 British Standard conventional drawings of worm gears

Figure 23 shows British Standard conventional drawings of a worm and a worm gear.

Figure 24 A Meccano worm gear system

Bevel gears

The drive shafts of bevel gears are usually at right angles, but may be at other angles. Bevel gears are commonly used in the driving mechanisms of the axles of rear wheel drive cars.

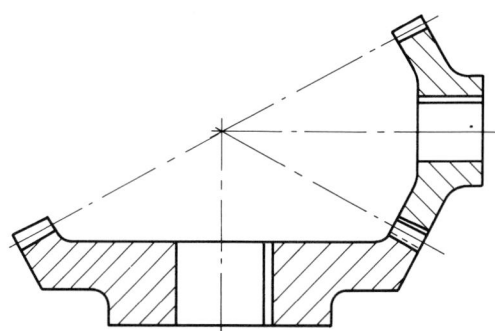

Figure 25 British Standard conventional drawing of bevel gears

Rack and pinion

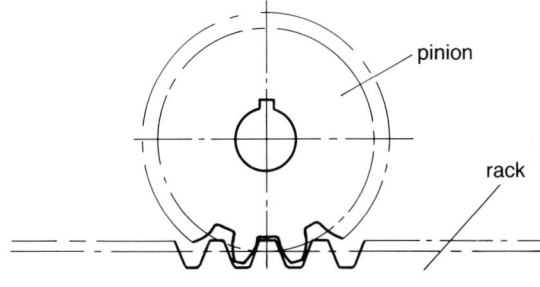

Figure 26 A rack and pinion system

In rack and pinion gears, either the gear rotates and drives the rack, or the rack moves along a straight line and drives the gear.

Changing speeds

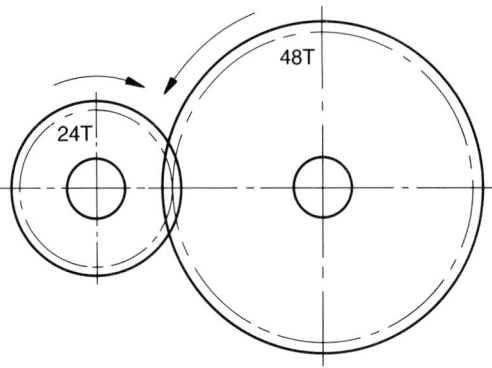

Figure 27 Gear speed change

Project 1

Construct the gear train shown in Fig. 27, with the driven gear having twice as many teeth as the driving gear.
1 For one revolution of the driven gear, how many revolutions does the driving gear make?
2 Now divide the number of teeth on the driven gear by the number of teeth on the driving gear.

The results of 1 and 2 are the same because either method can be used to find the velocity ratio (V.R.) of the gear system.

$$\text{V.R. (1)} = \frac{\text{distance moved by effort}}{\text{distance moved by load}}$$

$$= \frac{\text{number of revolutions of driving gear}}{\text{number of revolutions of driven gear}}$$

$$= 2$$

$$\text{V.R. (2)} = \frac{\text{number of teeth on driven gear}}{\text{number of teeth on driving gear}}$$

$$= \frac{48}{24}$$

$$= 2$$

The **gear ratio** is the same as the velocity ratio.

Project 2

Now make the smaller gear the driven gear. What is the gear ratio now?

$$\text{V.R.} = \frac{\text{distance moved by effort}}{\text{distance moved by load}}$$

$$= \frac{\text{number of revolutions of driving gear}}{\text{number of revolutions of driven gear}}$$

$$= \frac{1}{2}$$

$$\text{Gear ratio} = \frac{\text{no. of teeth on driven gear}}{\text{no. of teeth on driving gear}} = 1:2$$

Torque

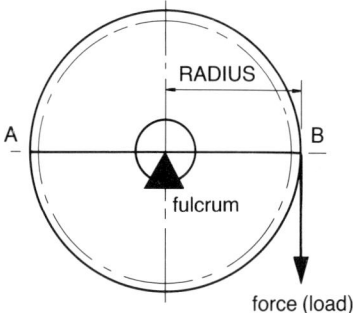

Figure 28 Torque

$$\text{torque} = \text{force} \times \text{radius}$$

It is measured in newton metres (Nm).

In Fig. 28, AB forms a lever with its fulcrum at the centre of the gear and the ends the gear teeth. The load is at one end and the effort at the other. If the force remains constant:

if the radius decreases then the torque decreases;

if the radius increases then the torque increases.

In gear trains, the loads where the two gear teeth mesh are opposite and equal. To calculate the M.A. of the gear train shown in Fig. 29(a), if the driving gear has 25 teeth and the driven gear 50 teeth:

$$\text{M.A.} = \frac{\text{load torque}}{\text{effort torque}}$$

$$= \frac{\text{force} \times \text{RADIUS}}{\text{force} \times \text{radius}}$$

$$= \frac{\text{RADIUS}}{\text{radius}}$$

$$= \frac{\text{number of teeth on driven gear}}{\text{number of teeth on driving gear}}$$

$$= \frac{50}{25}$$

$$= 2$$

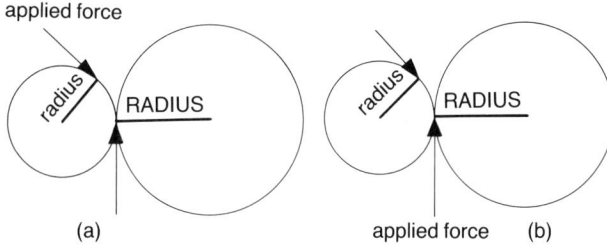

Figure 29 Mechanical Advantage (M.A.) of gear systems

Simple gear train with idler gears

An **idler gear**:
1 adds distance between two spindles;
2 reverses the direction of the driven gear;
3 does not affect the final gear ratio of driving: driven, even when several idler gears are in the gear train.

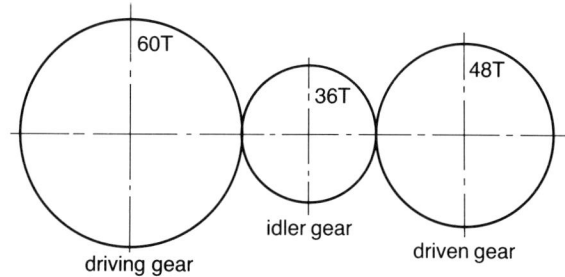

Figure 30 Simple gear train with idler gear

Compound gear trains

In compound gear trains, the middle gears share the same shaft. They do alter the final gear ratio.

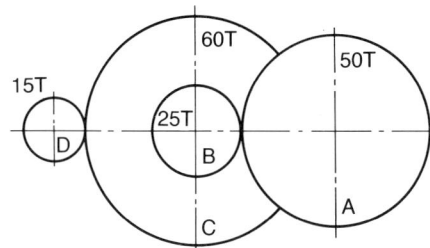

Figure 31 Compound gear train

In the gear train of Fig. 31, the overall gear ratio (A:D) is:

$$V.R. = \text{gear ratio} = \text{ratio of A:B} \times \text{ratio of C:D}$$
$$= \frac{50}{25} \times \frac{60}{15} = 8:1$$

or eight revolutions of the driving gear to one revolution of the driven gear.

Exercise

If D is the driving gear and rotates at 800 RPM what is the RPM of gear A?

$$\text{Gear ratio} = V.R.$$

The gear ratio reduction is 8:1. Velocity of driving gear is 800 rpm.

$$\text{Velocity of driven gear} = \frac{1 \times 800}{8} = 100 \, \text{rpm}$$

Wedges

The smaller the angle between the two faces of a wedge the greater is the force that it can exert for a given applied force, although the distance through which it can cause movement will be less. A wedge can be regarded as an inclined plane.

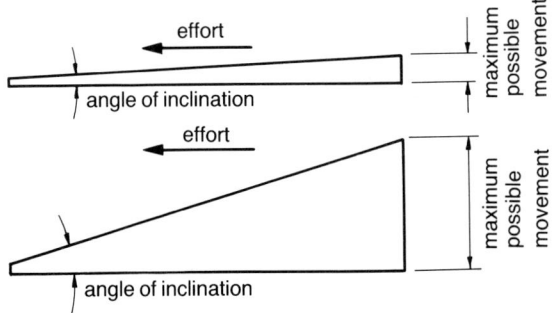

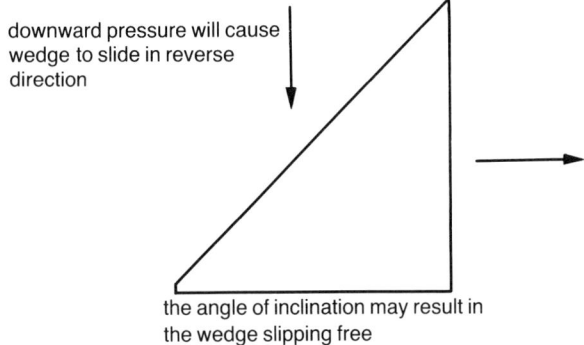

downward pressure will cause wedge to slide in reverse direction

the angle of inclination may result in the wedge slipping free

Figure 32 The action of wedges

Screw threads

Screw threads work in the same way as wedges. An inclined plane wrapped round a cylinder is a helix.

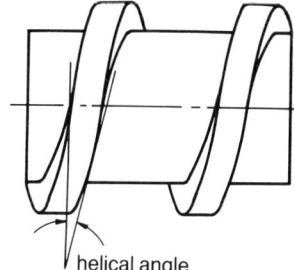

Figure 33 Helical angle · helical angle

A small helical angle produces threads in which the teeth are close together: the screw thread is said to have a fine **pitch**. Screw threads made in weak materials, such as cast iron or chipboard, require coarse thread (in which the teeth are more widely spaced).

Exercise: Examine and compare different forms of screws, such as:
1 woodscrews compared with chipboard screws;
2 steel bolts compared with screwed studs;
3 other screws.

Forms of screw thread

Figure 34 shows some common forms of screw thread.

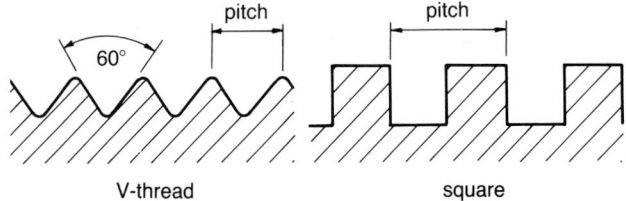

V-thread · square

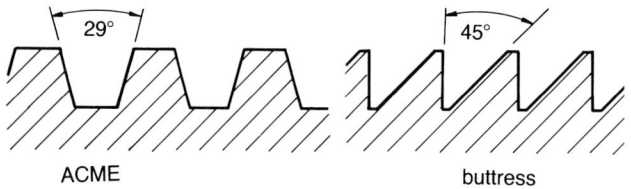

Figure 34 Types of screw threads

☐ **V-thread screws:** used for general construction purposes.

☐ **Acme and square threads:** G-cramps, lathe lead screws.

☐ **Buttress threads:** vice and press screws.

The screw jack

If the handle of a screw jack is turned through one complete revolution, the jack will lift a load the height of the lead of the screw.

In a *single-start* screw, lead and pitch are equal. In a screw with more than one start:

$$lead = pitch \times number\ of\ starts$$

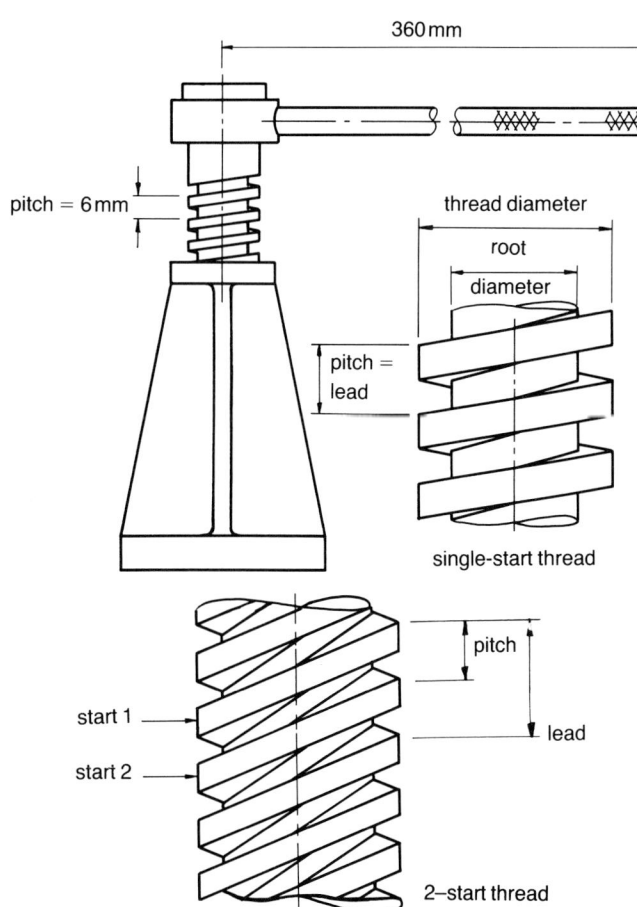

Figure 35 Screw jack

In the example in Fig. 35, pitch = 6 mm. If it is a 2-start screw:

$$lead = 6 \times 2 = 12\,mm$$

Example

Calculate the V.R. of the jack screw in Fig. 35. The handle is 360 mm long. In one revolution the end of the handle moves $2\pi \times$ radius

$$= 2 \times 3.142 \times 360$$

$$= 2262.2\,mm$$

The distance moved by load = 6 mm

$$V.R. = \frac{distance\ moved\ by\ effort}{distance\ moved\ by\ load} = \frac{2262.2}{6} = 377$$

Exercise: If the jack has an efficiency of 50% and is to lift a load of 1000 N, calculate:

■ the M.A.;
■ the effort needed to raise the load.

Friction force

To make a block move, you must apply a force. Equal and opposite forces act on the block at right angles to the surfaces in contact.

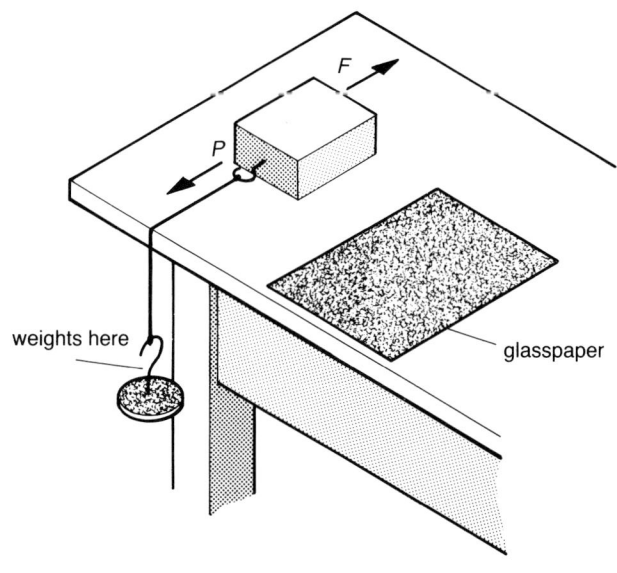

Figure 36 Experiment – friction force

Project

Using the equipment shown in Fig. 36 conduct the following experiment. The wooden block has one side polished and one side covered with glasspaper. Make notes of your findings and of the conclusions you draw.
1 Place the block of wood polished side down.
2 Add masses one by one until the block just begins to move and will continue moving slowly.
3 Turn the block glasspaper side down onto the sheet of glasspaper. Add masses until the block just begins to move and will continue moving slowly.

The **frictional force** F, prevents the block from moving in the early stages of the experiment.

F acts in a direction opposite to the pulling force P. If perfectly smooth surfaces were possible there would be no friction (F would be 0). The block would move when the smallest possible force P was applied.

Since no surface is perfectly smooth, frictional forces are always greater than zero. With very rough surfaces, P must be very much greater than F if the block is to move.

Lubricants (such as oil) reduce friction by keeping surfaces slightly apart from each other.

Lubrication

Under a microscope even very smooth surfaces appear to be rough. So friction is always a problem when surfaces move over each other.

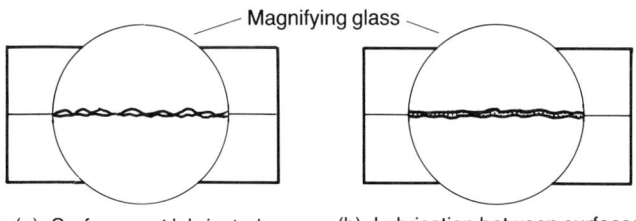

(a) Surfaces not lubricated   (b) Lubrication between surfaces

Figure 37 The action of a lubricant

Lubricating oils and greases have additives mixed with them to make them suitable for:
1 low or high temperature applications;
2 use under water;
3 application to parts of machines which move against each other.

The purpose of lubrication is to reduce friction by applying a thin film of oil or other lubricant (less than 0.0075 mm thick) between two surfaces. A well-chosen lubricant will maintain the film when under load, without creating more resistance.

Methods of applying lubrication
Gravity feed

☐ **Oil can:** Used to apply thin oils to machines such as bicycles. The oil is dripped in through an oil cap.

☐ **Oilcup and screw-down greaser:** The cup transfers oil or grease to a bearing gradually over a period of time.

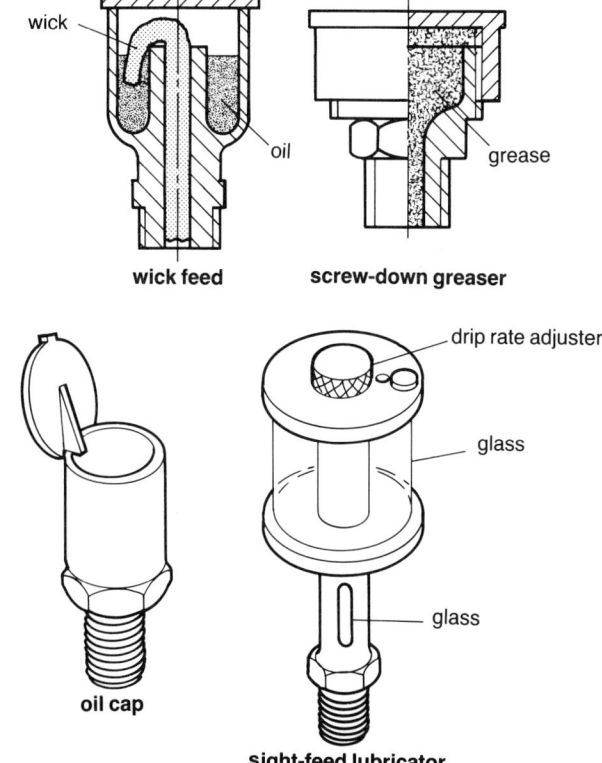

Figure 38 Methods of applying lubricants

☐ **Wick-feed lubricators:** A cotton wick soaks oil to the working area as it is needed.

☐ **Sight-feed lubricator.**

Pressure feed

☐ **Oil or grease gun:** Lubricant is fed under pressure from the gun to an oil/grease nipple.

☐ **Oil pump:** A continuous supply of oil is supplied by a pump. The oil drains back down to a sump to be pumped round again.

Splash feed

Part of the moving machinery is always in a reservoir of oil. As the machinery turns, oil splashes up and over all moving parts. This method is used in the gearboxes of cars.

Bearings
Plain bearings

☐ **Journal:** a bearing to support a radial load.

☐ **Thrust bearing:** a bearing to take a sideways or axial load.

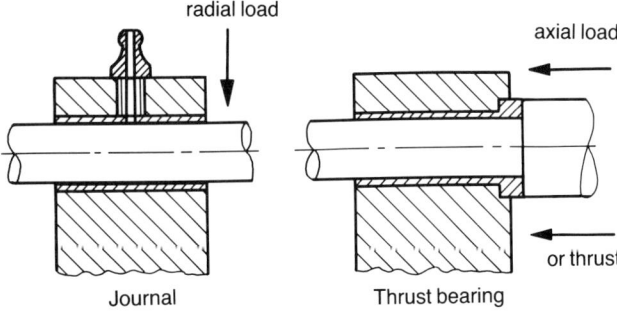

Figure 39 Plain bearings

Materials for plain bearings

Usually materials for bearings are softer than the shafts running in them. Thus they:
  1 will 'give' slightly so that a badly aligned bearing is less likely to seize up;
  2 wear slightly faster than the shaft, forming slight recesses – oil will be held in these recesses.

☐ **PTFE:** This is a plastic which has a very smooth, almost friction-free surface; no lubrication needed.

☐ **Nylon:** This is another plastic. An extremely good bearing material for use in dry conditions. Swells when damp. No lubrication needed. Cheap to produce, does not corrode. Light load use only.

☐ **White metal:** These are alloys of tin or lead. Used for bearings which do not have to withstand high pressures or speeds.

☐ **Bronze:** Alloys of either phosphor/tin or copper/tin bronze are used where bearing loads are high and speeds are medium to low.

☐ **Oilite:** This is a 'sintered' material (made from compressed powdered metals). It is very porous, and can be impregnated with lubricant. Used in the motors of audio cassette players, disc drives, etc.

Ball and roller bearings

Project

Without using undue force, and keeping the forces equal, compare the distances moved by an empty soft drinks can when it is pushed in both the positions shown in Fig. 40.

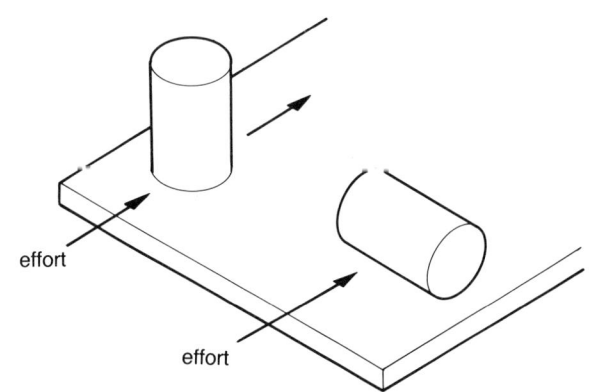

Figure 40 Experiment – bearings

You should find that the can will roll much further than it will slide. The action of ball or roller bearings is similar – they roll, keeping friction to a minimum.

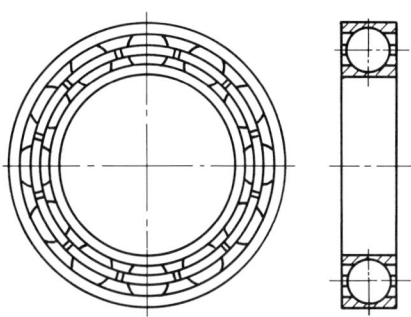

Figure 41 Front and sectional end views of a ball bearing race

Types

☐ **Self-aligning:** These allow some aligning.

☐ **Roller and needle roller bearings:** These have a larger contact area than ball bearings. They can take much heavier loads.

☐ **Taper roller bearings:** These are sold in pairs. They are designed to accept large thrust and radial forces.

Shafts in alignment

Coupling shafts

A **coupling** connects two shafts.

☐ **Muff couplings:** join two shafts.

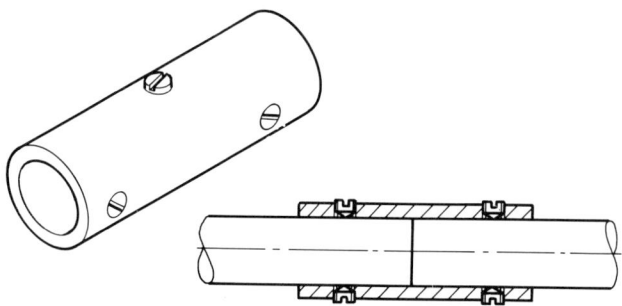

Figure 42 A muff coupling

☐ **Flange couplings:** These are used to bolt shafts together.

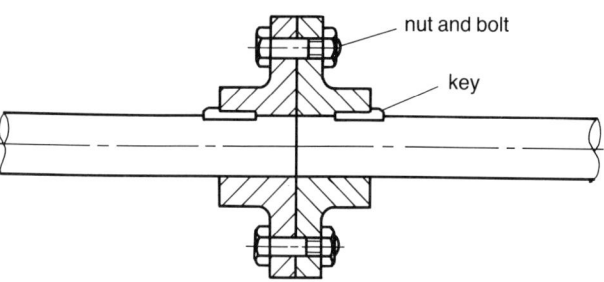

Figure 43 A flange coupling

☐ **Flexible joints:** These join shafts when some flexing is needed to absorb minor shocks when stopping and starting a motor.

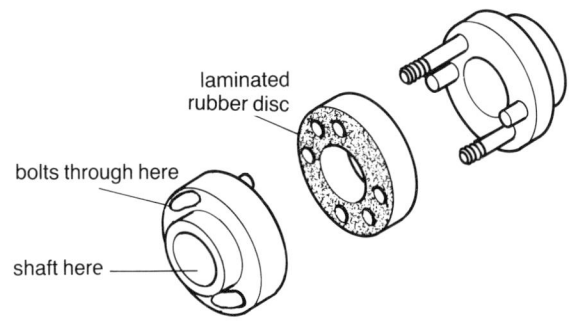

Figure 44 A flexible joint coupling

☐ **Universal joints:** These are joints which will allow swivelling in any direction.

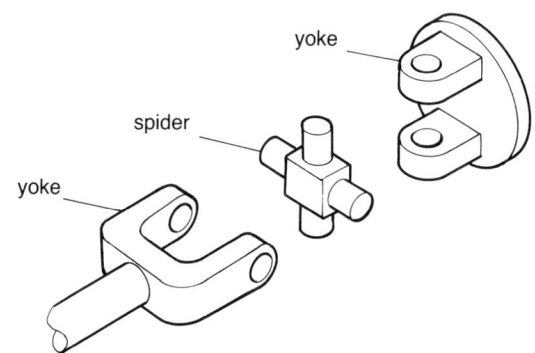

Figure 45 A universal joint

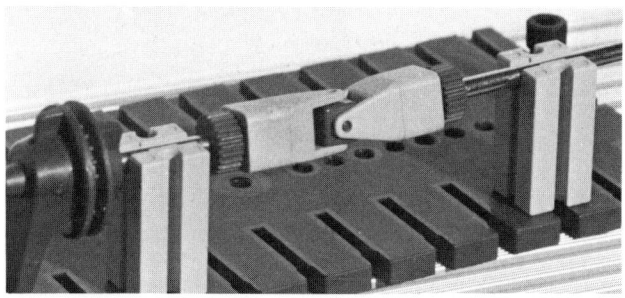

Figure 46 A model of a universal joint

☐ **Sliding coupling:** This allows sliding movement between shafts.

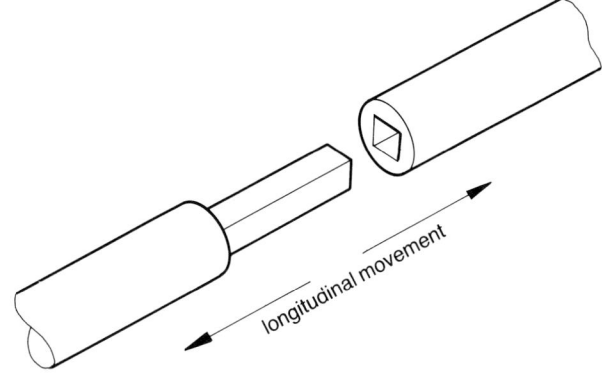

Figure 47 A sliding coupling

Securing to shafts

Pulleys and gears may be fixed to shafts using keys or splines, or be held by friction.

☐ **Friction fit:** The diameters of the shaft and its hole are so close that they have to be fitted together with force.

☐ **Grub screw:** Grub screws rely on friction between the screw threads holding the pulley against the shaft.

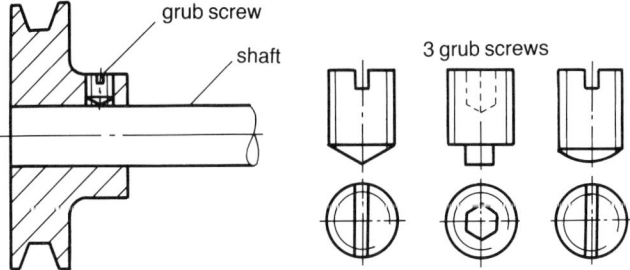

Figure 48 Grub screws

☐ **Cotter pins:** A number of different types of cotter pins can be used.

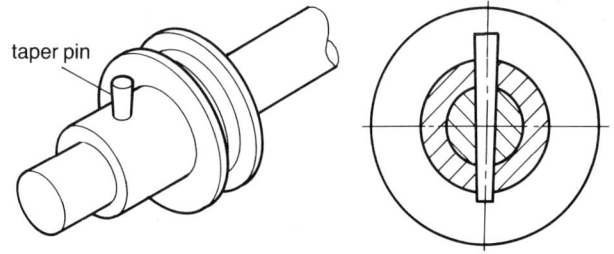

Figure 49 Cotter pin

Research

Make notes and sketches of different forms of cotter pins. Examine your bicycle for one source, and try vices in the workshop as another.

☐ **Keys and keyways:** These may be fixed rigidly or may allow some freedom of movement.

Research

Investigate Woodruff, feather and taper keys. Make sketches showing how they can be used for securing wheels to shafts.

☐ **Splines:** Splined shafts and gears/pulleys are often found in gear boxes.

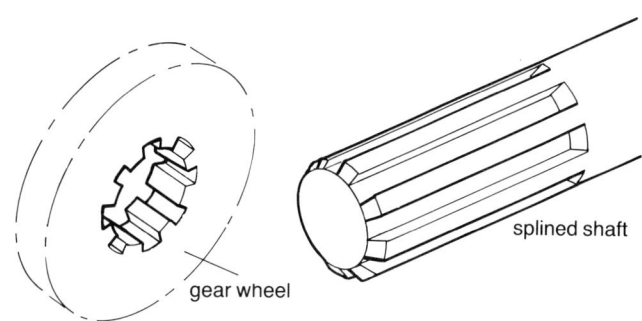

Figure 50 A spline

Types of motion

1 **Linear:** motion in a straight line.
2 **Reciprocating:** straight line motion backwards and forwards.
3 **Oscillating:** e.g. the pendulum of a clock.
4 **Rotary:** motion in a circle.

Conversion of motion from one type to another
Rotary to linear

☐ **Rack and pinion:** Some drilling machine tables are raised or lowered by rotating a handle to move a pinion along a rack. The saddle of a screw-cutting lathe is moved along its bed by a rack and pinion system. The steering sytems of many motor cars are based on a rack and pinion system.

☐ **Cams:**   There are several types of cams, three of which are shown in Fig. 51. Cams are devices for converting one form of motion into another – for example – horizontal to vertical; rotary to vertical.

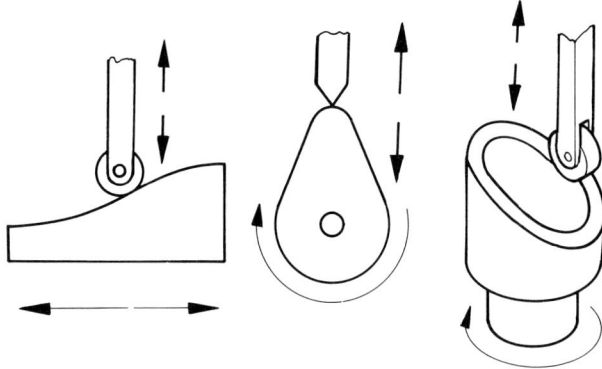

Figure 51 Cams

Linear to rotary

☐ **The internal combustion engine:**   The up-and-down linear reciprocating motion of the piston in an internal combustion engine is converted into rotary motion in the crankshaft to which it is attached, by a connecting rod.

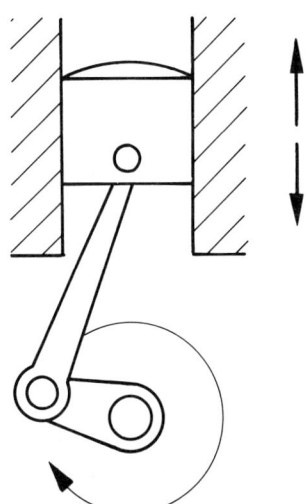

Figure 52 Action of internal combustion engine cylinder

Note: The rotary motion of the crankshaft is converted into linear motion of a vehicle along a road by the wheels of the vehicle.

Exercises

Find other methods by which linear, rotary, reciprocating and oscillating motions are converted into other forms. Sketch and write notes on the systems that you have observed.

The winch

If the crank radius is 280 mm, the drum diameter 75 mm, and the gears have 15 and 90 teeth, respectively, calculate the V.R. of the winch shown in Fig. 53.

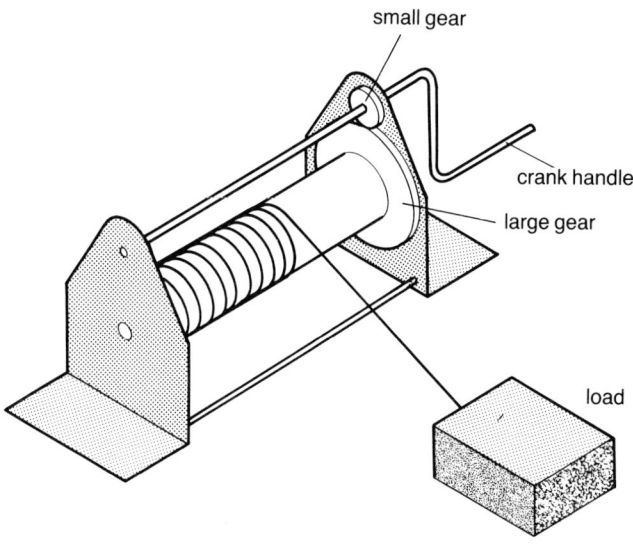

Figure 53 A winch

For one revolution of the drum: the cable is pulled in $\pi \times 75 = 3.142 \times 75 = 235.65$ mm

and the handle must rotate $\dfrac{90}{15} = 6$ times.

The hand turning the handle must move
$$6 \times 2 \times \pi \times 280 = 10557.12 \,\text{mm}$$

$$\text{V.R.} = \frac{\text{distance moved by effort}}{\text{distance moved by load}}$$

$$= \frac{10557.12}{235.65}$$

$$= 44.8$$

Project

Using a cotton reel for a drum, design and make a winch which will lift a load of 20N. The efficiency of the machine must be 60% at least. Show all calculations to 'prove' your answer.

Exercises

1 Figure 54 shows a pulley and belt drive as used in some reel-to-reel tape-recorders.
■ What is the purpose of pulley C?
■ Describe a method of fixing the pulley to the shaft.
■ Use arrows to show the direction of rotation of pulleys B and C
■ At how many rpm will the motor pulley B have to run to produce 8 rpm in output pulley A?
■ To add a digital counter it is necessary to extend one of the shafts. Sketch a suitable coupling.

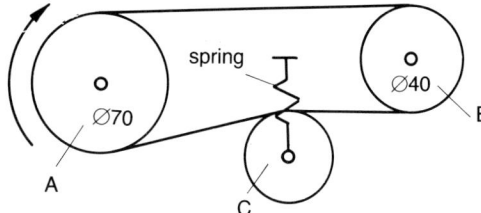

Figure 54 Exercise 1

2 Figure 55 shows a hydraulic jack for a car. For every 200 mm movement of the handle, the load moves 12.5 mm.
■ Calculate the velocity ratio of the jack.
To lift a load of 1575 N an effort of 175 N is required.
■ What is the mechanical advantage of the system?
■ What is the overall efficiency of the system?

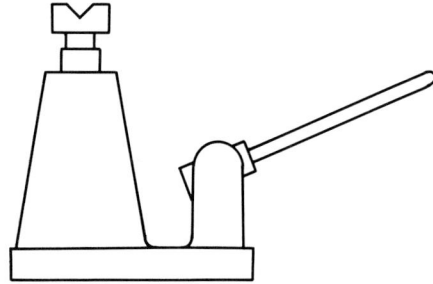

Figure 55 Exercise 2

3 Figure 56 shows a lift shaft and electric motor.
■ Sketch a gear system which has a 'fail-safe' element should the motor fail.
■ Sketch a mechanism to be fitted to the top of the shaft to make sure the cable runs freely.
■ If the lift were badly loaded it might tilt and jam in the shaft. Sketch a method for correcting tilt.

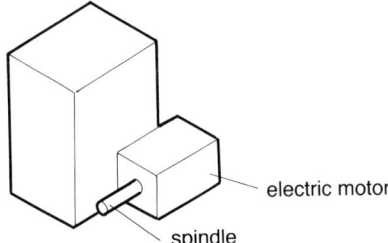

Figure 56 Exercise 3

4 Name mechanisms suitable for use in each of the following circumstances:
■ a mountain railway;
■ the vertical movement of a fork-lift;
■ a department store escalator;
■ the drive from motor to chuck in a pedestal drill;
■ moving the endless line of buckets on a dredger;
■ wagging the tail of a pull-along toy.

5 In each of the drawings in Fig. 57, gear A turns at 42 rpm. For each drawing:
■ show the direction of rotation of each gear;
■ name the type of gear in use;
■ calculate the gear ratio and rpm of the final gear.

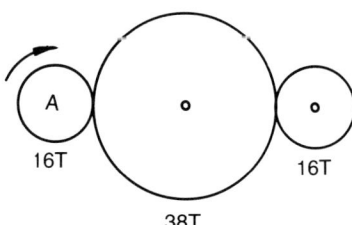

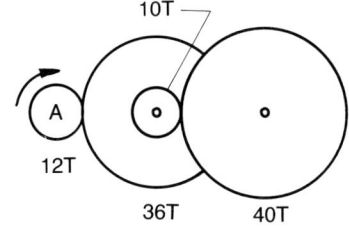

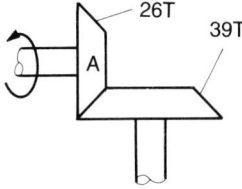

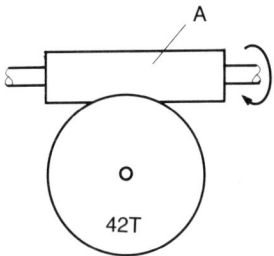

Figure 57 Exercise 5

6 Figure 58 shows two non-aligned drive shafts and two driving wheels.

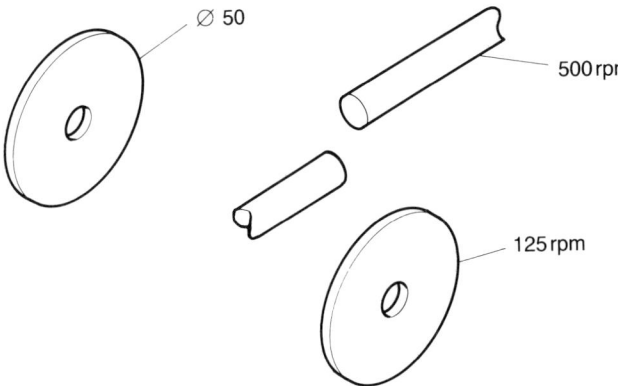

Ø 50

500 rpm

125 rpm

Figure 58 Exercise 6

■ Sketch and name a suitable mechanism for joining together the two non-aligned shafts.
■ Sketch a simple mechanism to transfer the power to the rear wheels.
a State the final gear ratio required.
b Calculate the distance the wheels will travel in one minute.
7 Keyways, splines and grub screws may be used to fix gears to shafts.
■ Sketch an example of each.
■ Give an example where each would be used.
8 Figure 59 shows a 'black box' where a crank is used to produce the movements shown. Make three sketches of mechanisms that would produce the required movement and label the appropriate parts.

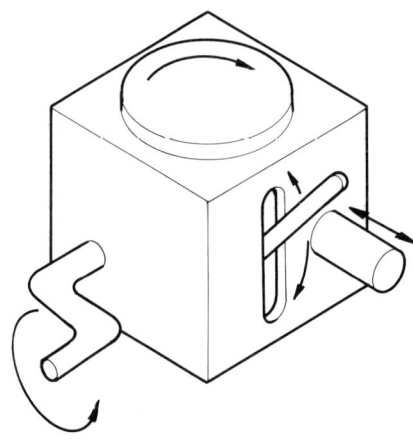

Figure 59 Exercise 8

9 Explain why a machine is never 100% efficient.
10 There are three classes of lever. Sketch an example of each one.

8 Structures

Introduction

This area of knowledge is concerned with framed structures. Some examples are: pylons for electricity cables; TV and radio masts; bridges; roof trusses; scaffolding; the frameworks of cold frames, greenhouses and large buildings such as aircraft hangers and barns; storage shelving in factories; cranes and hoists; climbing frames, ladders and stepladders; the gantries supporting electric cables and the signal systems on rail lines; the frames supporting rocket launchers.

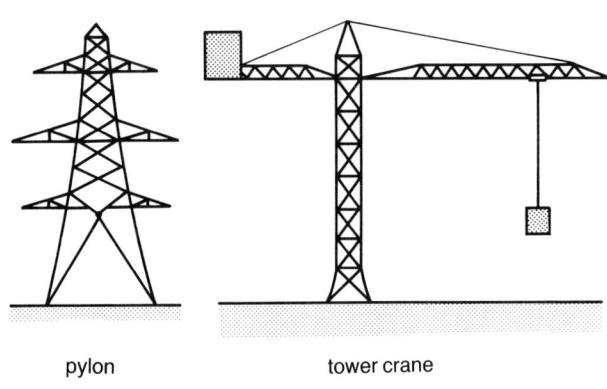

pylon   tower crane

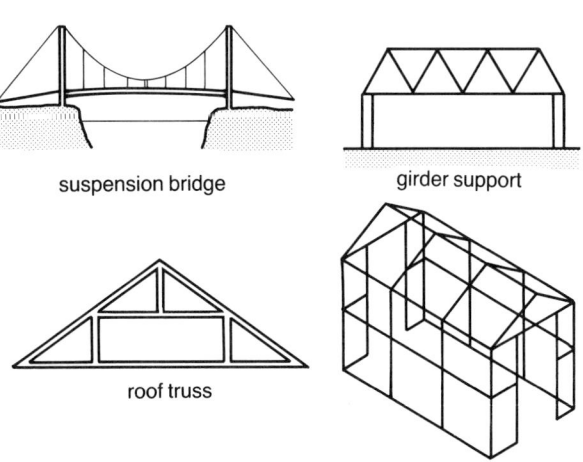

suspension bridge   girder support

roof truss   greenhouse frame

Figure I Examples of structures

Exercise 1: List any other framed structures you can think of. Houses, furniture, vehicles and other items which are part of our environment, are also structures.

Exercise 2: List some structures which are not framed structures.

Forces

All structures are subjected to forces acting upon them – the loads that they are designed to carry; weather, e.g. winds; their own weight. These forces tend to be destructive and can cause distortion of the structure. Structures should be designed to withstand these distorting forces. The forces can be classified as:
1 tension forces;
2 compression forces;
3 shear forces;
4 torsion forces.
These four types of force can cause:
1 stretching – Fig. 2;
2 buckling – Fig. 3;
3 bending – Fig. 4;
4 flexing – Fig. 5;
5 separation – Fig. 6;
6 twisting – Fig. 7.

Tension forces

Materials tend to stretch under tension forces.

Exercise 3: Figure 2 shows four examples of the action of tension forces. List some other examples.

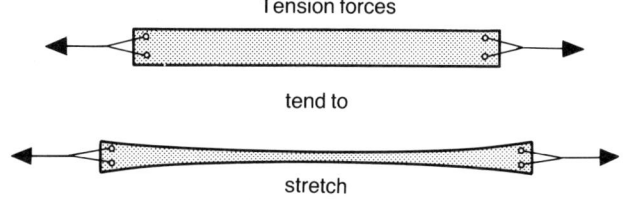

Tension forces

tend to

stretch

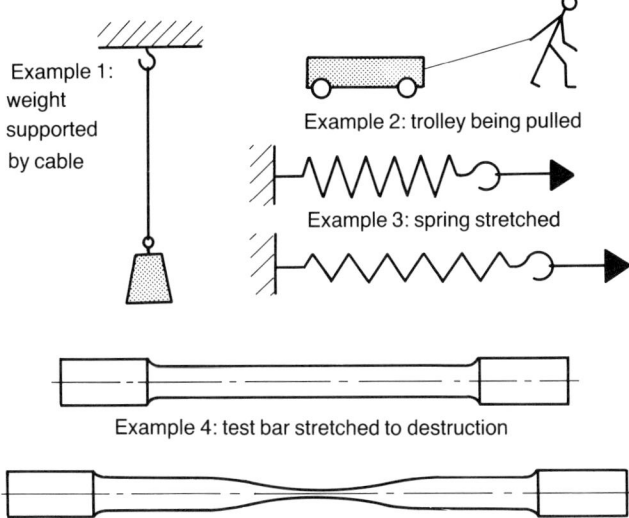

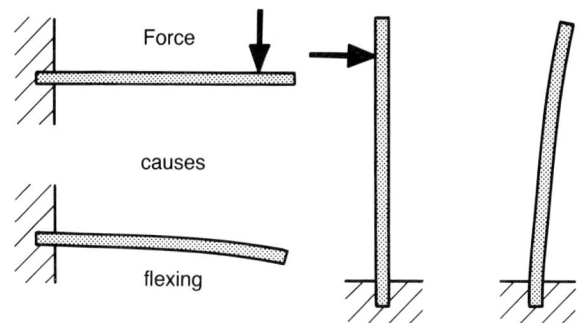

Figure 5 A force causing flexing

Shear forces

Materials under shear force tend to separate.

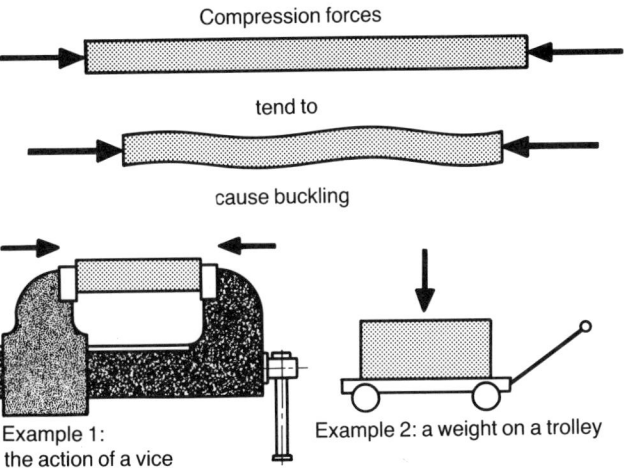

Figure 2 The action of tension forces

Compression forces

Compression forces tend to cause buckling.

Exercise 4: Figure 3 shows two examples of the action of compression forces. List some other examples of the action of compression forces.

Exercise 5: Figure 6 shows two examples of shear forces. List further examples of the action of shear forces.

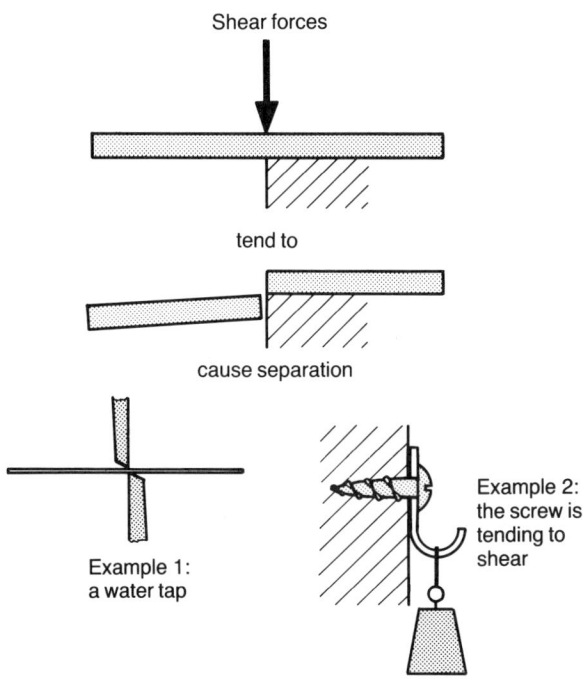

Figure 6 The action of shear forces

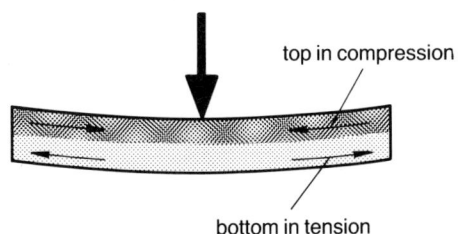

Figure 4 A force causing both compression and tension – upper part of the object is under compression, lower part is under tension

Torsion forces

Materials under torsion force tend to become twisted. Torsion forces apply *torque* (see page 84).

Exercise 6: Figure 7 shows two examples of torsion forces. List some other examples.

The design of structures depends in part on the methods used to resist the forces acting upon them. Figure 8 shows examples of methods by which the action of forces can be resisted.

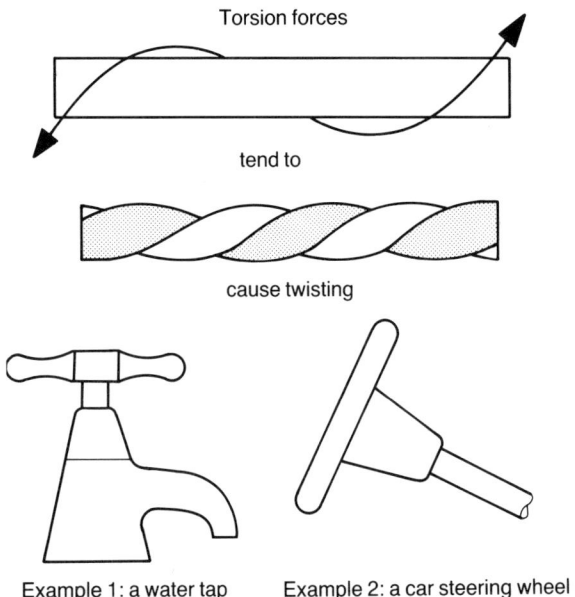

Example 1: a water tap   Example 2: a car steering wheel

Figure 7 The action of torsion forces

Exercise 7: Make a number of freehand sketches showing examples of other construction methods you have seen for resisting the action of tension, compression, shear and torsion forces.

Figure 8 A test piece mounted ready to be placed in a tension testing machine

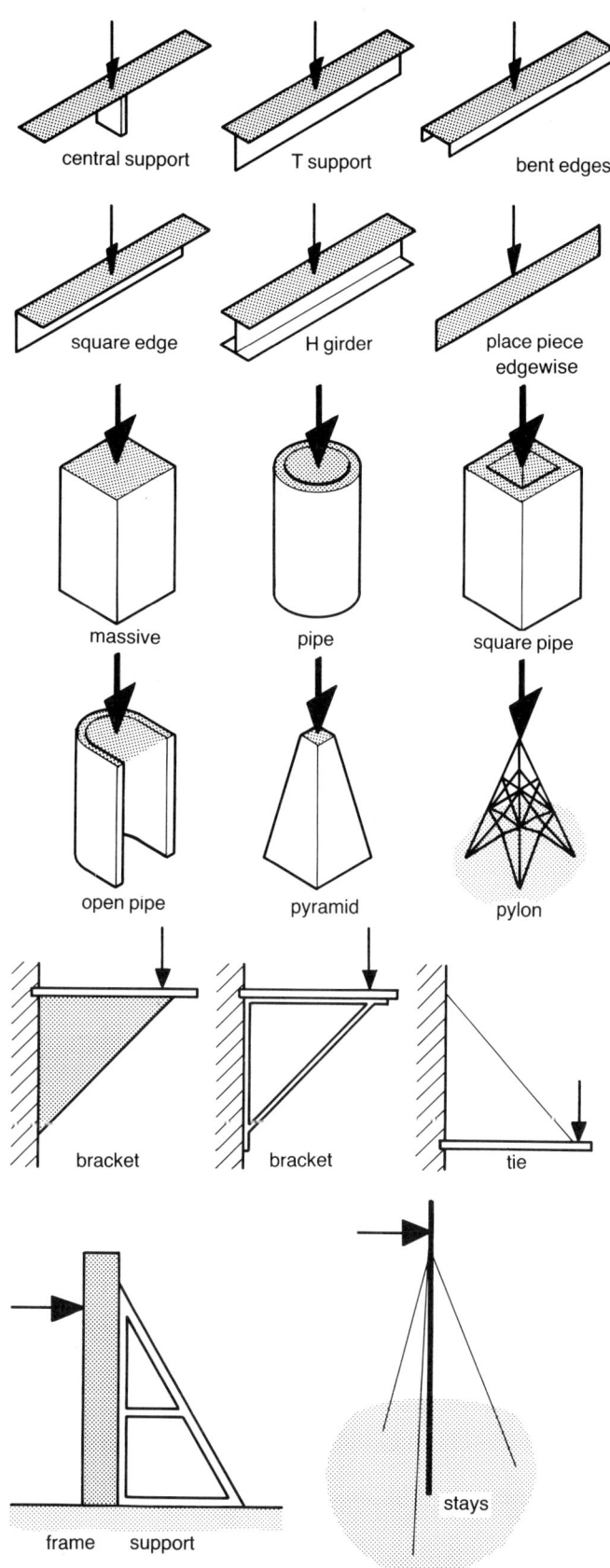

Figure 9 Some construction methods of counteracting the effects of forces

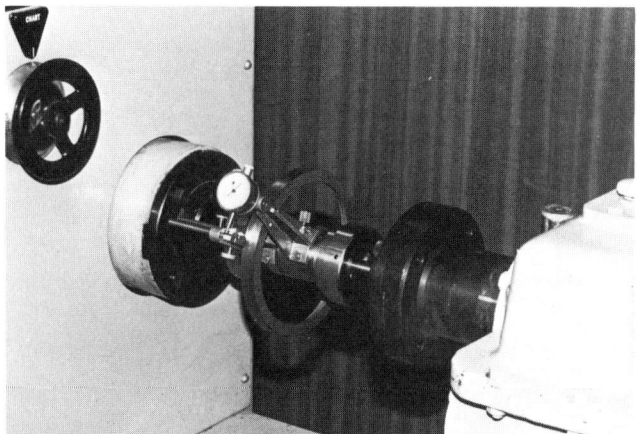

Figure 10 The test piece mounted in the test machine

Force-vector diagrams

Force-vector diagrams are used in designing to help to calculate the forces acting on and within a structure. This helps to ensure that a structure is strong enough for its intended use.

A vector has both *magnitude* and *direction*. It can be represented in a diagram by a line, its length giving the scaled magnitude and its angle and position giving its direction. Vectors can represent forces, velocity, acceleration and other such quantities.

In the following force-vector diagrams, the forces are all acting in the same plane.

Project 1

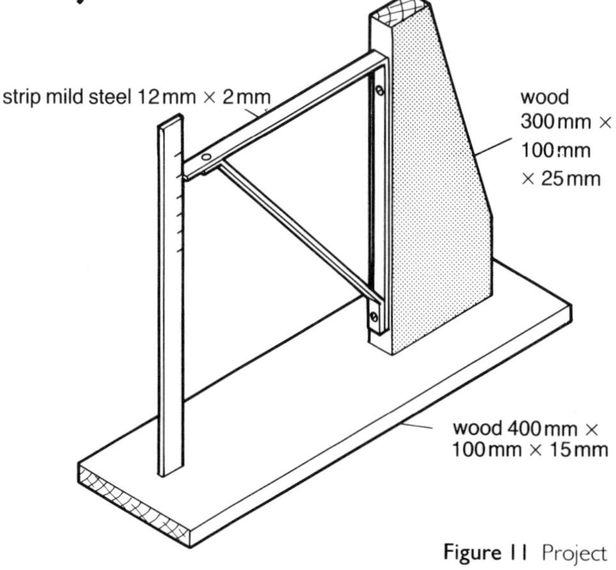

strip mild steel 12 mm × 2 mm

wood
300 mm ×
100 mm
× 25 mm

wood 400 mm ×
100 mm × 15 mm

Figure 11 Project 1

The apparatus shown in Fig. 11 demonstrates the forces acting in triangular structures.

1 Fig. 12. Remove the brace. Hang a 200 g mass (2 N weight) from A. Measure the downward movement of A.

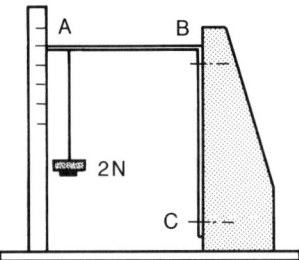

Figure 12 Project 1 with 200 g mass

2 Change the 200 g mass for a 500 g mass. What is the movement of A now?
3 Remove the screw at B. What happens when a 200 g mass is hung from A? What happens when a 500 g mass is hung from A?

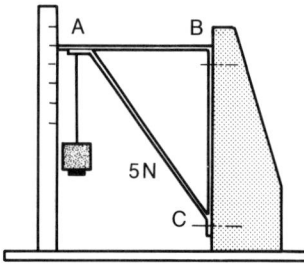

Figure 13 Project 2 with 500 g mass

4 Fig. 13. Replace the screw at B, remove the screw at C and insert the brace into the framework, by bolting to the frame at A and re-inserting the screw at C.
5 Hang a 200 g, then a 500 g mass from A. What happens?

It can be seen that forces are acting in all the three members (parts) of the triangle ABC of the apparatus in Project 1. Figure 14 shows these forces in a diagrammatic form.

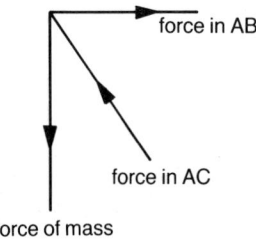

force in AB

force in AC

force of mass

Figure 14 Forces acting in the bracket

Forces in equilibrium

When the brace is added to the triangular frame in Fig. 13, all the members of the bracket remain stationary, even when masses much heavier than 500 g are hung from A. The forces acting in the three members of the frame are then said to be in **equilibrium**.

Triangle of forces

The forces acting in the bracket can be shown by a **triangle of forces**. Figure 15 is a triangle of forces diagram for the bracket, with the brace at an angle of 40° to the vertical and with masses of 200 g and 500 g hung from A. The 200 g mass is equivalent to approximately 2 newtons (2 N) force and the 500 g mass to 5 N.

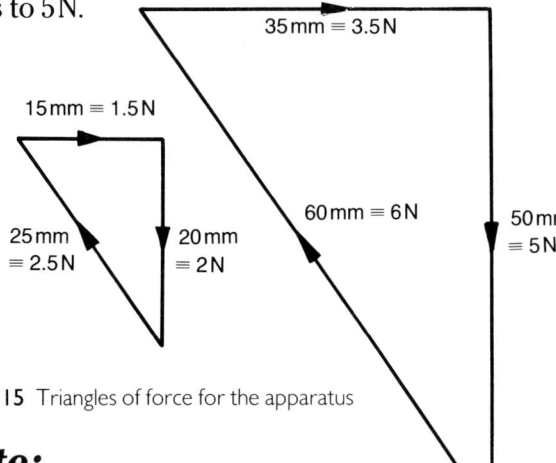

Figure 15 Triangles of force for the apparatus

Note:

1 The triangles are scaled, with the vertical members drawn 20 and 50 mm long to represent 2 and 5 newton forces respectively. 1 N ≡ 10 mm.
2 The forces in the frames are in equilibrium.
3 The values of the forces acting in members AB and AC are found by measuring the lengths of the triangle sides in millimetres and then using the scale to find the sizes of the forces acting in them.
4 In any structure in which the forces are in equilibrium, the forces in the triangle will all be in the same direction around the triangle – either all clockwise, or all anti-clockwise.
5 The triangles are complete – their sides meet.
6 The sides of the triangle represent *forces*, not the dimensions of the structure.

Project 2

Figure 16 gives another method of showing forces in equilibrium and their triangles of forces.

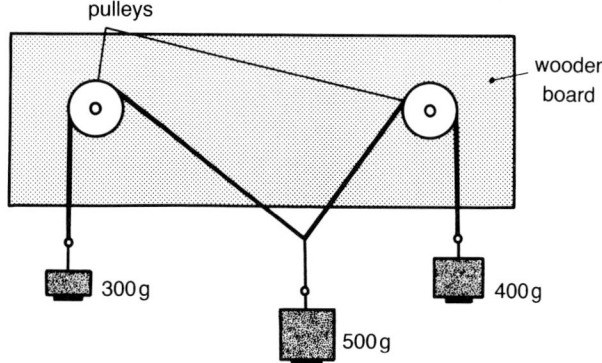

Figure 16 Second experiment

1 The board can be any size and of any material. The pulleys must run freely.
2 Place masses of 300 g, 400 g and 500 g on the carriers. Wait until the system is quite still, i.e. in equilibrium.
3 Draw a diagram like Fig. 17. The lines should be parallel to the strings holding the weights on the board.
4 Letter the drawing as shown – this lettering of a force diagram is called **Bow's notation**. The vertical force of 5 N is called force AB, that of 4 N, force BC and that of 3 N force AC – after the letters each side of their respective lines of force;
5 Now draw the triangle of forces for the three forces (Fig. 17), starting with the 5 N vertical force. Use a scale of 10 mm ≡ 1 N.

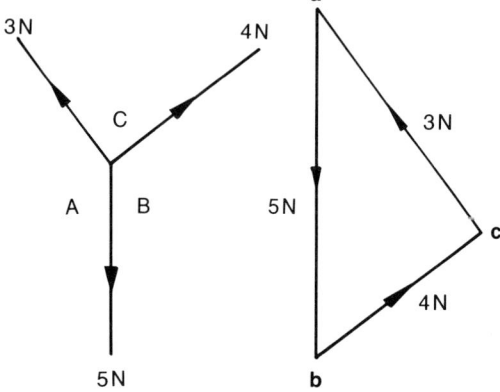

Figure 17 Bow's notation and triangle of forces for second project

6 Label the angles of this triangle as shown, with lower case letters. In this triangle, force AB is now called force **ab**, force BC is force **bc** and force AC is force **ac**.
7 Note how the lines of the two forces bc and ac in the triangle of forces are parallel to the strings on the board of the apparatus and that the angles at which they operate are the same.

Resultant of forces

In Fig. 18, a block resting on a rough plane can only just be moved by the two forces shown. A single force could produce the same movement.

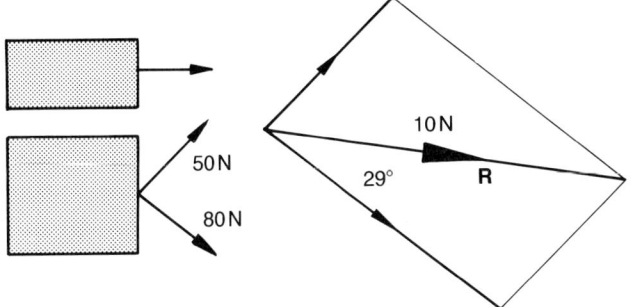

Figure 18 To find a resultant

To find this single force; proceed as follows.
1 Draw:
  a 50 N vector parallel to the 50 N force;
  a 80 N vector parallel to the 80 N force.
2 Draw lines parallel to the two vectors to complete a **parallelogram of forces** (Fig. 18).
3 The diagonal **R** is the **resultant** of the two forces, giving both the magnitude and the direction of the required single force.
4 By measurement, **R** is 10 N acting at 29° to the 80 N force.

Exercises

1 Fig. 19(a). By constructing a triangle of forces, find the magnitude and direction of the forces acting in the three members of the bracket.

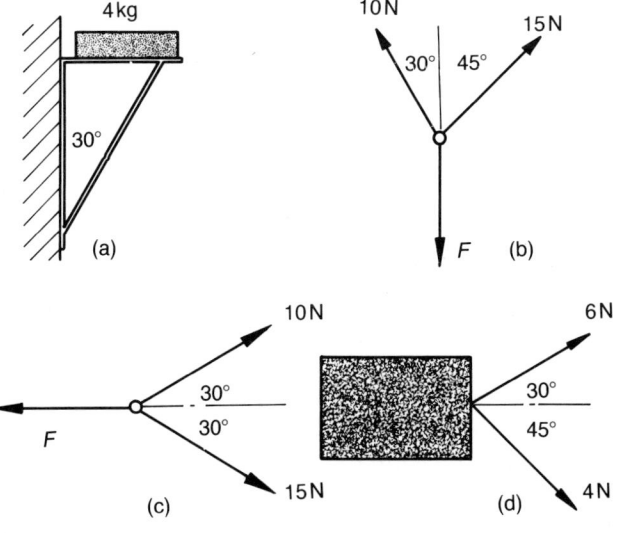

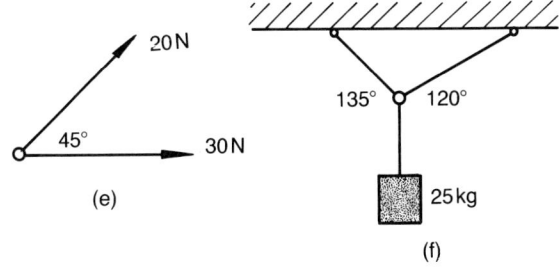

Figure 19 Exercises

2 Fig. 19(b). The ring is held stationary by the three forces acting on it as shown. Construct a triangle of forces and find from your drawing the value of force *F*.
3 Fig. 19(c). Find the value of *F*.
4 Fig. 19(d). Find the size and direction of one force which would replace the two given forces.
5 Fig. 19(e). Find the size and direction of one force which could replace the given two.
6 Fig. 19(f). The 25 kg mass is suspended from a ceiling. Calculate the forces in the two ropes.

Polygon of forces

A polygon can be regarded as a series of touching triangles. Because of this, the theory of the triangle of forces can be used for a system of more than three forces.

Example 1: Fig. 20. The ring is attached to a wall by a rope. What is the force exerted in the rope and its angle to the wall when the three forces are applied?

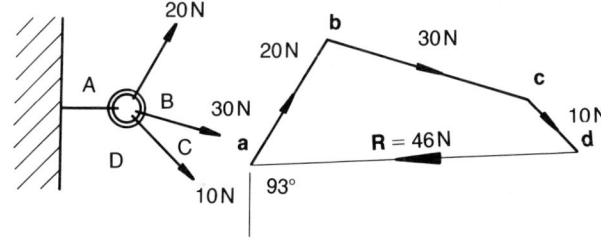

Figure 20 Polygon of forces – Example 1

1 Copy the diagram and label it using Bow's notation.
2 To a suitable scale, draw **ab** parallel to AB, **bc** parallel to BC and **cd** parallel to CD.
3 Join **da** – the resultant of the three forces. Measure **da** – you should find the resultant is 46 N at 93° to the wall.

Example 2: Find the magnitudes and directions of the two unknown forces D and E in Fig. 21.

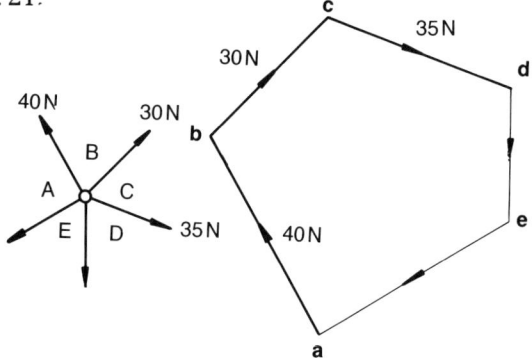

Figure 21 Polygon of forces – Example 2

1 Copy the diagram and label it using Bow's notation.
2 Working to a suitable scale, draw **ab**, **bc** and **cd** parallel to AB, BC and CD.
3 From d, draw a line parallel to DE.
4 From a, draw a line parallel to AE, completing the polygon of forces.
5 Measure ae and de to give the values of the two unknown forces. You should find AE = 40N and DE = 24N.

Note: In both examples, the arrows showing the directions of the actions of the forces are all in the same direction around the polygons. This is because the systems are in equilibrium. No movement is taking place.

Exercises

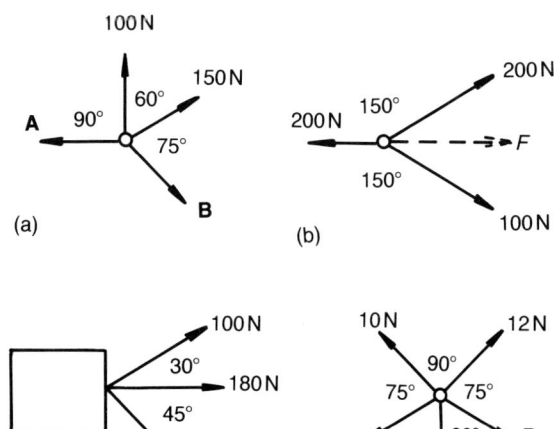

Figure 22 Exercises

1 Fig. 22(a). Find the magnitude of forces A and B.
2 Fig. 22(b). What is the magnitude of the force F needed to ensure equilibrium?
3 Fig. 22(c). What single force, acting at which angle, would replace the three forces?
4 Fig. 22(d). Find the values of the forces C and D which hold the system in equilibrium.

Simply firm frames

Project

Construct the square frame shown in Fig. 23. It can be made from strips of wood, pinned at the corners with panel pins; from strips of mild steel, jointed with small nuts and bolts; or from strips of Meccano, etc.

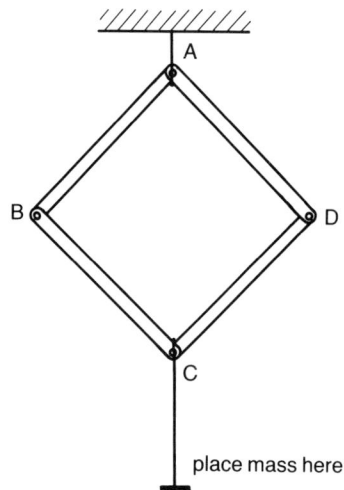

Figure 23 Experiment – a simply firm frame

1 Hang the frame from corner A.
2 Hang any weight at C. What happens?
3 How can you prevent the frame from sagging, to make it **simply firm**?

You should find that one diagonal strip added in either direction (A to C, or B to D) prevents sagging. A second diagonal offers no advantage over a single one.

Try the same project with frames of each of the shapes shown in Fig. 24. Add strips until each frame is simply firm.

You should then be able to work out that to make a frame simply firm:

number of members $= 2n - 3$

where $n =$ number of members in the original frame.

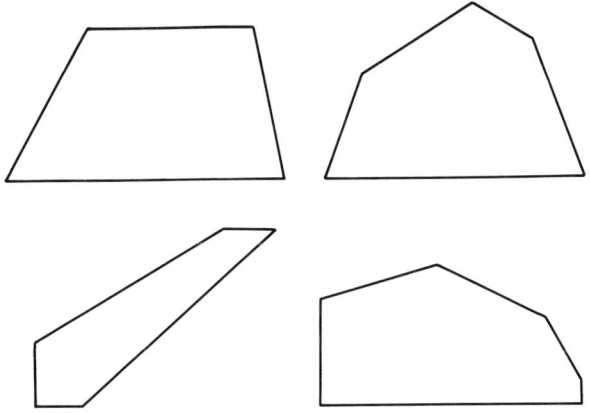

Figure 24 Experiments to confirm simply firm frames

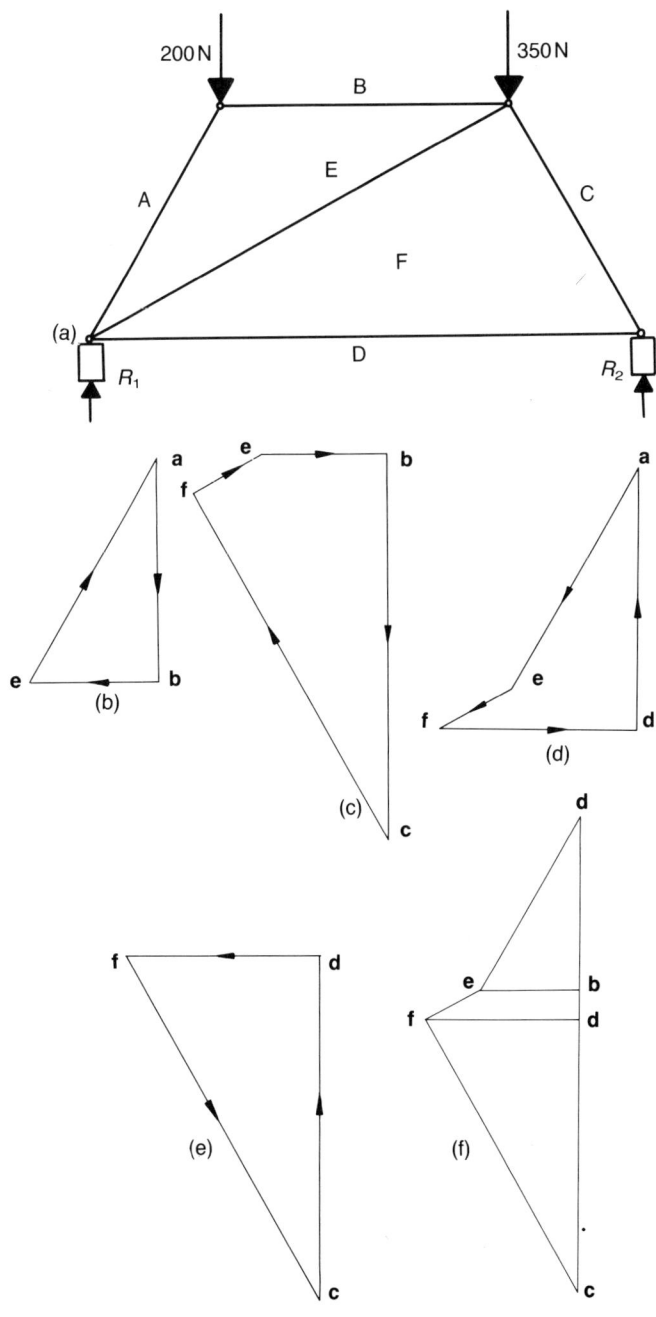

Figure 25 Stress diagram for a simply firm pin-jointed frame

Pin-jointed frames

Figure 25(a) is a line drawing of a simply firm pin-jointed frame, with vertical forces acting on two of the pin joints. The frame is resting on two supports. The sum of the two reactions at the support points must be equal to the sum of the two forces acting downward at the upper pin joints.

To find the forces acting in the members of this frame:

1 Make a scale drawing of the frame (Fig. 25(a)).

2 Label the frame according to Bow's notation.

3 Now split the frame into either triangles or polygons. Always start with a triangle.

4 The 200N force acting at the top left corner is acting on the triangle abe – Fig. 25(b). From this triangle measure the value of forces **ab** and **be**.

5 With the value of the force acting in BE and the vertical force of 350N, the values of the forces in CF and EF can be found – Fig. 25(c).

6 The value of force **cf** can now be found. This enables us to draw Fig. 25(d), from which R_2 can be found.

7 To find R_1, add the two forces of 250 and 350N and take R_2 from the total – Fig. 25(e).

8 Fig. 25(f) combines all the force diagrams into one **stress diagram**. The usual practice is to draw this single diagram, from which all forces acting in the frame can be found.

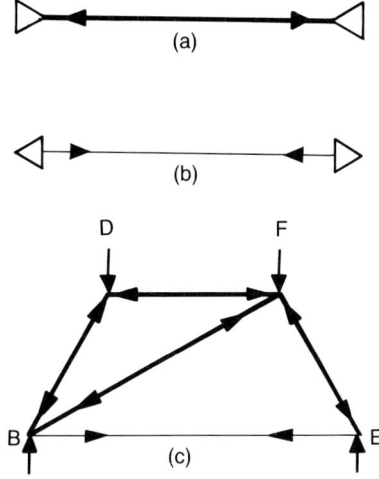

Figure 26 Compression and tension forces in the frame

Compression and tension in frames

At each joint in a simply firm pin-jointed frame, there must be two forces acting on each of the members at the joint – the force acting along the member and an equal and opposite force keeping it in equilibrium. In Fig. 26(a) arrows show the directions of all forces acting in all members of the frame from Fig. 25.

Taking the two members BE and DF, Fig. 26(a) and (b) show the external, equal and opposite forces acting on them. From Fig. 26(b) it can be seen that BE is in compression and DF is in tension. The diagram for the frame can now be re-drawn with thick lines (compression) and thin lines (tension) – Fig. 26(c).

□ **Results:** By measuring the scaled lengths of the appropriate lines in the stress diagram, the forces in the frame are found to be:

ae: 235N – in compression;
be: 120N – in compression;
ef: 75N – in compression;
cf: 360N – in compression
df: 185N – in tension;
R_1:   235N;
R_2:   315N.

All figures have been taken to the nearest 5N.

Shear and bending forces in a beam

If a simply supported beam is loaded, the beam tends to shear and bend (Fig. 27).

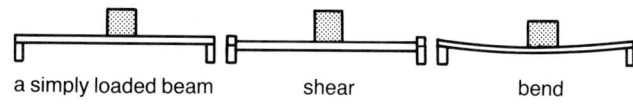

a simply loaded beam   shear   bend

Figure 27 Shear and bend in a beam

Shear force diagrams

Shear forces in the beam can be found by taking moments (page 78) about the point at which the load is positioned (Fig. 28).

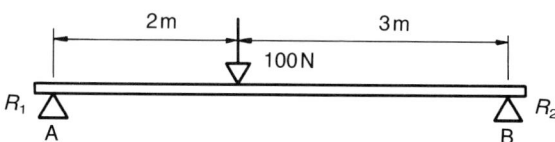

Figure 28 Example for shear force diagram – simply loaded beam

Simply loaded beam

As the beam is in equilibrium, clockwise moments about A = anti-clockwise moments about A.

Clockwise moment $= 100N \times 2m = 200Nm$
Anti-clockwise moment $= R_2 \times 5m$

$$\text{Thus:} R_2 = \frac{200Nm}{5m} = 40N$$

Similarly, taking moments about B,

Clockwise moment $= R_1 \times 5m$
Anti-clockwise moment $= 100N \times 3m = 300Nm$

$$\text{Thus:} R_1 = \frac{300Nm}{5m} = 60N$$

$$\textit{Check:} R_1 + R_2 = \text{load on beam}$$
$$60N + 40N = 100N$$

Positive and negative shear

Shear forces are either negative (−ve) or positive (+ve) (Fig. 29). With this in mind, the shear force diagram for the beam in Fig. 28 can be constructed, with $R_1 = 60\,\text{N}$ and $R_2 = 40\,\text{N}$.

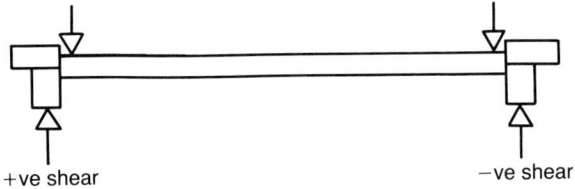

+ve shear   −ve shear

Figure 29 Positive and negative shear

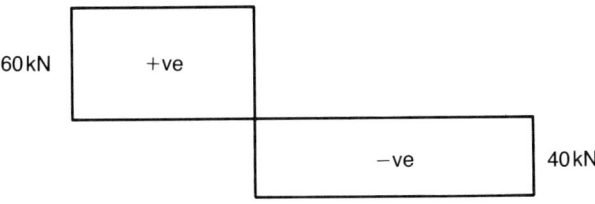

60 kN   +ve

−ve   40 kN

Figure 30 Shear force diagram – simply loaded beam

Shear force diagram for a cantilever

Load a cantilevered beam with 250 N (Fig. 31).

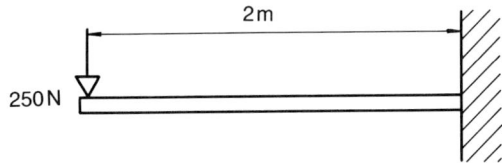

2 m

250 N

Figure 31 Example – shear force in a cantilevered beam

The beam is in equilibrium, thus the reaction to the 250 N load at the wall must be 250 N acting vertically upwards. This reaction would tend to make the right-hand end of the beam move up – see Fig. 29. Thus the shear force diagram, Fig. 32, will show all negative shear.

−ve   500 Nm

Figure 32 Shear force diagram for cantilevered beam

Shear force diagram for a beam carrying more than one load

Construct the shear force diagram for each load as if other loads were not present. Then combine the several diagrams to form the shear force diagram

for the beam. Figure 33 shows a beam with two loads, together with its individual and combined shear force diagrams.

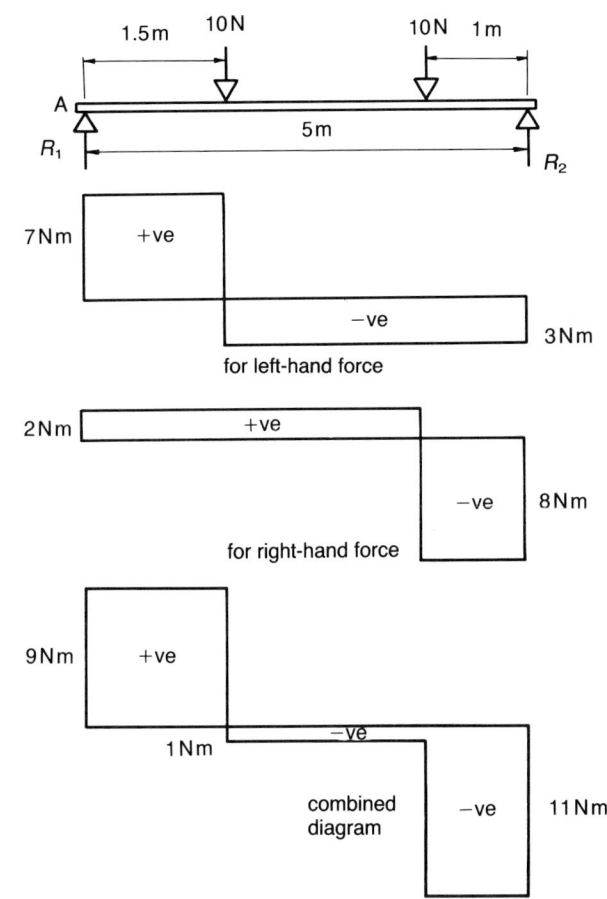

1.5 m   10 N   10 N   1 m

A

5 m

R_1 R_2

7 Nm   +ve

−ve   3 Nm

for left-hand force

2 Nm   +ve

−ve   8 Nm

for right-hand force

9 Nm   +ve

1 Nm   −ve

combined diagram   −ve   11 Nm

Figure 33 Shear force diagram for beam with two loads

Exercises

1 Fig. 34. Draw the shear force diagram.
2 Fig. 35. Construct a shear force diagram.
3 Fig. 36. Draw the shear force diagram.
4 Fig. 37. Construct a shear force diagram.

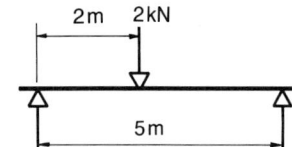

2 m   2 kN

5 m

Figure 34 Exercise 1

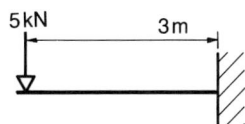

5 kN   3 m

Figure 35 Exercise 2

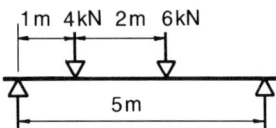

Figure 36 Exercise 3

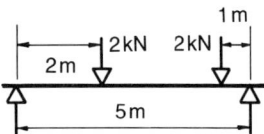

Figure 37 Exercise 4

Bending moment diagrams

Positive and negative bending moment

If a beam tends to sag as a result of the bending force, it is said to have had a positive (+ve) bending moment applied. If it tends to bow upwards (to *hog*), it is said to have had negative (−ve) bending moment applied.

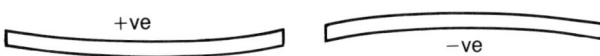

Figure 38 Positive and negative bending moment

Example of a bending moment diagram

Figure 39 shows a simply loaded beam. To find R_1 and R_2, take moments about A.

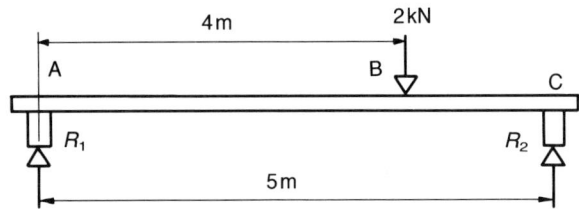

Figure 39 Example – bending moment, simply loaded beam

$$\text{Clockwise moment} = 4\text{m} \times 2\text{kN}$$
$$= 8\text{kNm}$$
$$\text{Anti-clockwise moment} = 5\text{m} \times R_2$$

Thus: $5R_2 R_2 = 1.6\text{kN}\,(1600\text{N}) = 8\text{kN}$

$R_2 \qquad\qquad\qquad\qquad = 1.6\text{kN}\,(1600\text{N})$

Also $R_1 = 2 - R_2$
$= 2 - 1.6$
$= 0.4\text{kN}\,(400\text{N})$

Taking moments about B, the bending moment of the 2 kN force is found by taking the reaction of 0.4 kN at A and multiplying it by the distance AB. Thus:

$$0.4 \times 4 = 1.6\text{kNm}$$

or taking the reaction at C and multiplying it by the distance BC. Thus:

$$1.6 \times 1 = 1.6\text{kNm}$$

or use the formula:

$$\frac{\text{force} \times \text{AB} \times \text{BC}}{\text{length of beam}}$$

This gives

$$\frac{2 \times 4 \times 1}{5} = 1.6 \text{ as before.}$$

The bending moment diagram for the beam is shown in Fig. 40.
Note: The beam is tending to bow under the action of the 2 kN force – the bending moment diagram is all positive.

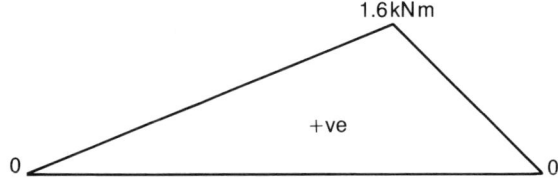

Figure 40 Bending moment diagram for example

Bending moment diagram – cantilevered beam (Fig. 41)

Taking anticlockwise moments about A, the bending moment of the 2 kN force is $4\text{m} \times 2\text{kN} = 8\text{kNm}$.

The bending moment diagram for this beam is shown in Fig. 42.
Note: (**1**) the beam is tending to hog – the bending moment diagram is all negative.
(**2**) A bending moment diagram can be used in a similar manner to a line graph. The bending moment for any position on a beam can be *read* by measuring the scaled height of the line above the base line.

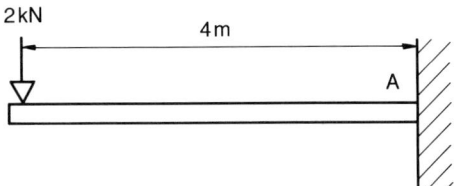

Figure 41 Example – bending moment, cantilevered beam

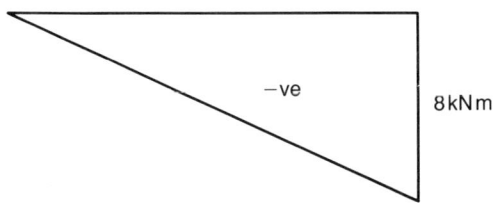

Figure 42 Bending moment diagram for cantilevered beam

Bending moment diagram for a beam with more than one load

(Fig. 43)

As with the shear force diagram, Fig. 33, a combination of bending moment diagrams of single forces can be constructed when more than a single force is involved. See Fig. 43.

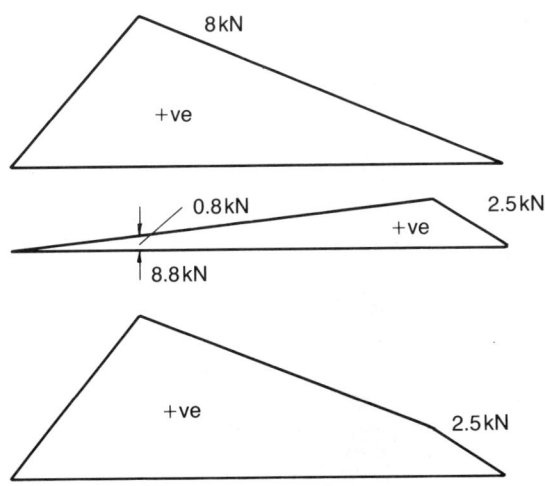

Figure 43 Example – beam with two loads and its bending moment diagram

Exercises

1 Fig. 44. Draw a bending moment diagram.
2 Fig. 45. Construct the bending moment diagram.
3 Fig. 46. Construct the bending force diagram.
4 Fig. 47. Draw a bending force diagram.

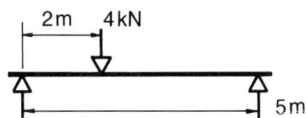

Figure 44 Exercise 1

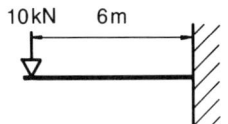

Figure 45 Exercise 2

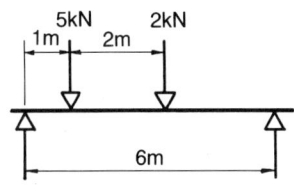

Figure 46 Exercise 3

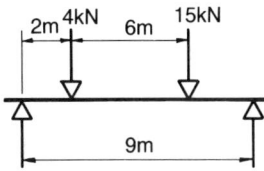

Figure 47 Exercise 4

Stress and strain
Strain gauges

The principle upon which strain gauges work is that when a metal wire is stretched (strained), its resistance to electrical current increases.

Most strain gauges are made from thin wire cemented to insulating paper (Fig. 48).

—Fig 8.48—

Resistance in the strain gauge is measured by:

$$\text{resistance} = \frac{\text{length}}{\text{area}} \quad \text{or} \quad R = \frac{l}{A}$$

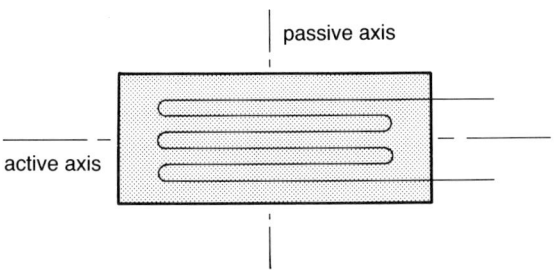

Figure 48 Diagram of the wire in a strain gauge

As l increases, so does the resistance. The changes in resistance are very small, and so strain gauges are connected to amplifying circuits, which show the changes on a dial test indicator. Strain gauges are cemented on the surfaces along which strain is to be measured, with their active axes in line with the direction of strain. In making up test rigs, such as that shown in Fig. 49, many (sometimes several hundreds) of gauges are employed.

In modern industrial design, the use of CAD (computer aided design) specialist software programs allows the forces and resulting stresses to be measured by computer. Such software programs require a great deal of computer memory and so they are not suitable for school use. This type of program is known as a **finite element analysis** program.

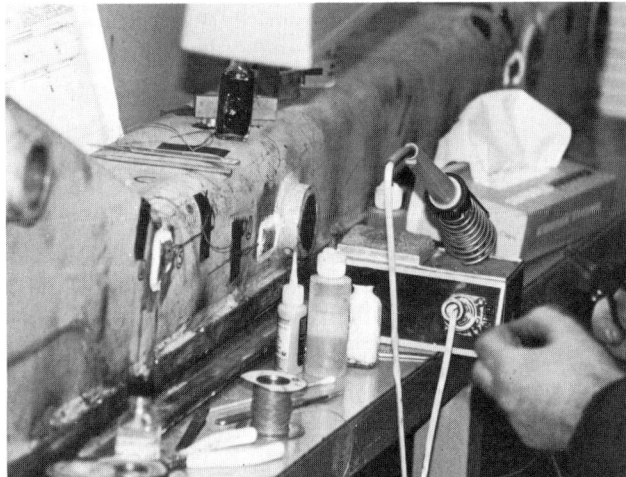

Figure 49 Strain gauges being applied to a test rig

Modulus of elasticity

When a material is loaded with a gradually increasing force, it becomes stressed and stretches. As the applied force is increased, the

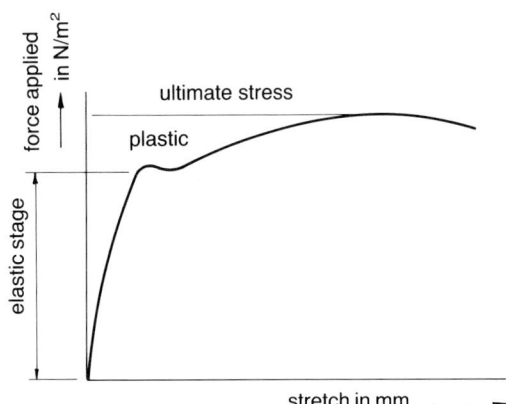

Figure 50 Stress–strain graph for mild steel

stress reaches a point, known as **ultimate stress**. If the applied force is continued beyond the point of ultimate stress, the material will become plastic and will eventually break. Figure 50 shows a stress–strain graph for mild steel.

If the material is not stretched beyond its elastic stage, i.e. the applied force is between O and A on the graph, the material will return to its original length when the force is removed. If the material is stretched beyond its elastic stage, it will be permanently stretched, even when the applied force is removed.

The force required to reach the elastic limit for any given material is proportional to the length by which the material is stretched. This proportion is known as the **modulus of elasticity** of the material, often referred to as **Young's modulus**. The modulus of elasticity is different for each material.

$$\text{Young's modulus} = \frac{\text{stress}}{\text{strain}} \text{ in newtons per mm}^2 \text{ (N/mm}^2)$$

Factor of safety

When building a structure, a designer will calculate the forces acting upon it and the stresses acting on the materials from which it is made. The shapes and sizes of the parts of the structure can then be determined. When the structure is put to use, other forces may act upon it – overloading, wind action, the wearing of parts, etc. These additional forces may make the structure less safe. The designer will therefore add another factor into the calculations: the **factor of safety**, measured by:

$$\text{factor of safety} = \frac{\text{ultimate stress}}{\text{working stress}}$$

The value of the factor of safety is usually taken as either 4 or 5 to work out the *ultimate stress*.

Note: Figures 51 to 55 show examples of models made as parts of projects on structure design. A variety of materials can be used – Meccano, LEGO® or Fischer Technik kits; wood strips; paper or card. Some experiments may be carried out with structures made from drinking straws, paper, cardboard or strips of plastic material.

Figure 51 A lifting device

Figure 53 A support frame

Figure 52 Part of a lifting device

Figure 54 Bridge girder frames made from 6 mm square softwood strips

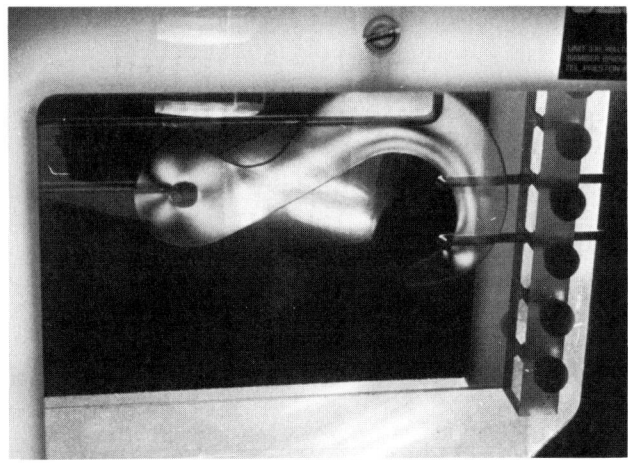

Figure 55 A plastic hook under a polarising screen, showing lines of stress

Exercises

1 ■ Figure 56 shows part of a see-saw. Copy the diagram and mark the fulcrum point if the two figures are of equal weight.
■ Is the underside of the board subject to tension, compression or torsion?
■ Person A, weighing 72 kg, is seated 1.5 m from the fulcrum. In order to balance the see-saw, how far from the fulcrum must person B, who weighs 56 kg, sit?

Figure 56 Exercise 1

2 A banquet table with eight legs can support a load of 2400 N if it is evenly spread. How much force will each leg support?

3 Figure 57 shows a frame for supporting stage scenery. Calculate the number of redundant members.

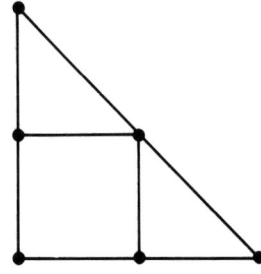

Figure 57 Exercise 3

4 ■ Figure 58 shows an outer framework used in a bridge construction. Add lines to show the least number of extra members required to make the framework simply rigid.
■ Draw a cross-sectional shape for member D which will cause the structure to be stiff.

Figure 58 Exercise 4

5 Figure 59 shows the centre section for a cantilever bridge. Complete the figure to show the cantilever principle.

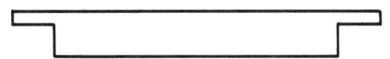

Figure 59 Exercise 5

6 ■ In a model suspension bridge a support cord 500 mm long has a diameter of 14 mm and supports a load of 6160 g. Calculate the stress in the cord.
■ Calculate the strain on the cord if it extends 20 mm when under load.
■ What is 'Young's modulus'?
Apply Young's modulus, using the information from the first two parts of this question.

7 Why is it necessary to recognise a 'factor of safety' when designing a structure?

8 ■ An electricity pylon is kept rigid by the triangular arrangement of its members.
Sketch a typical pylon.
■ Gussets and ribs also provide strength and rigidity for a structure. Search for and sketch one example each of a gusset and a rib.

9 Figure 60 shows a spanner, to be used to undo a nut. Calculate the moment of force about point A.

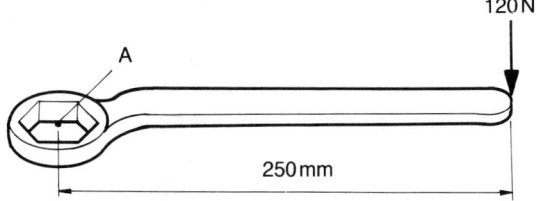

Figure 60 Exercise 9

10 ■ Figure 61 shows the brake pedal on a car. Calculate the force exerted on piston P.

■ Calculate the force on the brake pedal when the force on the brake piston is 600N.

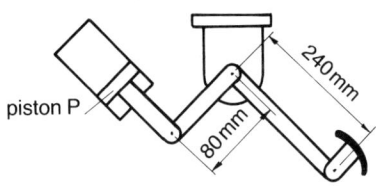

Figure 61 Exercise 10

11 ■ By using calculations or scale drawings, determine the forces present in each member of the hoist shown in Fig. 62.
■ State which member(s) is under compression.

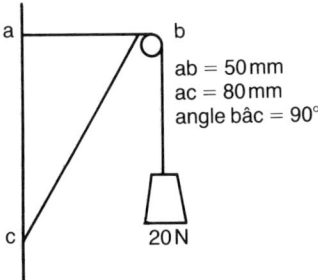

ab = 50mm
ac = 80mm
angle bâc = 90°

20N

Figure 62 Exercise 11

12 Figure 63 shows a support for a hanging flower basket and the cross-sectional shape of the material available for making the strut.

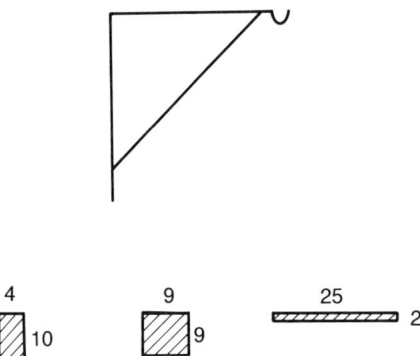

4   10   9   9   25   2

Figure 63 Exercise 12

■ Copy the diagram and label the strut L.
■ The material has to be used with the depth as shown. Show by calculation which cross-section will give maximum support.

13 The arrows in Fig. 64 show the direction in which two men are attempting to pull at a stone block in order to move it. From the information given, find the resultant of the force – i.e. the direction in which the block will tend to move over an even surface.

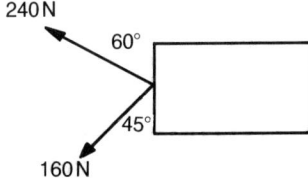

240N   60°
45°
160N

Figure 64 Exercise 13

9 Materials

Introduction

When choosing materials for a design it may be necessary to consider some (or all) of the following points.

1 Specific gravity; tensile strength; hardness; percentage elongation under strain; yield stress.

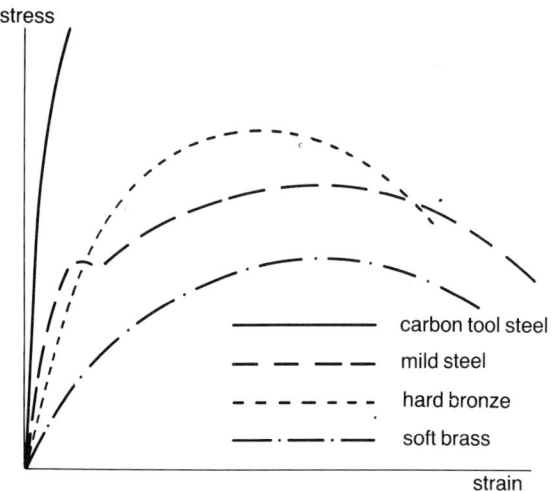

Figure 1 Stress-strain graph comparing mild steel, carbon tool steel, a hard bronze and a soft brass

2 Fatigue strength.
3 Resistance to corrosion.
4 Are the materials malleable and/or ductile?
5 Can they be worked with the tools at hand?
6 Reaction to other materials in the same design.

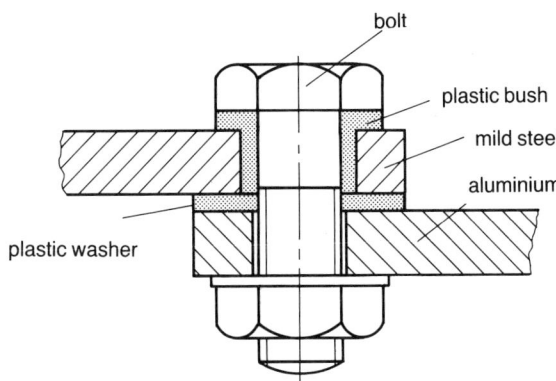

Figure 2 Bolting dissimilar metals to avoid electrical corrosion

e.g. when some metals are joined to others, electrical activity may be set up between them. This may cause corrosion.

7 Cost.
8 Availability.

You should also consider possible material failures. If a breakdown occurs through material failure:

1 It may be inconvenient.
2 Parts may have to be repaired and/or replaced.
3 Replacement and/or repair can be expensive.
4 Other parts of a design may be damaged.
5 People may be injured.

Failure may be caused by:
1 Creeping, e.g. in materials such as those used for hauling, lifting or suspending.
2 Fatigue: materials will eventually break down under repeated and continuous stressing up and down or in and out.

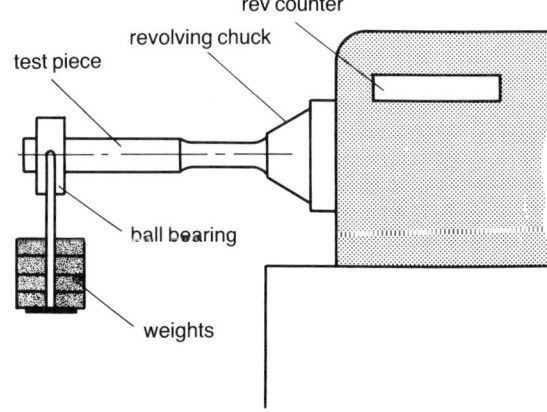

Figure 3 Diagrammatic drawing of a rotating bend test to test fatigue limits

3 Corrosion – by water, atmosphere, chemicals – and wear.
4 Exceeding the safe working loads of the design.
5 Unnecessarily heavy parts, themselves causing excessive wear and strain.
6 Poor positioning of holes and notches and their size, may weaken parts of a design.

Factor of safety

The factor of safety for materials is given by the equation:

$$\text{factor of safety} = \frac{\text{ultimate stress}}{\text{working stress}}$$

If a design is to be used under favourable conditions, a factor as low as 3 may be suitable. If failure could endanger life, then a factor as high as 25 may be necessary.

Metals

Metals can be divided into two broad groups, **ferrous** metals (containing iron) and **non-ferrous** (not containing iron). Both groups may be mixed with other materials to form **alloys**. Ferrous metals include the many **steel alloys**. Other alloys in common use are **brasses** (alloys of copper and zinc) and **soft solders** (alloys of lead and tin). All metals, whether ferrous, non-ferrous or alloys are **crystalline** in structure.

Ferrous metals

Pure iron is soft and ductile. When alloyed with other substances, iron becomes a strong structural material. Steels are alloys of iron with substances such as carbon, manganese, molybdenum, nickel, chromium and tungsten.
    Depending on the alloy, steels are:

1 strong;
2 tough;
3 corrosion resistant;
4 resistant to wear.

Carbon steels

Common carbon steels are:

☐ **Dead mild steel:**   0.1% carbon. Used for soft wires, tubing, etc. A soft metal, easily worked. Very ductile.

☐ **Mild steels:**   From 0.1% to 0.3% carbon. The most common. Used extensively in school CDT projects.

☐ **Medium mild steel:**   From 0.3% to 0.6% carbon. High tensile strength steel. Tough.

☐ **Carbon tool steel:**   1% carbon. Hard, tough steels for making tools. Carbon tool steels can be hardened and tempered, to take and retain a sharp edge.

Other steel alloys

☐ **Stainless steels:**   Alloys mainly of iron and chromium. They contain at least 13% chromium to make sure the steel is stainless, but some contain as much as 30%.

☐ **High speed steels:**   Tungsten steel alloys which are used for cutting tools which can operate at high speeds and high temperatures.

☐ **Nickel alloy steels:**   These are tough and stainless.

☐ **Silver steel:**   This is high-carbon-content steel ground to good dimensional accuracy. The grinding results in its silver colour – there is no silver metal in the steel.

☐ **Cast iron:**   Molten iron poured into moulds to form castings. Used where massive compression strength is needed. As much as 4% carbon content.

☐ **Malleable iron:**   Cast iron which has been baked at high temperature for a long time and allowed to cool slowly. Cast iron tends to be brittle. Malleable iron is far less brittle.

Surface protection of mild steel sheet

☐ **Tinplate:**   Mild steel, coated with tin. Easily worked.

☐ **Galvanised iron:**   Zinc-coated mild steel.

☐ **Other metal coatings:**   Aluminium, chromium or nickel.

☐ **Plastic coatings:**   Plastic coating such as vinyls.

☐ **Case hardening:**   If mild steel is baked in carbon at high temperatures for a long time, the carbon is absorbed into the *skin* of the steel, and forms a thin layer of high-carbon-content steel. This surface can be hardened by heat treatment, after the steel has been machined to its final shape. The result is a mild steel component with a hard, tough skin. Spindles, bearings and components in which surfaces are subjected to wear are case hardened.

Non-ferrous metals

Copper

After steel, copper is the most commonly used metal. Copper is important when used in its almost pure state, or when alloyed with zinc or tin to form brasses or bronzes. Copper has several properties which make it such an important metal.

1 It is very resistant to corrosion by weather. A good anti-corrosion surface forms naturally when the metal is exposed to the atmosphere.

2 It is very ductile – can be easily worked to shape.

3 It is a very good conductor of heat and electricity.

☐ **Brasses:** They are alloys of copper and zinc. The proportions of copper to zinc vary widely according to use, but there must be at least 50% copper in the mixture. Examples are: gilding metal: about 95% copper to 5% zinc; yellow brass: about 65% copper to 35% zinc; brazing spelter: about 50% copper to 50% zinc.

Brasses are very corrosion resistant and good conductors of electricity. They are much harder than copper. They work easily and well by machine.

☐ **Bronzes:** These are alloys of copper and tin.

☐ **Phosphor bronze:** This is a copper, tin and phosphorus alloy – used for bearings in machinery, some springs and gears.

☐ **Gunmetal:** A form of brass.

☐ **Copper in aluminium alloys:** If a small percentage of copper (between 0.1% and 1%) is alloyed with aluminium, it becomes tougher and harder. The result is called an aluminium copper alloy. On a weight-to-weight ratio, aluminium copper alloys are stronger than some steels.

Aluminium

Aluminium is soft, malleable and easily worked. When exposed to air a corrosion-resistant surface film forms. This film can be coloured by the electrolytic process known as **anodising**. Aluminium is a good conductor of electricity. It is used for a huge variety of purposes because of its light weight and resistance to corrosion under poor conditions.

Zinc

Zinc is important because of its very good corrosion-resistant properties. It is used for casings of dry battery cells; roofing items such as weather station screens; galvanising; as a constituent of brasses. As a die-casting material, zinc is alloyed with copper and aluminium in ratios such as:

zinc:copper:aluminium   95:4:1.

It can be die cast to give very accurate castings, requiring a minimum of machining.

Tin

Tin is a coating material for mild steel; a constituent of soft solders; tin foils; tinned copper pipes; fuse wires.

Other non-ferrous metals

☐ **Lead:** is a very *heavy* metal. Because of the dangers of lead poisoning, lead is not now used as part of paints and its addition to petrol to prevent 'pinking' is gradually dying out. Lead is used for the plates of accumulator batteries, as shielding in X-ray apparatus and as noise-deadening sheet material.

☐ **Gold:** is a *precious* metal: very expensive. It is used for coating some electrical circuit contacts. Very dense and heavy. Very high resistance to corrosion. Pure gold is soft and ductile. In use it is alloyed with small percentages of copper or silver, which increase its hardness.

☐ **Silver:** is ductile and easily worked. Used in some electrical components because of its good electrical conductivity. *Sterling silver,* for silversmiths' work, is about 93% silver and 7%

copper. *Hard silver solder* is alloyed – silver: copper 80:20.

☐ **Platinum:** is also a precious metal (more expensive than gold). Very resistant to corrosion and extremely ductile.

Heat treatments of metals

When metals are being formed to shape, they **work harden**. This may cause the metal to break up or fracture. A work hardened metal can usually be softened by heating the metal until red hot and allowing it to cool. Some metals can be quickly cooled by plunging the red hot metal in water or in oil, e.g. copper. Other metals must be allowed to cool slowly, e.g. steels and brasses. This process is known as **annealing**.

If carbon tool steel (1% carbon steel) is made red hot and suddenly cooled by plunging it into water or oil, the metal becomes dead hard and very brittle. If the hardened steel is polished and slowly re-heated its surface changes colour, from a straw colour, to brown, then purple, then blue and finally black. If the steel is suddenly cooled at selected surface colours it is said to be **tempered**. The colours show varying degrees of hardness, making the steel suitable for uses as follows:

Colour	Uses
pale straw colour	hammer heads, lathe tools for brass
dark straw	lathe tools for mild steel, milling cutters
brown	punches, taps and dies
brown/purple	screwdrivers, wood chisels
blue	steel rulers, springs, spanners

Plastics

Plastics are **polymers** – chemicals which have long-chain molecules. They are made from crude oil, coal and natural gas. Some plastics include small quantities of limestone, salt or water.

There are two main stages in their production:
1 Monomers (small-molecule chemicals) are manufactured from the raw materials.
2 The monomers are **polymerised** (joined together) to form long-chain polymers.

Monomer molecules are small. Polymer molecule chains may contain thousands of atoms. These long molecular chains give plastics the properties which make them such useful design materials.

There are three main groups of plastics: thermoplastics, thermosetting plastics and elastomers.

Thermoplastics

Thermoplastics become soft and pliable when heated. They can be heat-moulded to shape. On cooling, they assume the shape to which they have been formed. Examples are:

☐ **Polyethene:** Waxy feel; good electrical insulator; good resistance to corrosion by chemicals; very low water absorption, obtainable in powder form for heat and pressure moulding.

☐ **Polyvinyl chloride:** *pvc* or *vinyl*. Rigid and stiff. If plasticisers are included, pvc becomes flexible. Very good chemical and electrical resistance. Sheet pvc can be heat moulded under pressure.

☐ **Polypropylene:** Its unique flexing qualities make it suitable for designs where hinging actions are required.

☐ **Foam plastics:** Foam plastics in common use are *polypropylene* and *polyurethane*. Polypropylene foam is the white foam material used for packaging. Polyurethane foam is used for upholstery. At the moment, a law is being passed through Parliament forbidding the use of polyurethane foams in furniture unless they have been treated to make them fire resistant.

☐ **Acrylics:** *Perspex* and *Oroglas* are acrylics. They are hard, stiff materials, readily worked to shape. Can be shaped under pressure at temperatures between 130°C and 175°C. Very good optical qualities.

☐ **Polyester films:** Used for tapes for audio and visual tape recorders. Good electrical and chemical resistance.

☐ **Polytetrafluoroethene:** *PTFE*. Very low coefficient of friction, making it suitable for bearings, bearing linings and for non-stick coatings on kitchen cooking equipment. One of the best of all electrical and chemical insulators.

☐ **Nylon:** Used for small bearings, gears, door catches, hinges and lock parts. White and waxy. Machines well.

Thermosetting plastics

Once they have been polymerised, their shapes cannot be altered. Thermosetting plastics can be formed to shape under the action of pressure and heat, before they are fully polymerised (set). Examples are:

☐ **Formaldehydes:** *Phenol formaldehyde* was one of the first plastics to be invented, being the material previously known as *Bakelite. Urea formaldehyde* is of a light colour. *Melamine formaldehyde* is used as the top layer of *Formica* plastic sheet.

☐ **Epoxy resins:** These are an important group of plastics. As they set (polymerise) they do not shrink and they will adhere to most materials in common use. Because of these properties, epoxy resins are used as adhesives in a great range of situations, from jointing in electrical circuit boards to the jointing of metals in cars and other vehicles. Epoxy resins with added *fillers* are used to repair damaged vehicles (among other uses).

☐ **Polyester resins:** These are the plastic bases of GRP (glass reinforced plastic) and other resin-based materials such as carbon-fibre reinforced plastic. They are purchased as thick syrup-like resins. When hardeners and catalysts are added, they set hard at room temperatures.

Elastomers

Elastomers are materials with elastic properties. The most common elastomer is rubber. Examples are:

☐ **Natural rubber:** Plastics are usually thought of as man-made substances. Polymers also exist in nature, however. Rubber (from latex) is one of these.

☐ **Neoprene:** Similar to natural rubber. Good resistance to oils and chemicals, to natural light, to ozone and to permeation by gases.

☐ **Isoprene:** More elastic than natural rubbers, but with similar properties.

☐ **Butyl rubber:** Good resistance to tearing, flexing and abrasion, and good resistance to permeation by gases. For these reasons it is used in the manufacture of car tyres and in hoses of all types.

☐ **Silicone rubbers:** Used for moulds. Poor mechanical properties. Found in encapsulations for electronic components and circuits.

☐ **Urethane:** Strongest and hardest elastomer. Good resistance to abrasion and to oils and fuels such as petrol.

Woods

Wood is the most widely used material in the world. Some of the reasons for this are:

1 Wood could be an inexhaustible resource, if replanting to replace the trees felled for timber is carried out.
2 Woods are strong and tough for their weight.
3 Woods are easy to work.
4 Woods are comparatively cheap materials.
5 Large constructions can be built from wood, even without the use of heavy equipment.
6 Woods have a beauty of grain and colour which other materials do not have.

Two classes of woods

Woods can be classified as softwoods, from coniferous trees, or hardwoods, from broad-leaved trees. (The terms 'soft' and 'hard' do not always describe the strength of the wood, as you will see below.)

Softwoods

These are generally soft, but a few are comparatively hard. Softwoods have a *fibrous* structure. A cut surface will be seen to be smooth.

☐ **Examples of softwoods:** Redwood (from Scots pine), larch, spruce, western red cedar, yew (one of the hard softwoods).

Hardwoods

These are generally hard compared with softwoods, but a few hardwoods are quite soft –

the best being *balsa*, which is quite soft yet, because it comes from a broad-leaved tree, is classed as a hardwood.

Hardwoods are *cellular* in structure. When a cut surface of a hardwood is examined its open cells can be seen.

☐ **Examples of hardwoods:** oak, beech, mahogany, ash, elm, teak.

Timber from trees

The process of producing usable timber from growing trees is as follows:

1 The trees are felled.
2 Branches are removed and the logs taken to a sawmill.
3 the logs are sawn into planks and boards.
4 Newly felled wood is wet. The planks must be dried by **seasoning**. Open-air seasoning takes a year or more. Kiln seasoning can be carried out in a few weeks.
5 Seasoned planks are sawn to usable dimensions.

Manufactured boards

Much of the wood used today is in the form of boards manufactured from a variety of woods. **Examples are**:

☐ **Plywoods:** Layers of wood with their grains running at right angles to each other (Fig. 4).

☐ **Blockboards:** Plywoods in which the central layer is thicker than the outer layers (Fig. 5).

☐ **Chipboards:** Boards made from chips held together by resins (Fig. 6).

☐ **Hardboards:** Another form of board made from chips bound together with resins (Fig. 7).

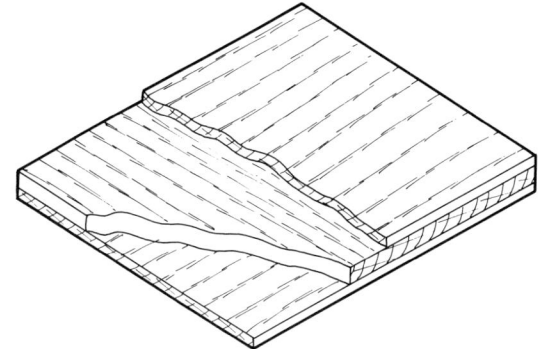

Figure 4 The construction of three-plywood

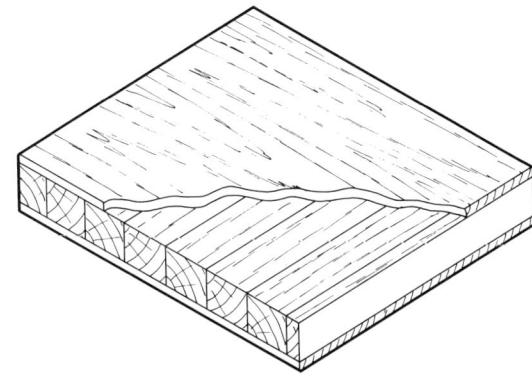

Figure 5 The construction of blockboard

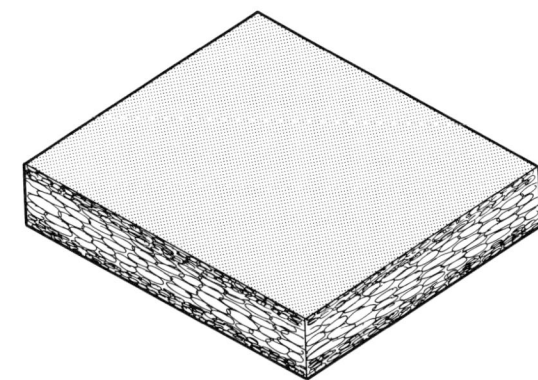

Figure 6 Chipboard

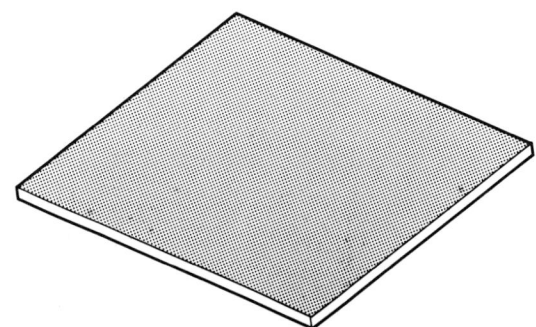

Figure 7 Hardboard

Other materials

Portland cement

This is a mixture of clay and chalk (or limestone), heated to high temperature in a kiln. It is then ground to a powder and some gypsum is added.

When mixed with water, chemical reactions cause the cement to harden gradually. The cement appears to set hard in hours, but it continues to harden over much longer periods, sometimes taking a year or more to become fully hardened.

Rapid setting cement is made by grinding the clay/chalk clinker to a finer powder. Cements which are water repellent, heat resisting and have other special properties are made by adding a variety of chemicals to the basic mixture.

For building, cement is mixed with sand or gravel. The cement is first mixed dry with either 2 or 3 parts of soft sand to 1 part of cement, before adding water. It can then be used for brickwork.

For paths, roads or reinforced concrete structures, mixes of 1 part cement to 4, 5 or more parts of sand/gravel are suitable.

Ceramic materials

Pottery is made from clays which have been fired in a kiln at high temperatures – from 950 °C to 1150 °C. Red clays, grey clays and white clays (kaolin) are used to make earthenware, stoneware and china.

All pottery is made by similar processes (except that most earthenware is not glazed).

1 The clay is kneaded to remove all air.
2 It is shaped to the required design.
3 The design is fired in a kiln to remove moisture.
4 It is removed from the kiln and allowed to cool.
5 Glaze is applied to the surface of the design.
6 The design is placed back in the kiln to fire the glaze.

Jointing of materials

The jointing of materials cannot be covered in a book of this size. A few examples are given here, but the reader is advised to turn to textbooks dealing with methods for each material for further information.

Jointing of metals

Metal surfaces can be joined together using epoxy resin adhesives (see page 114). The resulting joints are very strong and for some purposes adhesive jointing is ideal. This is particularly so when the jointing areas are large and when the design is not being subjected to heavy loads and stresses. Other methods of jointing metals are:
1 Bolting parts with nuts and bolts – Fig. 8.

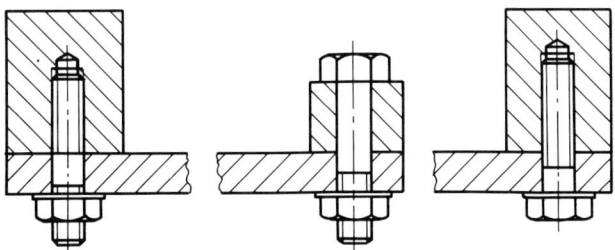

Figure 8 Bolting of parts

2 Joining parts with rivets – Fig. 9.

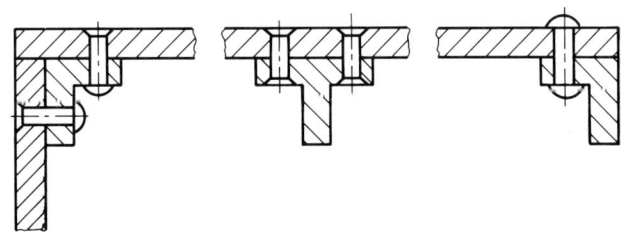

Figure 9 Riveting of parts

3 Soldering with soft solders, e.g. for containers made from tinplate or for jointing components in electrical and electronic circuits. Only suitable where no load or stress is placed on the join.
4 Brazing or silver soldering. Brazing is more suitable for metals such as mild steel and brasses; silver soldering for copper, gilding metal and silver. *Hard* soldering forms very strong joints.
5 Welding requires oxy-acetylene or electric welding equipment. It is suitable for mild steel

constructions. Properly welded joints using the correct welding materials form very strong joints, which for practical purposes are as strong as the metal itself.

6 Soldering/welding of aluminium. Because an oxide film forms rapidly on aluminium when it is heated, specially developed fluxes and solders must be used to heat join aluminium and its alloys.

Reinforcing materials

Three example of reinforcing are given here to show the principle involved.
1 Reinforcing wood by laminating.
2 Glass reinforced plastic (GRP).
3 Reinforcing concrete with steel rods.

Reinforcing wood

If layers of thin wood are joined by gluing (**laminated**), the resulting structure is very strong. The adhesives used are usually polymer glues such as urea formaldehyde, phenol formaldehyde, pva and, where great strength is necessary, epoxy resins.

GRP

GRP, glass fibre reinforced plastic, is made of glass fibres bonded into polyester resins. When fully polymerised, the resulting glass reinforced fibre item is flexible and strong. Carbon fibres are used for the same purpose, but are much more expensive than glass fibres.

Reinforced concrete

Steel rods are placed in concrete while it is still in its wet, plastic state and the concrete is then allowed to set. The resulting concrete mass is able to take much greater loads and stresses than concrete which is not so reinforced.

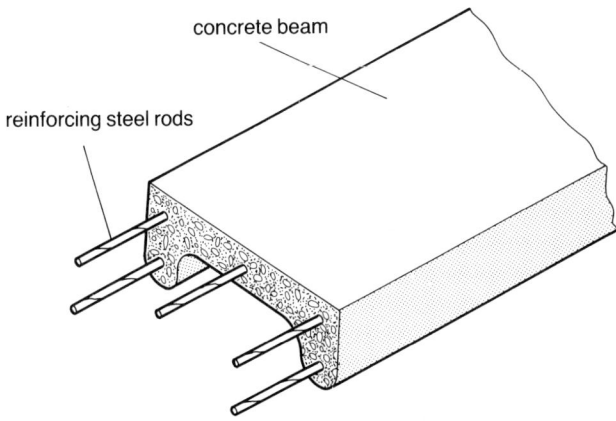

concrete beam

reinforcing steel rods

Figure 12 Reinforcing of concrete

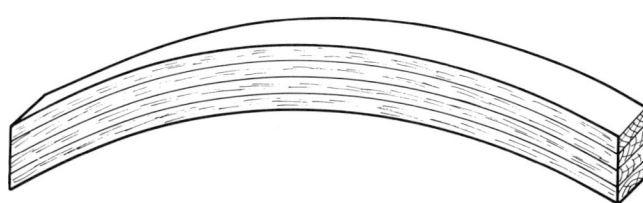

Figure 10 Reinforcing of wood by laminating

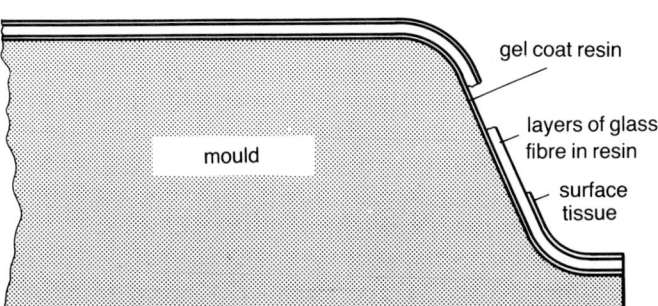

gel coat resin

layers of glass fibre in resin

surface tissue

mould

Figure 11 Glass reinforced plastic

Exercises

1 Name three factors which must be considered when choosing materials for a design.

2 When choosing a material for a design, why is it important to consider whether it might fail under the stresses imposed on the material by the functioning of the design?

3 What is meant by the *factor of safety* with regard to materials selected for a design? What would you suggest as a sensible factor of safety in a design where injury to people could result from failure of a material in the design?

4 What is meant by the term *non-ferrous*?

5 Describe the difference between *hardwood* and *softwood*.

6 Carbon steels are divided into several groups – *dead mild, mild, medium mild, tool*. Explain the differences between these groups and give a use for each.

7 Stainless steel is an alloy. What are the main constituents of this alloy?

8 Why is copper such an important metal?

9 What does the term *anodising* mean?

10 Name three methods of coating steel sheet to improve its rust resistance.

11 Explain briefly how *case hardening* is carried out.

12 There are three main groups or classes of *plastics*. What are the names given to these three groups? Name two plastics from each group.

13 What do you understand by the term *polymer*?

14 What is *manufactured board*?

15 How can the following materials be reinforced?
- concrete;
- polyester resin;
- wood.

16 Why are *epoxy resins* such an important group of adhesives?

17 What is meant by the term *seasoning* as applied to the treatment of wood?

10 Design Process

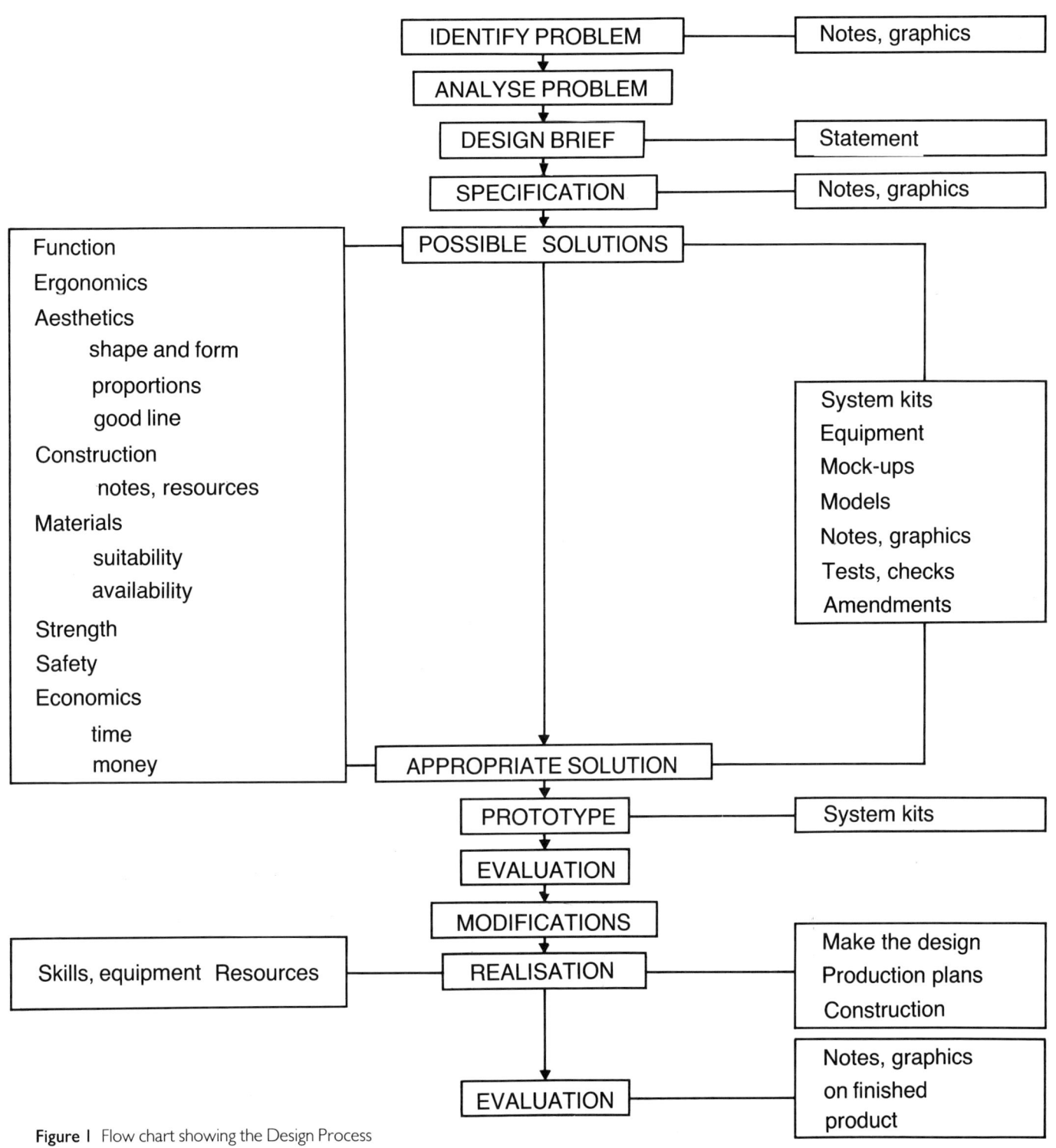

Figure I Flow chart showing the Design Process

A short description of a design process is given in the Introduction (pages 1 to 9). It is described at greater length in this chapter.

Identify problem

Any design results from attempts to find a solution to a problem which has arisen. The problem might be, for example, to satisfy the common human need to sit in comfort – resulting in the numerous designs of chairs in use today. There is also the modern problem that people wish to travel from place to place as quickly as possible, resulting in designs for vehicles of all types – bicycles, cars, buses, trains, ships, aeroplanes. Some examples of the types of problems which students at schools and colleges tackle are:

1 The problem of switching a lamp on and off from a remote position.
2 The problem of switching the lights in a house on and off at given times, without being in the house.
3 The problems arising when a person becomes bed-ridden such as:

■ the discomfort of lying in the same position for long periods of time;
■ finding difficulty in eating with comfort;
■ being unable to read in comfort.

4 How to help a disabled person to overcome the problem of not being able to move easily from place to place.
5 The problem of designing a robot which will lift a weight from one position to another.
6 The problem of designing a device allowing you to open the door of a garage while sitting in a car.

The problem to be overcome may not seem all that obvious at first. In the example given above of finding a way of switching lights on and off in a house, the original problem might have been that the owners of the house wanted to give the impression that the house was occupied when in fact it was not. Thus the problem of switching the lights was identified as a method by which the original problem could be overcome.

Analyse problem

When the problem has been identified, it is necessary to examine and analyse it. If, for example, one is trying to find a solution to the problem of fast travel from place to place, the analysis of the problem may take a form such as the following:

1 How many people are to be carried at a time?
2 How fast?
3 Wheeled vehicle? One, two, three, four or more wheels?
4 Are wheels in fact necessary?
5 Human-powered vehicle? Pulled, pushed, bicycle type?
6 Animal-powered vehicle? Which animal?
7 To travel on land, on water, in the air?
8 Wind powered? If so, is it to be a land, water or air vehicle?
9 Powered by some form of mechanical unit? Gas powered, oil or petrol powered, powered by electricity?
10 Any other form of power suitable?

Only after analysing such details arising from the original problem is it possible to arrive at a **design brief**.

Design Brief

When the problem has been clearly identified, a design brief must be stated. This usually takes the form of a written statement in which a clear description is given of what it is that is to be designed. Many design briefs take the form of a single sentence. Others may be more complicated.

Two examples of design briefs

1 Design a lighting unit, for temporary lighting in a bedroom, which is to be worked from its own power source.
Note the clear definitions in this brief – the lighting is to be temporary; it is to be suitable for a bedroom; it is to be powered from its own power source.
2 Design a timer, which can be adjusted to give timings between 0 and 5 minutes, accurate to 0.5 second.
Note again the clear definitions – the timer must be adjustable; it must range from 0 to 5 minutes; it must be graduated in 0.5 second intervals.

All designing should include a design brief. At any stage it can be referred to in order to check whether or not it is being followed. It is easy to be misled by outside areas when designing and to stray away from the brief. Note that amendments to an original design brief may be necessary if, during the process of design, it is found that the original brief does not state clearly what is to be designed.

Specification

A specification may accompany a design brief. This should specify details expected from the design. A specification to accompany the design brief for a temporary bedroom light given above might be:

- The lighting unit must be small enough to be placed on a bedside table.
- The power supply is to be two 1.5 V dry battery cells.
- The unit must incorporate an on/off switch.

Possible solutions

This is the stage in the design process when ideas are investigated, noted, drawn and, if necessary, tested.

Each idea which comes to mind should be drawn, with notes added to describe the drawings. Drawings may be freehand, drawn with instruments or drawn with the aid of a computer program (computer aided drawing).

Notes which accompany drawings may be written, typed, produced with the aid of a computer word processor, written in capitals or in lower case printed lettering, added to drawings with the aid of letter stencils, or added from dry transfer sheets.

Some ideas for solutions may need to be tested, using kits such as those developed for mechanics, structures, electrics, electronics and pneumatics. Possible solutions may be made up as models and then the models can be tested. Each idea for a solution may have to be **evaluated**, i.e. checked to test whether or not the idea is one which could be developed.

Consideration will have to be given to details such as:

1 Will the solution idea work as expected?
2 Are ergonomic considerations satisfied?
3 Will the idea be aesthetically pleasing, i.e. attractive to look at? Is its shape and form pleasing? Are its proportions good? In some examples, this consideration will not affect the idea – e.g. an idea for an electronics or pneumatics circuit need not necessarily be aesthetically pleasing.
4 Can the idea be constructed? There is no point in suggestions for solutions which cannot be made. Are the tools and equipment needed to make the design available for constructing a suggested solution?
5 Are materials for making the solution available, or can they be purchased?
6 Will the constructed solution be sufficiently strong?
7 Have all safety considerations been taken into account?
8 Is the solution economic? Not only whether the solution can be made economically in cash terms, but also whether its shape and form are economically designed, and whether there is time to construct the solution.

Appropriate solution

A time must come when one of the possible solutions can be chosen as an appropriate solution to the design brief. If none of the ideas for solutions are suitable, a modification of one or more of the ideas may be developed into an appropriate final solution. The selected, chosen solution should be accurately drawn and 'written up'. This chosen solution must be evaluated. All the details of function, ergonomics and so on should again be assessed with regard to the chosen solution.

Prototype

At this stage in the design process, consider whether a model (a prototype) should be made. The construction of a model could save considerable time and effort later. Errors, mistakes, poor constructions and other details needing attention, may show up in a prototype. The design can then be amended at this early stage, rather than at a later stage when the final design has been made and when it will be too late to remedy poor details in the design.

Evaluation

Evaluation ('finding the true value of') plays an important and repeatedly occurring part in any design process. Each and every idea for solving a design brief needs to be examined in order to assess whether it meets the requirements of its design brief. Evaluation of a chosen appropriate solution and its model (if made) is of particular importance, because this is the last stage before the design is made or realised. If an evaluation shows that modifications are necessary, these can be carried out at this stage, before starting to make the final design.

Realisation

This means making the design, Before making the design it may be necessary to draw up a production plan, listing the order in which parts of the design are to be made, together with the possible times that each part will take to complete.

The making of a design assumes that the maker has the necessary skills, resources and equipment and materials from which the design can be made.

Evaluation

This final evaluation of the completed design is of particular importance. An honest assessment of the design is essential. Even though you may have been completely and solely responsible for all stages of the design, this does not necessarily mean that you must just sing its praises. Try to look at the finished work with a truly critical judgement. If you find you cannot do this yourself, ask others to make the evaluation for you. Ask the following questions of the design:
1 Does the design meet the requirements of its design brief?
2 Does the design function as it should?
3 Does it meet ergonomic requirements?
4 Is it pleasing in its shape and form; its proportions; its colour; its surface finish; its general appearance?
5 Has it been successful economically – in terms of its financial cost, its appearance and the amount of time spent on making it?

6 Does it need modifying? Are the modifications worthwhile in terms of the time and money needed to carry them out?

Exercises

1 Answer all parts of this exercise.

Problem: As a keen cyclist and a camper, you wish to carry your camping kit by bicycle.

■ Analyse this problem. Make a list of ways in which the problem could be met.
■ One answer would be to design a trailer to be attached to the bicycle, which would carry your kit. This could cause further problems. Analyse these. Make a list of ways in which these could be met.
■ One problem in making a trailer is that of fixing lights to its rear – STOP, TURN LEFT, TURN RIGHT. Write a design brief, with a specification, for a design which would meet this problem.

2 Answer all parts of this exercise.

Problem: Two tags shaped as in Fig. 2 are to be made for each pupil in a school for tying on their shoes when changing for P.E. The tags are to be made from aluminium in the school workshop.

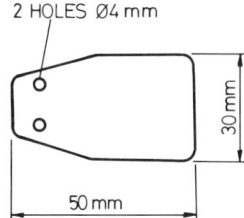

Figure 2 Exercise 2 – aluminium tags

■ Make a list of the problems arising from your being required to make the tags.
■ One problem is that so many tags are needed. List the methods by which this problem could be solved.
■ One of the items in your list above is that, after making the shaped aluminium pieces, two holes must be bored in each one. List methods by which this could be done.
■ Among your answers above is one involving a device which could be made to locate the position of the holes by a form of air bleed

mechanism and then operated the drill to drill the holes. Write a design brief with specifications for this mechanism.

Design briefs from examination papers

1 People who suffer from arthritis often find it difficult to carry out simple tasks such as turning the pages of a book which they are reading. Design a piece of equipment which, when placed on a table or other horizontal surface, will:
- support a book of maximum dimension 220 mm × 140 mm × 25 mm when closed;
- make it easier for a disabled person to turn the pages;
- be adjustable to different angles to suit different sitting positions;
- give the disabled person full control in turning over the pages.

LEAG

2 Many people are now actively interested in physical fitness and this in turn has given rise to an increase in the sale of home fitness equpment. Some of the equipment is very expensive and there is a need, therefore, for a fitness machine for home use which is inexpensive to make and which will enable the user to perform safely a wide variety of exercises.

Design a home fitness machine which will meet the above requirements. The machine which you design should also:
- allow a wide range of resistance to the exercises;
- be collapsible and portable;
- be easy for one person to assemble and take apart with tools or specialised equipment.

LEAG

Revision notes

The following is a brief summary of the design process.

1 What is the **problem** or **situation** to be solved?

2 Write a **Design Brief**. Include:
- what it is that is to be designed;
- a simple specification.

3 Make sketches and notes showing ideas for **possible solutions**. For each idea look at:
- areas requiring investigation:
- the need for research;
- do you need to experiment?

4 Examine difficulties arising from your possible solutions.

5 Choose one of your solutions as an **appropriate solution**, or select the best parts from several solutions to form an appropriate solution.

6 Is a **model** (a prototype) needed?

7 **Evaluate** your chosen solution.

8 Does it need to be **modified**?

9 In drawings and notes show how the design is to be made – **working drawings**.

10 **Realise** (make) the design.

11 Test the design. Modify it if necessasry.

12 **Evaluate** the completed design. Does it satisfy the design brief?

Note 1: The process of designing a technology project can only be successfully followed if good methods of communicating ideas, solutions and the necessary investigations and evaluations are used.

The best method of communicating these is by means of graphics and notes. All the drawings and notes connected with any one project should be kept in the folio or folder.

Note 2: Not all designing has to follow all the steps outlined in the flow chart, Fig. 1.
- Some of the steps may not be necessary in some circumstances, e.g. the problem for which you are designing may have already been stated.
- You may be able to arrive at a solution without much difficulty. This could mean that a long investigation is not needed.
- You may not need to make a model because all difficulties have been clearly understood in your notes and graphics.
- Some designing is the result of intuitive thinking, without the investigation which most of us need when designing.

On the whole, however, students at school and college are probably best advised to follow a procedure such as that outline in the given design process.

Note 3: Modern industrial and commercial designing is nearly always the result of team work. Only very rarely will modern designs be the result of a single person working on his/her own. Students at schools and colleges usually work at their own designing. This is good practice and helps them to understand the difficulties in producing well-designed work. However, it is advisable for students to have some practice in designing as part of small teams or groups.

11 Presentation

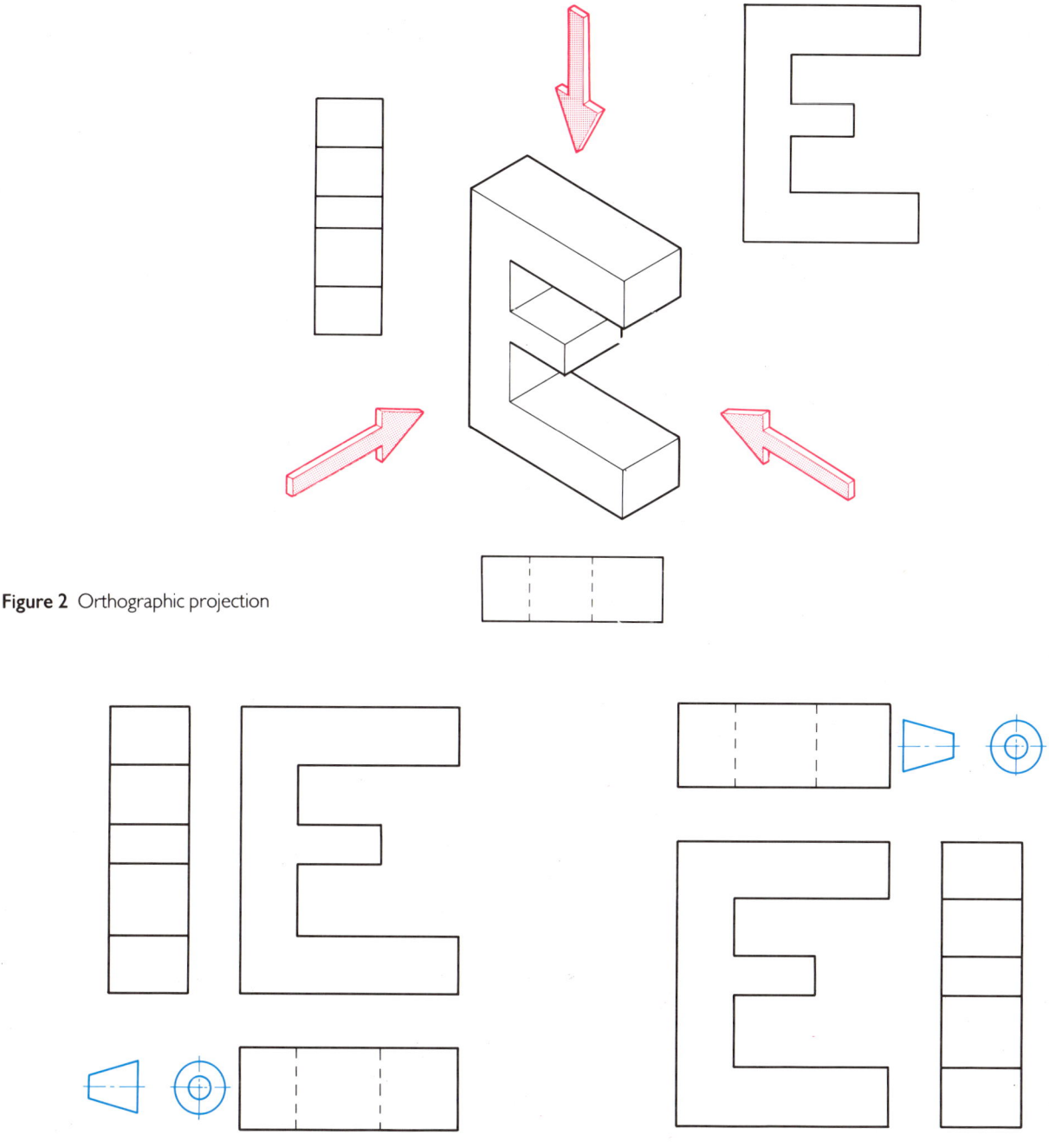

Figure 2 Orthographic projection

Figure 3 First Angle orthographic projection

Figure 4 Third Angle orthographic projection

Introduction

Good presentation of project notes, with good graphics, is important in designing. Many drawings for ideas, leading up to a final design solution, will be drawn freehand. Finished working drawings will probably be done with the aid of instruments. Colour can add to the effects of the presentation of graphics in all design work.

Figure 1 shows lines used in technology drawings. A thick line is about twice as thick as a thin line.

thick lines – all outlines

thin lines – dimension and projection lines, hatch lines

hidden detail – broken thin lines

centre line – dashed thin lines

break lines – thin lines

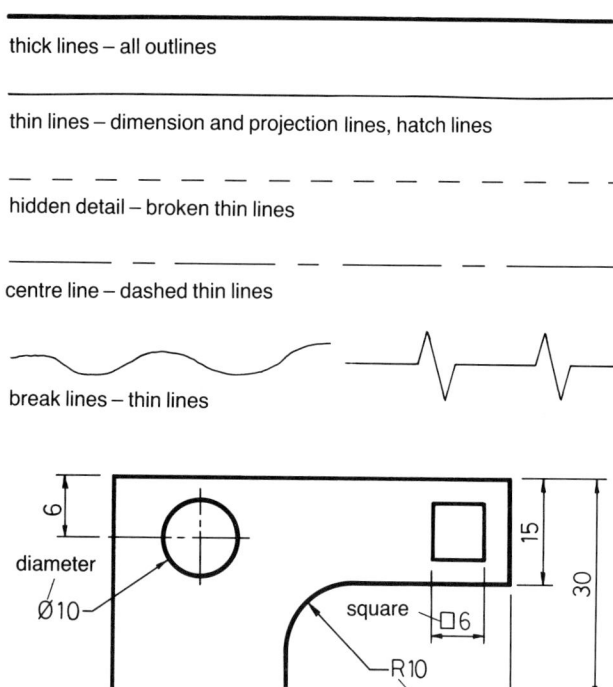

simple dimensioning

Figure 1 Lines and dimensions

Orthographic projection

Figure 2 shows the principles of orthographic views. The object is looked at from its front and what is seen is drawn as a FRONT VIEW. The object is then looked at from an end and what is seen is drawn as and END VIEW. The object is viewed from above and what is seen is drawn as a PLAN. Views are often called **elevations**.

The three views are drawn in either First Angle or Third Angle projection – Fig. 3 and Fig. 4.
Note:
1 End views and plans are in line and front views.
2 Any side of an object can be its *front*.
3 Either, or both, end views may be drawn.
4 An INVERTED PLAN can be drawn with the subject viewed from below.
5 In First Angle projection:
■ end views face outwards;
■ end views are drawn on the opposite side of the front view from the point they are viewed from;
■ plans face outwards;
■ plans are drawn *below* front views.
6 In Third Angle projection:
■ end views face inwards;
■ end views are drawn on the same side of the front view as the point from which they are viewed;
■ plans face inwards;
■ plans are drawn *above* front views.

Note: An orthographic projection can be a single front view.

Examples of orthographic projection

Figures 5 and 6 are First and Third Angle projections of part of the model for the sweets dispenser – see pages 147 to 155. Drawings such as this, which give enough information for another person to make the article, are called **working drawings**.

Pictorial drawing
Isometric drawing

Isometric drawings are made with 30°, 60° set squares. The method of constructing isometric drawing of straight-sided objects is shown in Fig. 7. The procedure for drawing a circle in isometric is shown in Fig. 8.
1 Draw the circle. Draw ordinates a, b and c.
2 Draw the isometric axes of the circle, at 60° to each other.
3 With a compass, transfer the ordinates a, b and c onto one isometric axis.
4 Draw a fair curve through a, b, c and d.
5 Line in the isometric ellipse.
6 Drawing circles in isometric can result in ellipses at these angles.

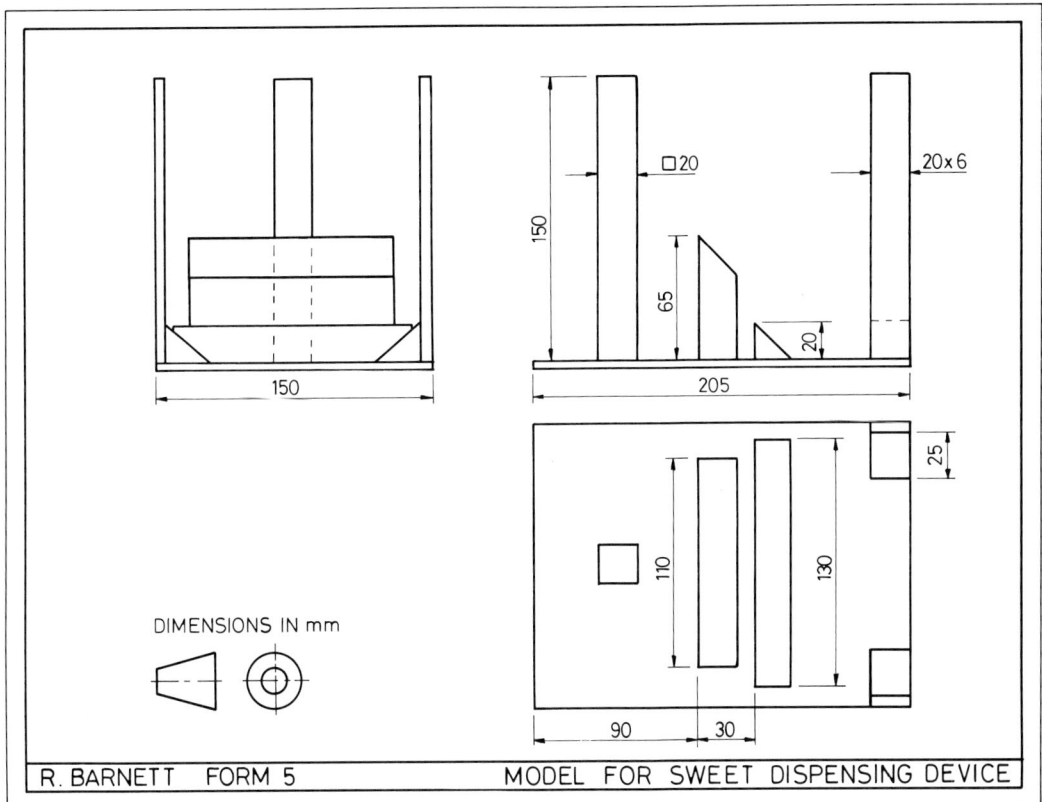

Figure 5 A First Angle orthographic projection

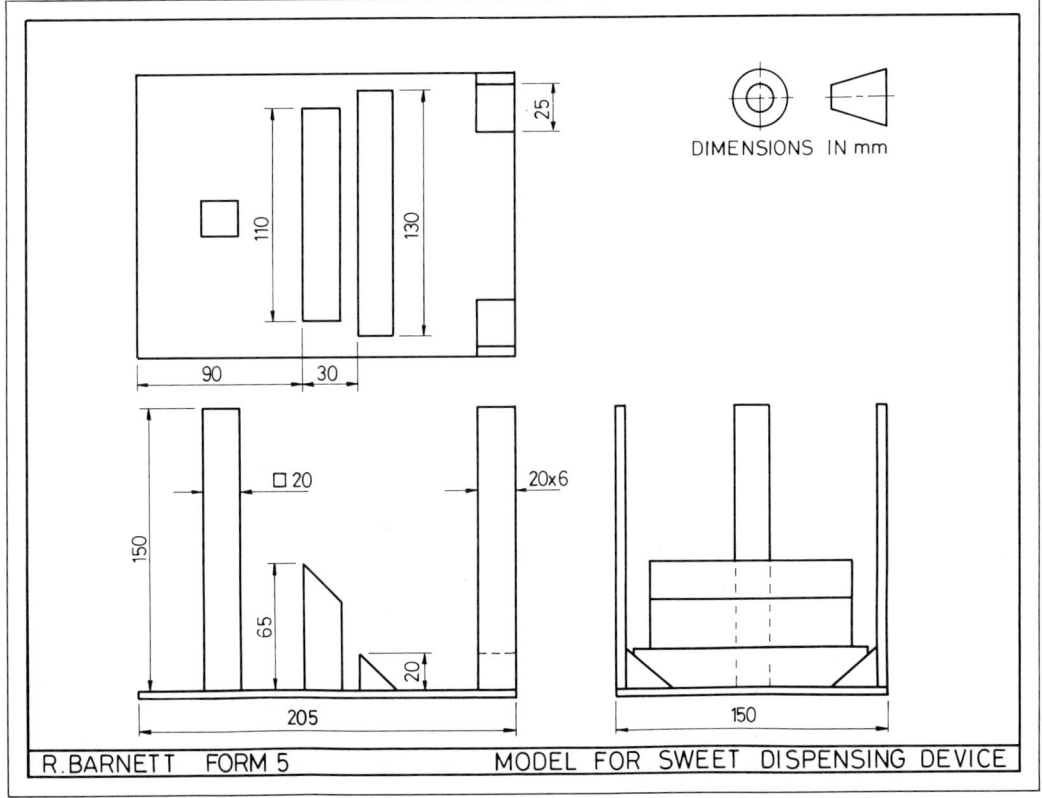

Figure 6 A Third Angle orthographic projection

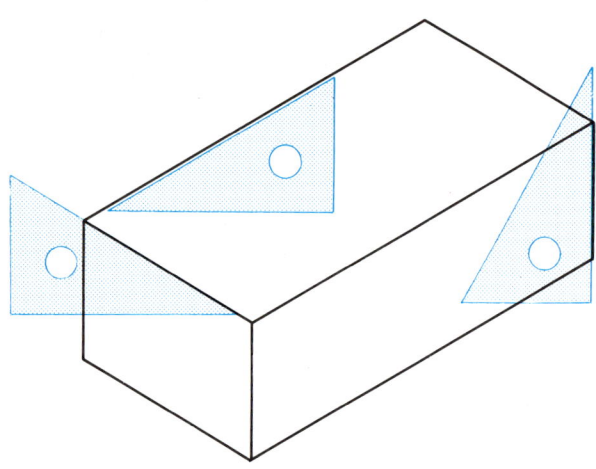

Figure 7 Method of isometric drawing

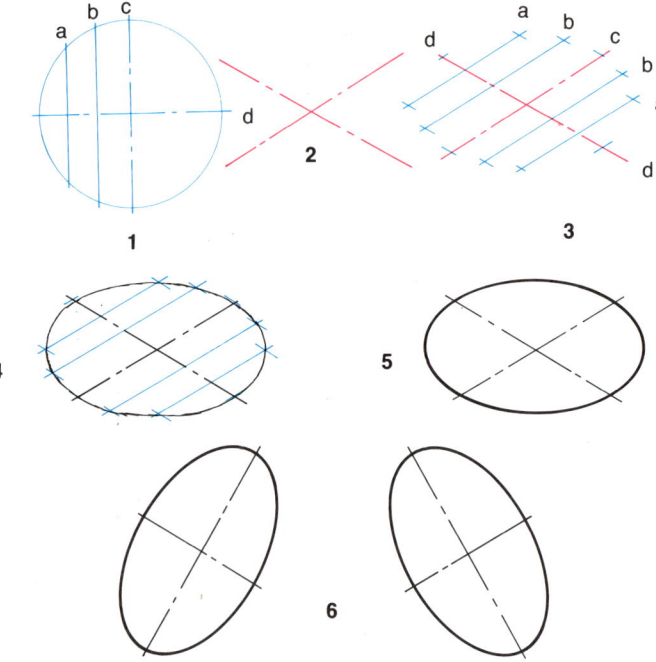

Figure 8 Method of drawing curves in isometric drawing

Cabinet drawing

A cabinet drawing starts with the front view. Then half size (scale 1:2) lines are drawn from points on this front view at 45° to produce a pictorial appearance.

Cabinet drawing is suitable for drawings which include complicated shapes in one view. Figure 9 shows how to draw a cabinet drawing:

1 Draw a front view of the object.
2 Draw lines at 45°.
3 Measure half size (scale 1:2) lengths along the 45° lines and complete the drawing.

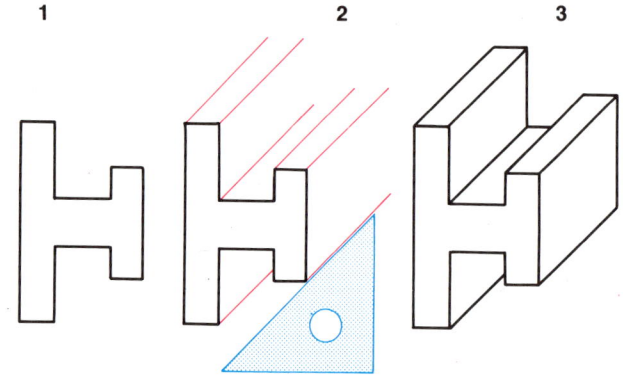

Figure 9 Method of cabinet drawing

Examples of isometric and cabinet drawing

Note: The coloured lines on the drawings of the pass-through switch and the diode show the best way of starting an isometric drawing: draw a box containing the whole article.

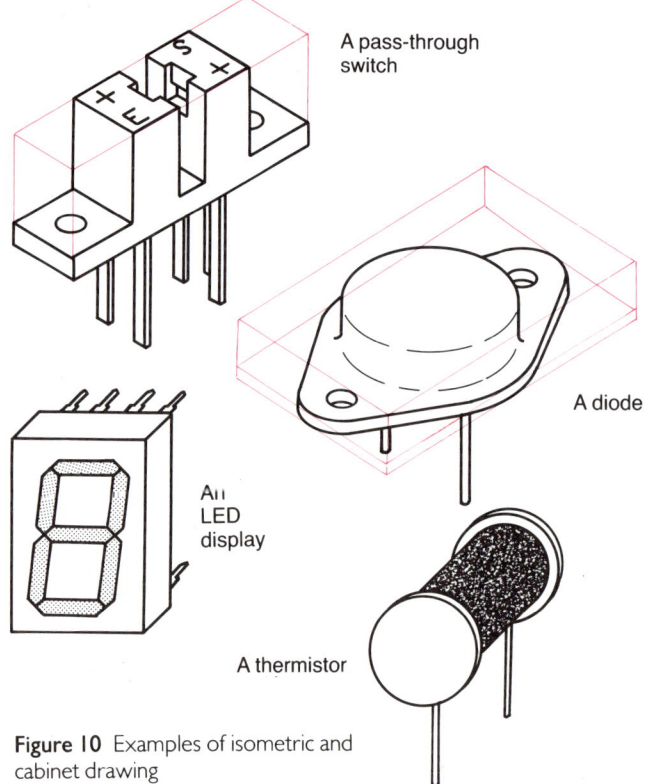

A pass-through switch

A diode

An LED display

A thermistor

Figure 10 Examples of isometric and cabinet drawing

Planometric drawing

Planometric drawings are made with 30°, 60° or 45° set squares. Start with a plan, and take vertical lines from points on it.

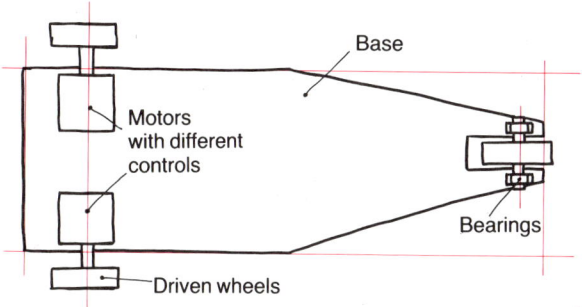

Figure 11 An example of freehand orthographic drawing

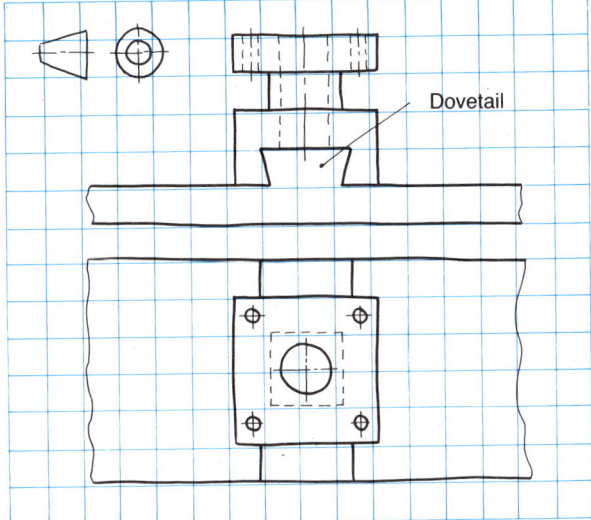

Figure 12 An example of freehand orthographic drawing on square grid paper

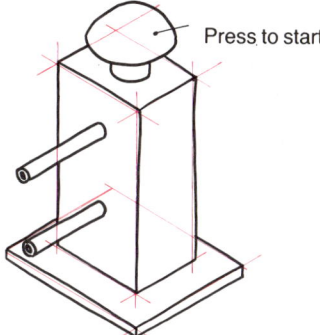

Figure 13 An example of freehand isometric drawing

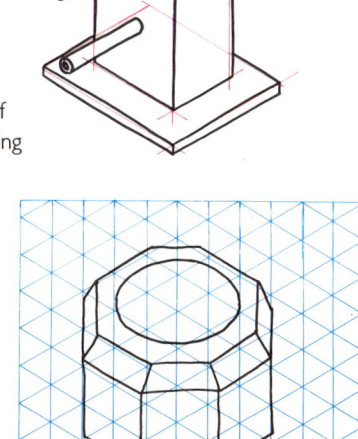

Figure 14 An example of freehand isometric drawing on isometric grid paper

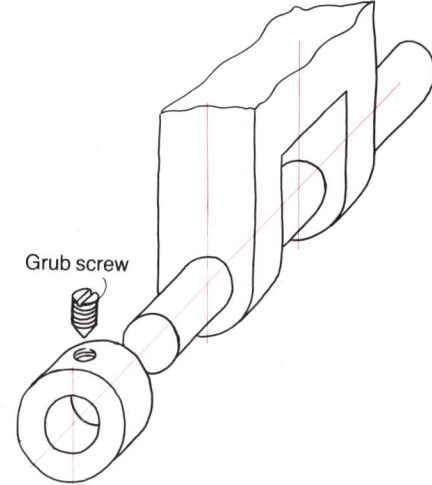

Figure 15 An example of freehand cabinet drawing

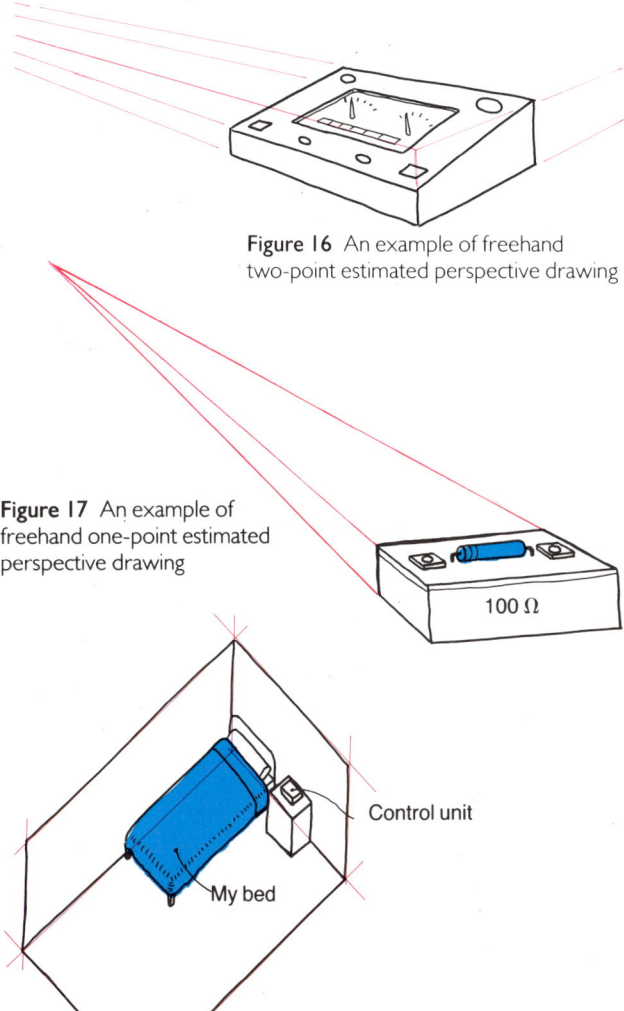

Figure 16 An example of freehand two-point estimated perspective drawing

Figure 17 An example of freehand one-point estimated perspective drawing

Figure 18 An example of freehand 45° planometric drawing

In 30°, 60° planometrics, verticals are drawn to the same scale as other lines in the drawing. In 45° planometrics, the drawing looks better if verticals are reduced to about 75% of the scale of other lines.

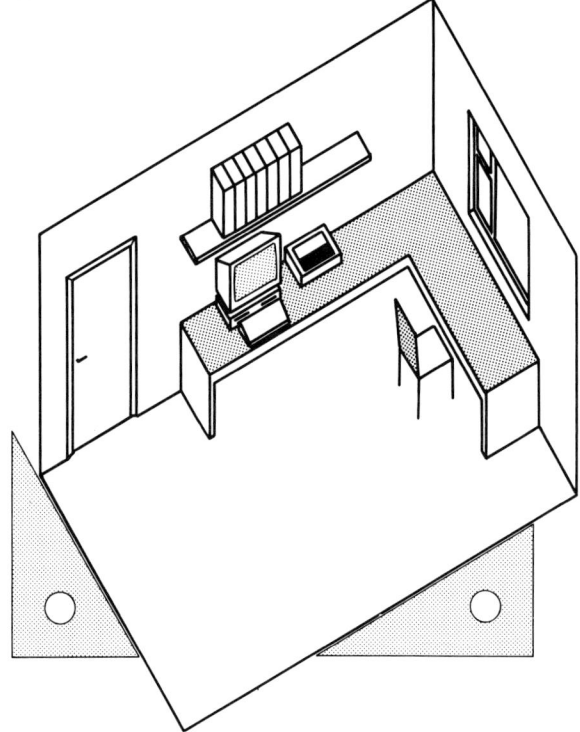

Figure 19 Example of a planometric drawing

Perspective drawing

Either **one-point** or **two-point** estimated perspective drawings are suitable.

All perspective drawing takes into account the fact that parallel lines appear to meet in the

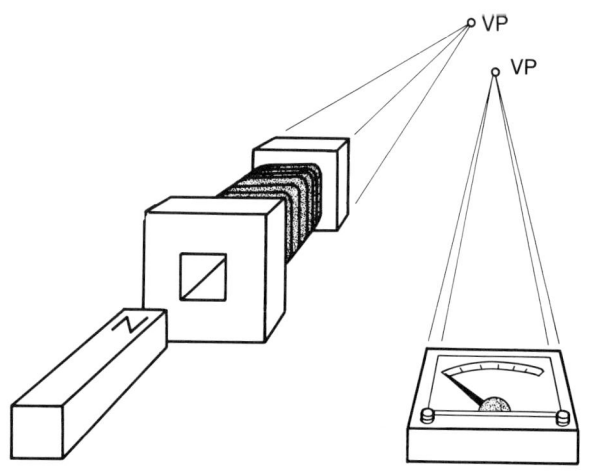

Figure 20 Examples of one-point estimated perspective drawing

distance. The meeting points are called **vanishing points** (VPs). In **estimated perspective**, VPs are estimated. Figures 20 and 21 are examples of the two forms of estimated perspective.

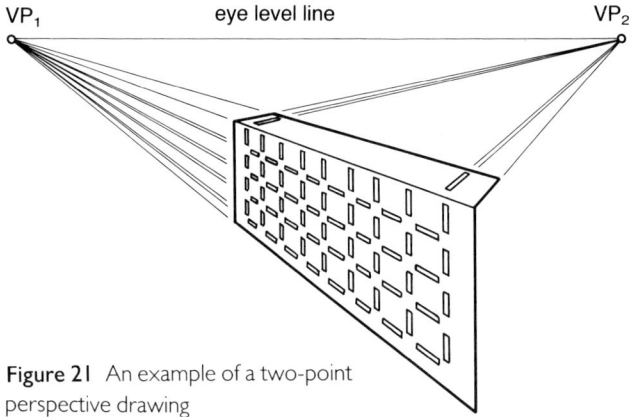

Figure 21 An example of a two-point perspective drawing

Note:
1 As equal spaces get nearer to the VPs, they appear narrower, in the same way as the width between parallel lines becomes narrower. This can be seen in Fig. 20 and Fig. 21.
2 The VPs are drawn at the height of an average person's eye level – about 1.7 m (scaled). This is not critical in estimated perspective drawing.
3 In two-point perspective, the two VPs should be on the same horizontal line.

Freehand drawing

When drawing ideas for solutions, most drawings will be freehand. Drawings with instruments are more likely to be for final solutions or for working drawings.

Any of the methods of drawing with the aid of instruments may be used for freehand drawings – orthographic, isometric, cabinet, planometric or estimated perspective.

When working freehand, it is often helpful to draw within guide lines, perhaps drawn with instruments, to help you to make sure that your drawing will have a good appearance, When a drawing has been completed the guide lines can easily be erased.

All the drawings on page 129 have been drawn freehand using the methods already briefly described. Where guide lines have been drawn to assist the final drawing, they are shown coloured.

☐ **Note:** If you find you can make better drawings with the aid of instruments, then produce the drawings for your design work with instruments. It is advisable, however, to practise freehand drawing, because drawings can then be produced much more quickly. If you are not confident of your skill at drawing, use instrument-assisted freehand drawing. But – the more you practise design drawing, the better will your graphics be.

Colour and shading

Colour can be added to drawings to:
1 make the meaning of the drawing clearer;
2 emphasise and draw attention to features;
3 show differences between parts of a drawing.

Colouring media

The two main types of colour for this form of graphics are colour pencils and colour pens. Other forms of colouring may be suitable, e.g. watercolour washes, marker pens, dry transfers, air-brushing. We will only concern ourselves here with colour pencils and pens.

Colour pens

Colour biro pens, colour Penstiks and other *throwaway* pens can produce good linework effects and/or good shading effects. Technical pens such as *Rotring* pens, which can be recharged with coloured inks, can also be used. If you do use technical pens with coloured inks:
1 clean the pen and its reservoir thoroughly after use;
2 keep the cap on the pen when not in use.

Colour pencils

Avoid wax crayon pencils – the resulting waxy finish can rub off onto other sheets of drawing and spoil them. Good quality pencil crayons are very suitable.

Colouring with pencils can produce a three-dimensional (3D) appearance on articles with flat surfaces, or it can show curved surfaces effectively.

Examples of colour and shading

☐ **Fig. 22:** 3D appearance achieved by making all outer lines thicker than other lines. A common graphic method, easy to apply.

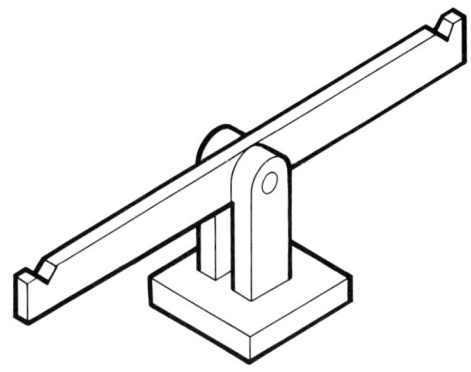

Figure 22 Thick outlines add to the 3D effect

☐ **Fig. 23:** single-tone shading, in colour or black pencil.

☐ **Fig. 24:** two-tone shading. Adding a second depth of tone adds to the 3D appearance of the drawing.

☐ **Fig. 25:** three-tone shading. A third depth of shade further increases the 3D appearance.

☐ **Fig. 26:** shading curved surfaces. The four examples show a gradually increasing depth of tone towards the outer part of the curve; curved or thin lines; and varied thickness of lines.

☐ **Fig. 27:** a freehand drawing emphasised by a coloured background.

☐ **Fig. 28:** The lines of the drawing are in two line thicknesses

☐ **Fig. 29:** a design for a project folder cover. Thick and thin coloured lines form borders to the cover and its photographs. Lettering added from Letraset dry transfer sheets.

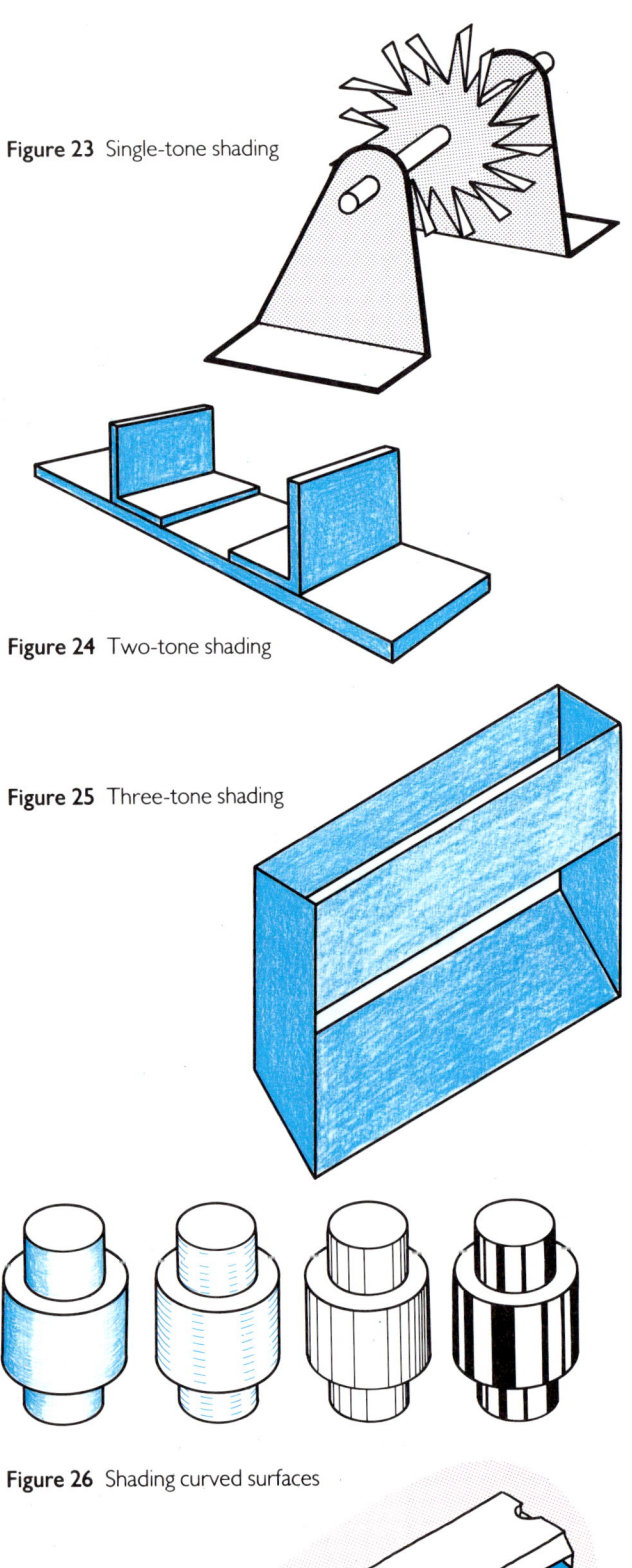

Figure 23 Single-tone shading

Figure 24 Two-tone shading

Figure 25 Three-tone shading

Figure 26 Shading curved surfaces

Figure 27
Colour background

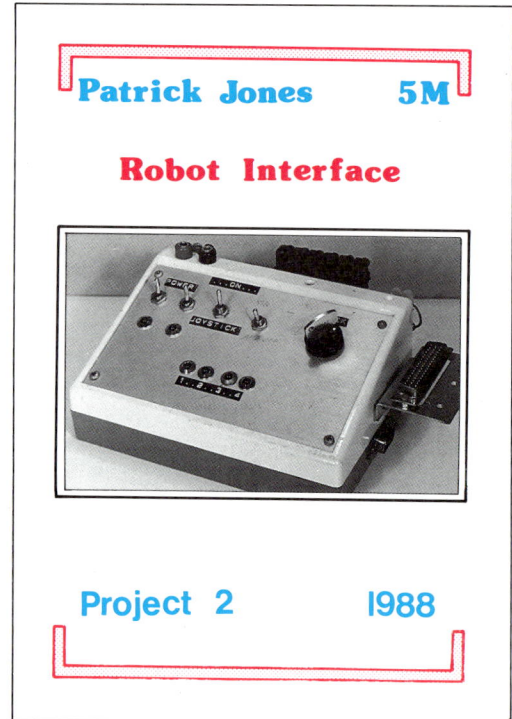

Figure 28 Thick line background outline

Patrick Jones   5M

Robot Interface

Project 2   1988

Figure 29 An example of a project folder cover

Note: Figures 30 to 33 appear on page 134.

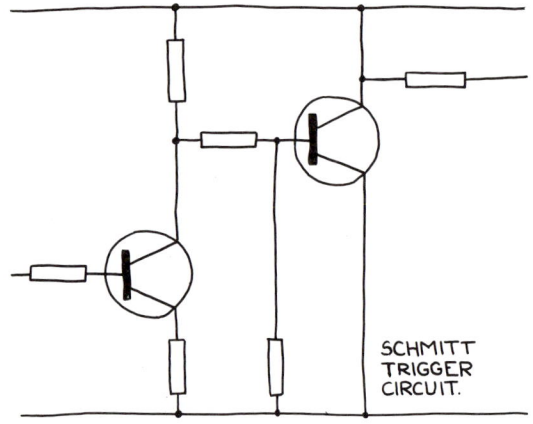

SCHMITT
TRIGGER
CIRCUIT.

Figure 34 Freehand electronics circuit

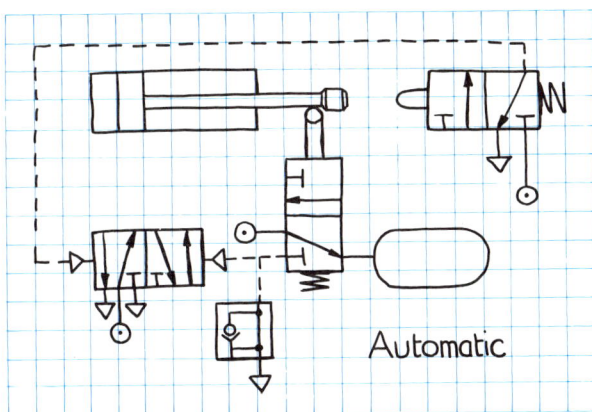

Figure 35 Freehand pneumatics circuit drawn on square grid paper

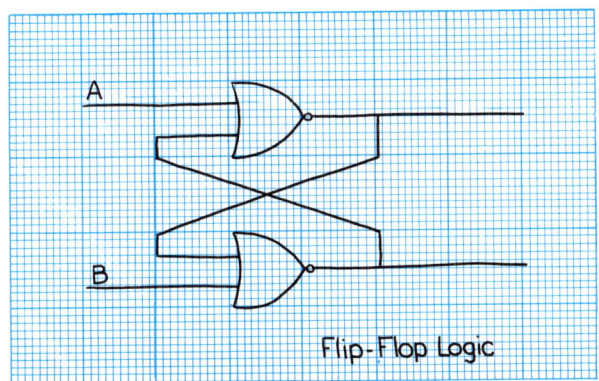

Figure 36 Freehand logic circuit drawn on graph paper

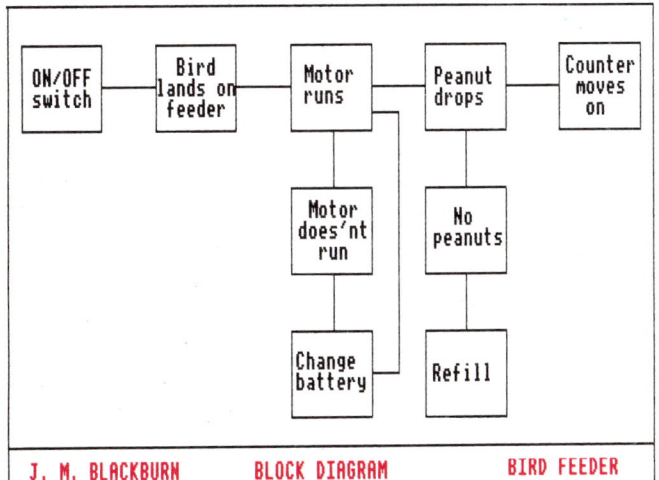

Figure 37 Computer printout from Bird Feeder project

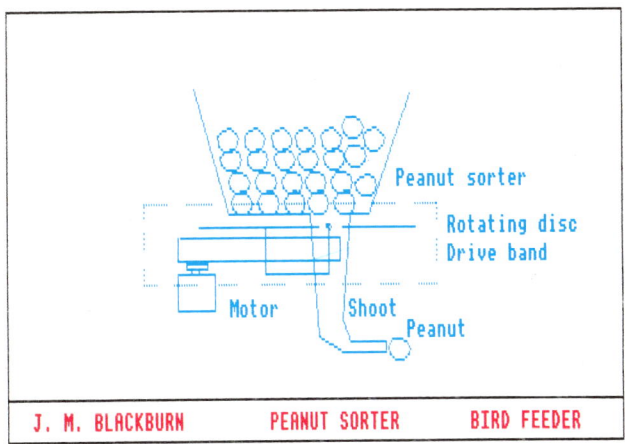

Figure 38 Computer printout from Bird Feeder project

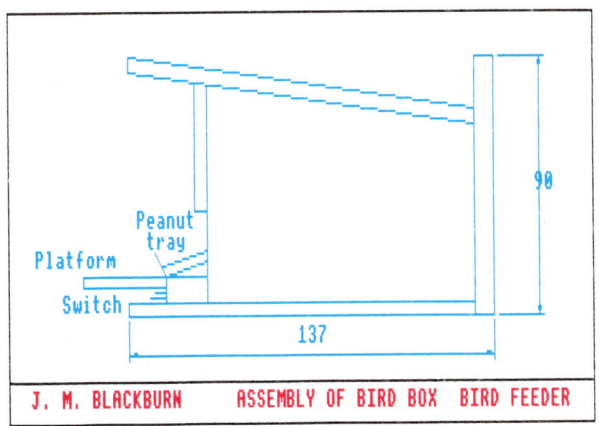

Figure 39 Computer printout from Bird Feeder project

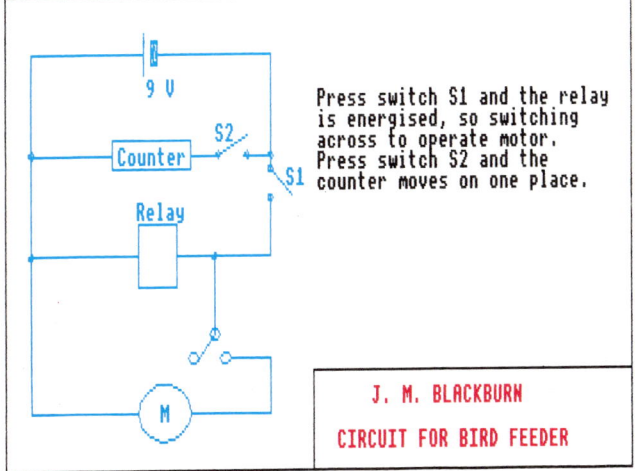

Figure 40 Computer printout from Bird Feeder project

Sectional views

Sectional views, or **sections** show the interior shape of an object. Imagine that the object has been cut through by a plane and the cut surfaces are then viewed at right angles.

Figure 30 shows the principles. Figure 31 is a two-view First Angle projection which includes a section.

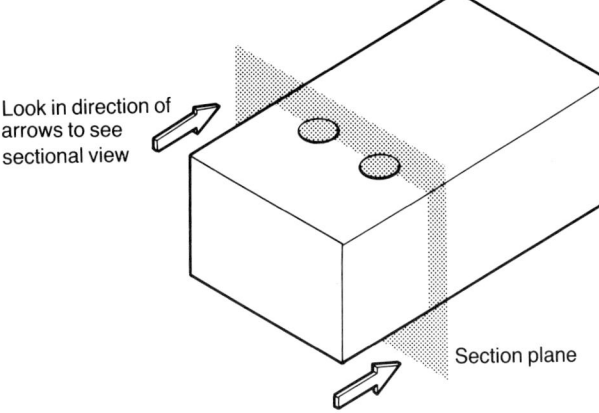

Look in direction of arrows to see sectional view

Section plane

Figure 30 The principle of sectioning

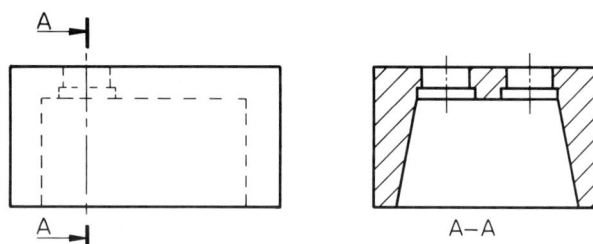

Figure 31 A simple sectional view

☐ **Note:**
1 The **section plane** is drawn as a thin centre line ending in short thick lines.
2 Arrows touch the thick lines of the section plane line to show the direction of viewing.
3 Letters label the section line.
4 The section is also labelled with these letters.
5 The cut parts of the section are hachured with thin lines at 45° about 4 mm apart.
6 All parts beyond the cut surface are shown.

Outside views within a sectional view

Some parts of sectional views are shown as outside views, e.g. shafts, spindles, nuts, bolts, keys, pins, cotters, web, ribs and similar features. Figure 32 is an example.

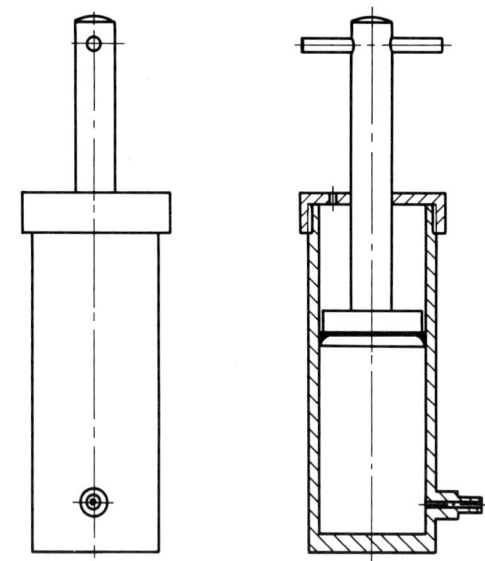

Figure 32 A sectional view containing outside views

Drawing aids

A number of drawing aids can be used to speed up drawing. A group is shown in Fig. 33. Aids are useful for drawing those details which have to be repeated over and over again, or which can only be accurately drawn using geometrical constructions.

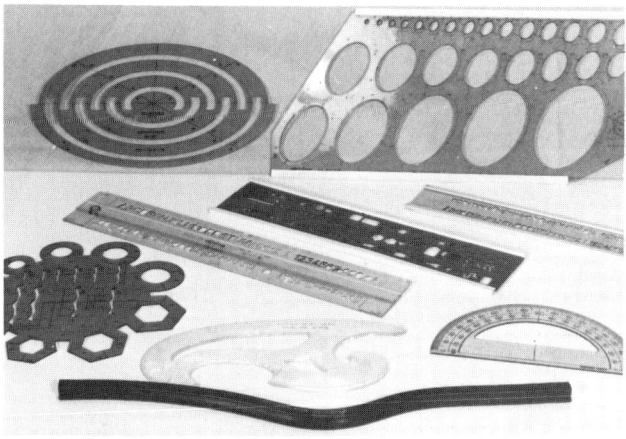

Figure 33 A variety of drawing aids

Circuit diagrams

Electrical, electronic and pneumatics, flow and logic circuit drawings may often be required in technology design work. Before a satisfactory circuit can be drawn it must be:

1 designed;

2 tested – often with circuit kits.

When you are satisfied that the circuit design works, you can draw its diagram. Designing of circuits is often carried out freehand. Three examples of freehand drawing of circuit diagrams are given:

☐ **Fig. 34:** an electronics circuit drawn freehand on plain paper, page 132.

☐ **Fig. 35:** a pneumatics circuit drawn freehand on square grid paper, page 133.

☐ **Fig. 36:** a logic circuit drawn freehand on graph paper, page 133.

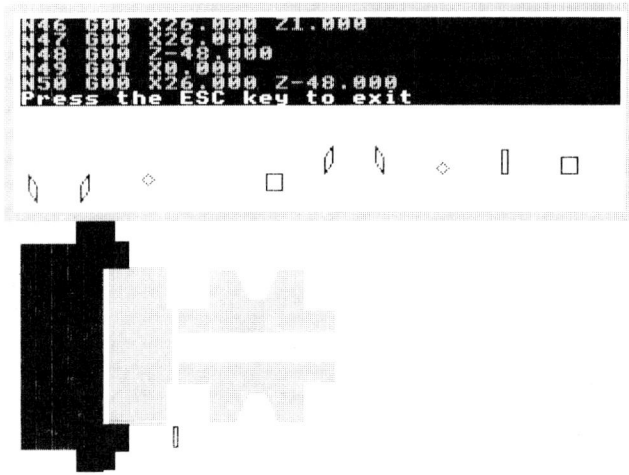

Figure 41 Computer printout from *Starturn* CNC program

Computer aided drawings

Figures 37 to 42 show printouts from pupils' technology projects. The drawings for these were produced on computers (BBC B and BBC Master) with the aid of CAD/CAM software (see page 73).

Figures 37 to 40 are part of a project dealing with the design of an automatic bird-feeding device. These were produced on a BBC Master computer using the CAD program *Techsoft Designer*. This is a 2D CAD, *menu-driven* software program worked either with the aid of a mouse, or from the keyboard or with a trackerball. The drawings were printed on an *Epson* dot matrix printer. They could equally well have been printed on a plotter.

Figures 41 and 42 are drawings made with the aid of Denford Machine Tool's *STARTURN* CNC program, working on a BBC B computer and printed on an Epson dot matrix printer.

The items shown in the drawings – a small pulley and a bolt – were turned on a Denford Starturn 4 lathe working from the computer program.

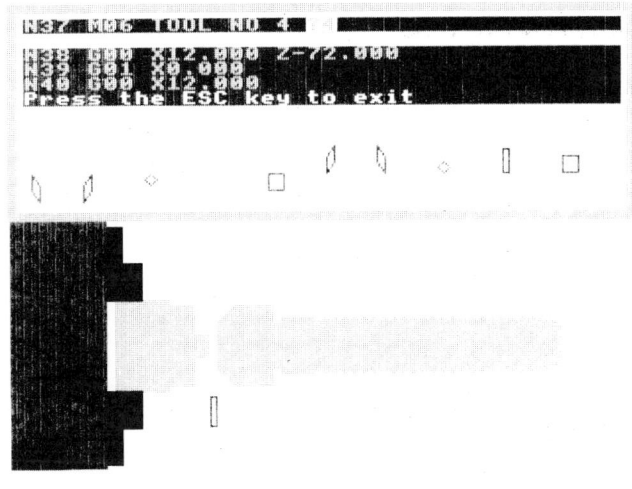

Figure 42 Computer printout from *Starturn* CNC program

12 Projects

Introduction

If you are entered for a GCSE CDT: Technology examination you will have to complete at least one project for marking by moderators appointed by an examination board. Projects for examinations are usually moderated by the end of the May of the year in which the examination is set.

You will have to submit either a major project or two mini-projects or a combination of major and mini projects. Projects in CDT: Technology are important because 50% of the marks for the whole examination can be given for project work.

How you present your projects to moderators is also important. If a project is well presented, a moderator will look on it much more favourably than if it is badly presented. This does not mean to say that a poor project which is well presented will gain much favour, but good project work which is well presented will gain higher marks than if it is poorly presented. Pay attention to the following details:

1 In general use either A3 or A4 sheets for graphics. A4 sheets are suitable for notes.
2 Gather all your graphics and notes together in a folder or folio, or in a manilla envelope.
3 Design a cover for your project folio which gives a clear idea of what it contains.
4 Include an index of the sheets and notes.
5 Number the sheets of drawings and notes so that they can be read in the order in which your design ideas have developed.
6 Notes may be written in longhand, printed in script, typed on a typewriter or typed on a word-processor. No matter which method you adopt, your notes must be easy to read.

Specimen projects

Details from a number of specimen CDT: Technology projects are included in the pages which follow. A variety of methods of drawing, of different forms of lettering for notes and various methods of presentation and use of colour are included in these. The specimen projects are:
1 pages 137 to 139 – *three-minute timer:* a project mainly concerned with an electronics circuit;
2 pages 139 to 142 – *model hovercraft;*
3 pages 142 to 147 – a group project: *an aero-generator;*
4 pages 148 to 149 – four sheets from an investigation into the design of a *mechanical robot arm,* which was part of a computer-driven robot project;
5 pages 147 to 155 – *sweets dispenser:* a project mainly dealing with pneumatics;
6 page 156 – the first sheet of a pupil's project: a *rock-a-baby-to-sleep* device.

Note:
1 The specimen projects given here are not complete. This is because of the lack of space in a book of this nature and the need to show a variety of methods.
2 Because the specimen projects were mainly derived from projects which have been submitted by pupils in schools, some of the details in them may not be correct.

Components needed for the three-minute timer project

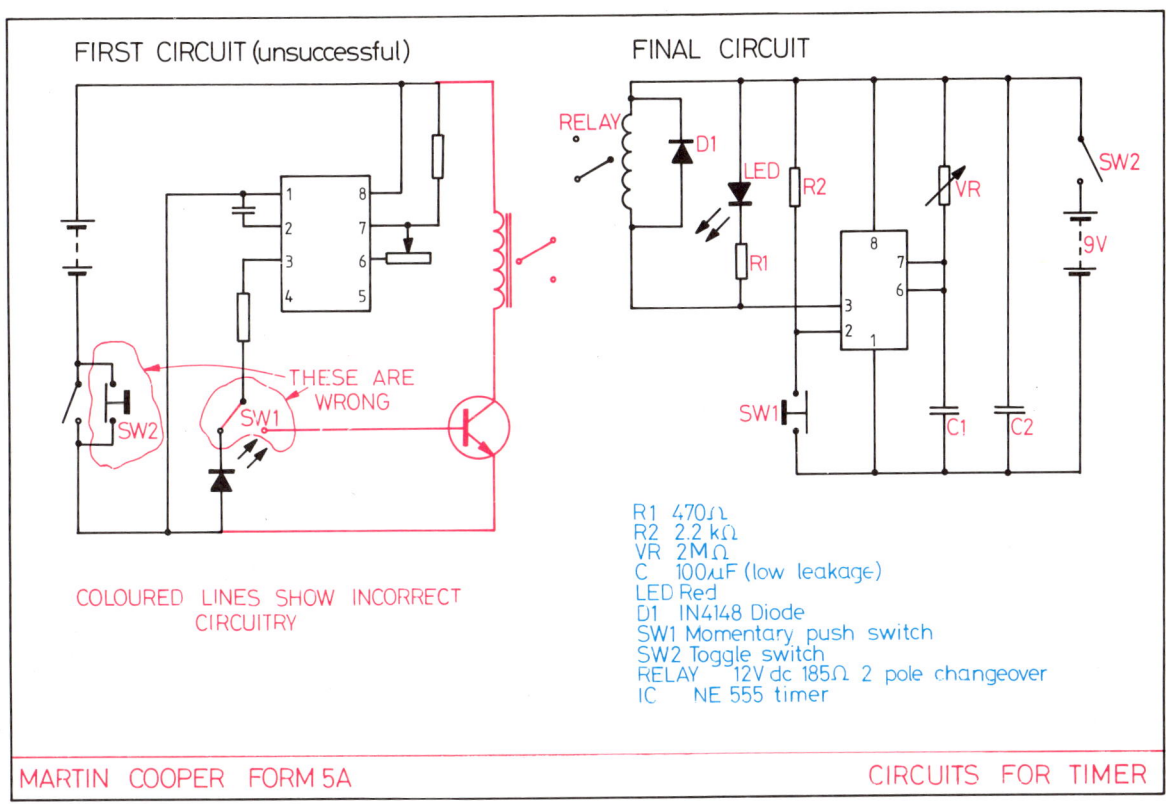

FIRST CIRCUIT (unsuccessful)

THESE ARE WRONG

SW2   SW1

COLOURED LINES SHOW INCORRECT CIRCUITRY

FINAL CIRCUIT

RELAY   D1
LED   R2
VR   SW2
R1   9V
8
7
6
SW1   3
2
1
C1   C2

R1 470Ω
R2 2.2 kΩ
VR 2MΩ
C   100μF (low leakage)
LED Red
D1 IN4148 Diode
SW1 Momentary push switch
SW2 Toggle switch
RELAY 12V dc 185Ω 2 pole changeover
IC   NE 555 timer

MARTIN COOPER FORM 5A   CIRCUITS FOR TIMER

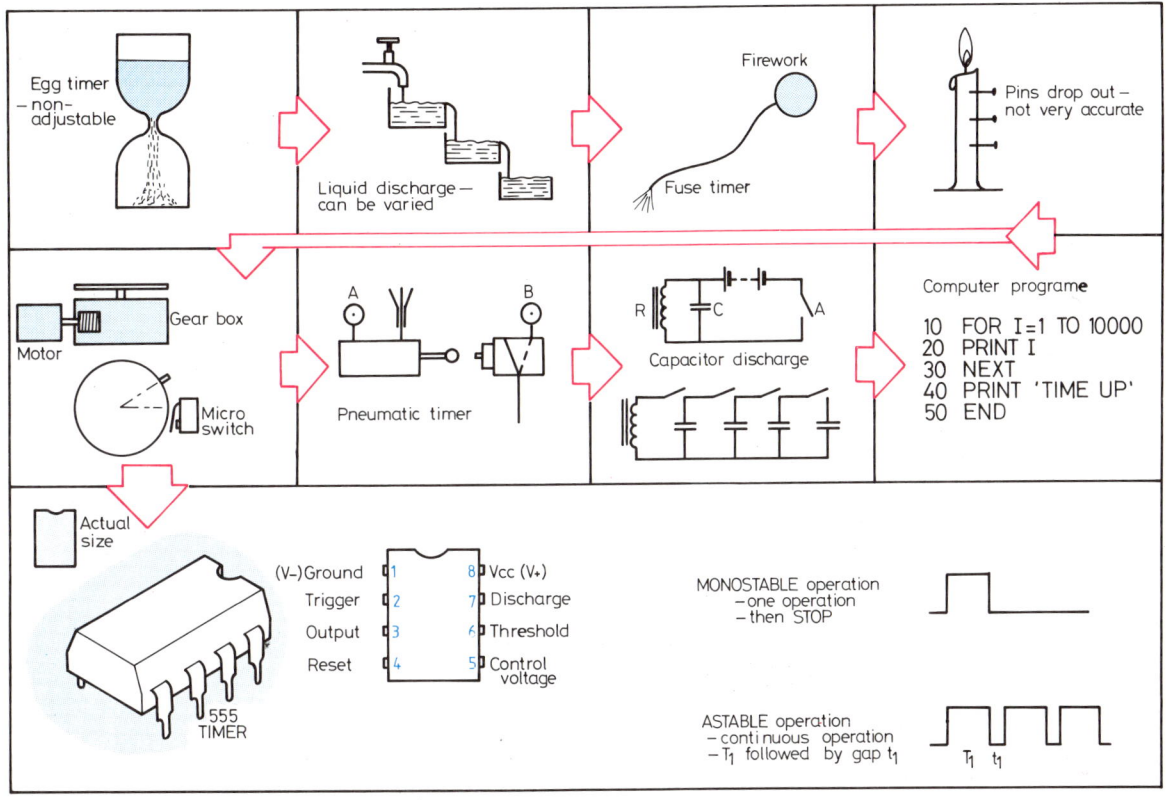

Egg timer – non-adjustable

Liquid discharge – can be varied

Firework
Fuse timer

Pins drop out – not very accurate

Motor   Gear box
Micro switch

A   B
Pneumatic timer

R   C   A
Capacitor discharge

Computer programe

10 FOR I=1 TO 10000
20 PRINT I
30 NEXT
40 PRINT 'TIME UP'
50 END

Actual size

(V–)Ground	1	8	Vcc (V+)
Trigger	2	7	Discharge
Output	3	6	Threshold
Reset	4	5	Control voltage

555 TIMER

MONOSTABLE operation – one operation – then STOP

ASTABLE operation – continuous operation – T_1 followed by gap t_1

T_1 t_1

Sheets from the folder of a student designing and making a three-minute timer

Project – three-minute timer

Problem

I am a keen photographer and I develop and print my own films and photographs. I need to be able to time the developing and printing processes accurately.

Design brief

Design a timing device, accurate to 1 second, which can be adjusted to give timings of between 0 and 3 minutes, in 1 second intervals.

Specification

1 A signal must be produced when a chosen time has elapsed. The signal must be easy to see, or to hear, or both.
2 The device must be easy to adjust in the darkness of a dark room.
3 The device needs to be reasonably small in order to avoid taking up too much space on the darkroom bench.

Investigation

□ **Shape and form:** The device should be of a shape and size to fit easily in the hands.

□ **Materials:** The box holding the device is to be made from a hardwood. I do understand that this may not be ideal, because wood is not a very good insulator, but I will have to use materials which are available. The electronic components can be mounted, or set in, transparent acrylic sheet.

□ **Shaping and forming:** The acrylic sheet will be heat formed to its required shape. The box containing the acrylic shape will be made up from pieces of hardwood.

□ **Jointing:** Heat sink soldering is required to join the electrical connections. The box can be jointed as simply as is possible.

□ **Strength:** The device requires to be sufficiently strong and robust in its construction to withstand the possibility of being knocked onto the floor while it is being used.

□ **Surface finish:** Smooth. Must be water and chemical resistant.

□ **Fittings and components:** Listed in the drawing shown on page 137.

□ **Safety:** Low power needed. No white light should escape as this could ruin exposures when printing or developing.

□ **Economics:** Not of real importance. The device will be much cheaper to make than a similar timer purchased from a photographic shop.

□ **Special factors:** The setting of the timing must be easy. Clear calibration is essential. The device could possibly switch on the enlarger at its light source.

Notes on drawings of possible solutions (page 137)

□ **Capacitor discharge drawings:** When switch A is closed, relay R is activated and capacitor C is charged. When A is opened, relay R continues to latch until capacitor C discharges. This electrical delay could be used as a timing device.
A 5000 µF capacitor will produce only a short delay – perhaps 4 or 5 seconds. To obtain longer delays, a number of capacitors will need to be connected in parallel.

□ **Computer program:** The program on Research Machines 380Z took 8 seconds to count up to 10 000, but by amending the program to give a read-out on a visual display unit (VDU) 3.04 minutes were required to count up to 10 000.

□ **555 Timer integrated circuit chip:** See data published by RS Components Limited. This IC timer is accurate to within $+/- 2\%$ in 3 minutes. It is capable of monostable or astable operation.
Monostable: one operation and then stop.
Astable: continues operating T_1, gap, T_2, etc.
The timing device requires monostable operation.
The formula for using the 555 timer IC is:

$$T = 1.1 \times VR \times C$$
$$= 1.1 \times 2 M\Omega \times 100 \, \mu F$$
$$= 1.1 \times 2\,000\,000 \times \frac{100}{1\,000\,000}$$

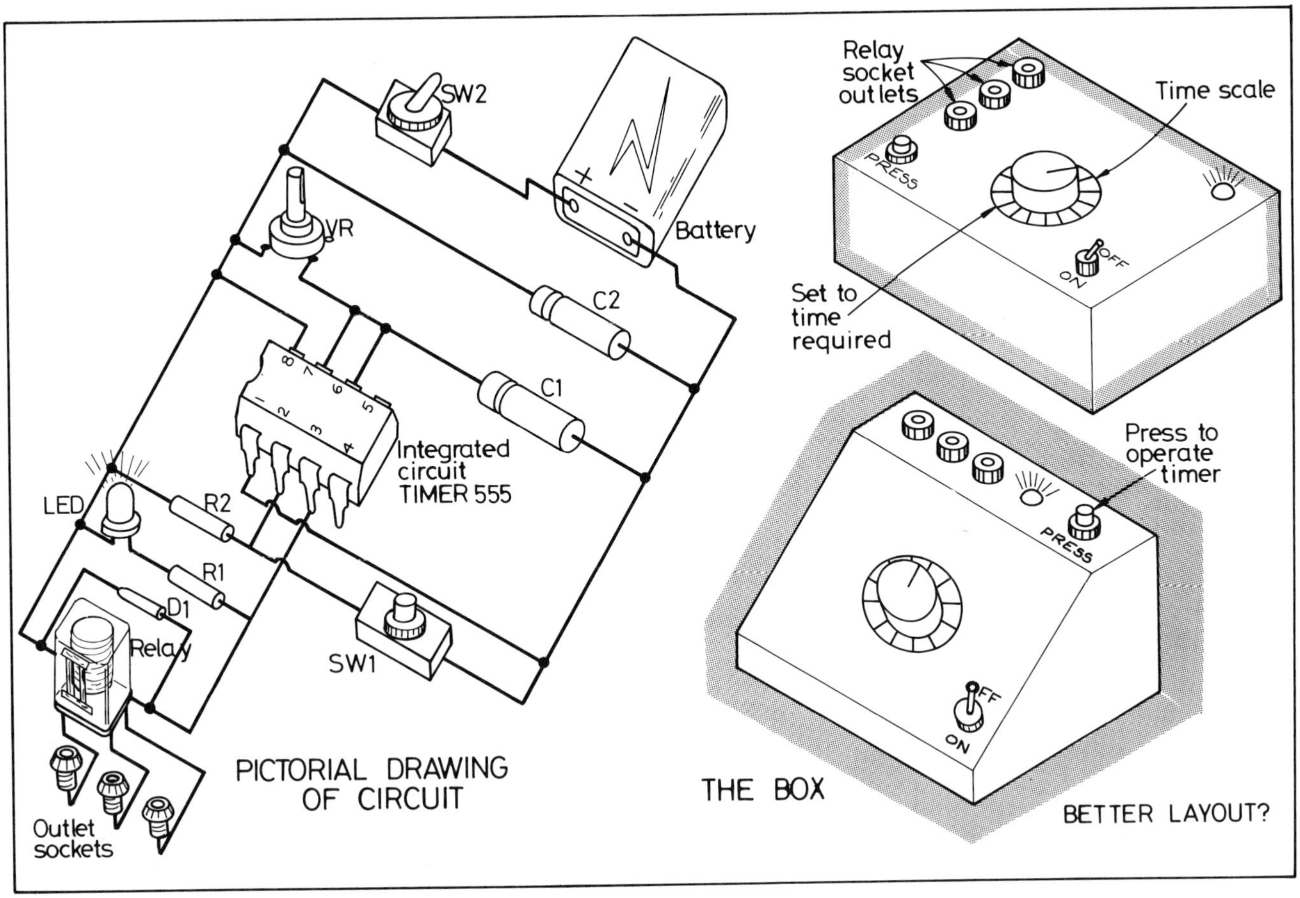

SW2

VR

C2

C1

Battery

Integrated
circuit
TIMER 555

LED

R2

R1

D1

Relay

SW1

PICTORIAL DRAWING
OF CIRCUIT

Outlet
sockets

Relay
socket
outlets

Time scale

PRESS

Set to
time
required

ON OFF

THE BOX

Press to
operate
timer

PRESS

OFF
ON

BETTER LAYOUT?

A student's design sheet for the three-minute timer project

$= 1.1 \times 2 \times 100$
$= 220$ seconds maximum.

Note: *VR* is a variable resistor.

The circuit operation

Turn the circuit on with switch SW2. Start the
timing period by pressing SW1.

Pressing SW1 causes pin 2 of the 555 chip to go
below one-third of the supply voltage. A
comparator inside the chip detects the change
and operates a flip-flop to make output (pin 3) go
to 9V. This turns off the LED and the relay. The
capacitor C now starts charging via R. The voltage
across the capacitor increases as the charge in the
capacitor increases. When this voltage reaches
two-thirds of the supply voltage, a further
comparator inside the chip causes the flip-flop to
re-set. This makes the output voltage at pin 3 fall
to nearly 0 volts – turning on the LED, operating
the relay and discharging the capacitor ready for
the next delay.

The finished timer

Project – Hovercraft Model

See pages 140–142.

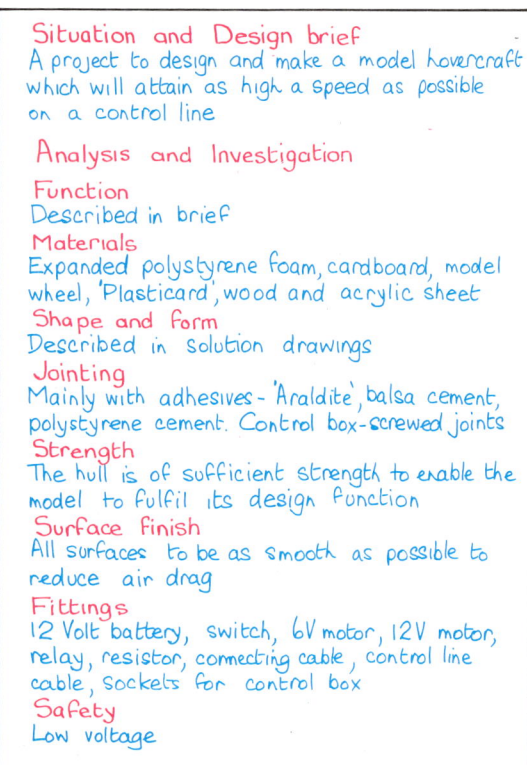

Situation and Design brief
A project to design and make a model hovercraft which will attain as high a speed as possible on a control line

Analysis and Investigation

Function
Described in brief
Materials
Expanded polystyrene foam, cardboard, model wheel, 'Plasticard', wood and acrylic sheet
Shape and form
Described in solution drawings
Jointing
Mainly with adhesives - 'Araldite', balsa cement, polystyrene cement. Control box-screwed joints
Strength
The hull is of sufficient strength to enable the model to fulfil its design function
Surface finish
All surfaces to be as smooth as possible to reduce air drag
Fittings
12 Volt battery, switch, 6V motor, 12V motor, relay, resistor, connecting cable, control line cable, sockets for control box
Safety
Low voltage

Hovercraft model

Nigel Heath

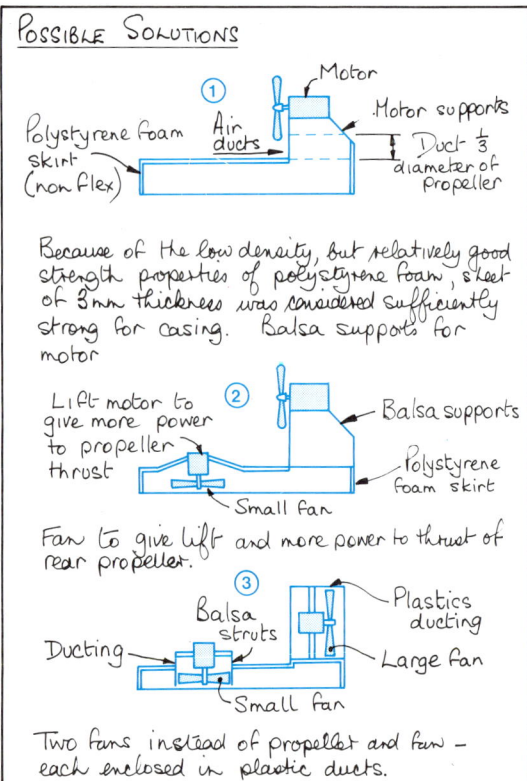

POSSIBLE SOLUTIONS

① Motor, Motor supports, Polystyrene foam skirt (non flex), Air ducts, Duct ⅓ diameter of propeller

Because of the low density, but relatively good strength properties of polystyrene foam, sheet of 3mm thickness was considered sufficiently strong for casing. Balsa supports for motor

② Lift motor to give more power to propeller thrust, Balsa supports, Polystyrene foam skirt, Small fan

Fan to give lift and more power to thrust of rear propeller.

③ Balsa struts, Plastics ducting, Ducting, Large fan, Small fan

Two fans instead of propeller and fan - each enclosed in plastic ducts.

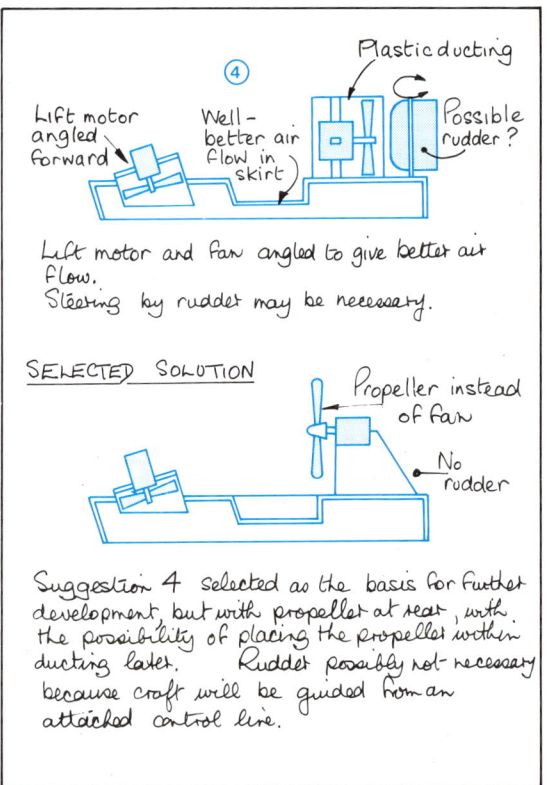

④ Plastic ducting, Lift motor angled forward, Well - better air flow in skirt, Possible rudder?

Lift motor and fan angled to give better air flow.
Steering by rudder may be necessary.

SELECTED SOLUTION

Propeller instead of fan, No rudder

Suggestion 4 selected as the basis for further development, but with propeller at rear, with the possibility of placing the propeller within ducting later. Rudder possibly not necessary because craft will be guided from an attached control line.

A student's design sheets for the hovercraft model

ELECTRIC CIRCUIT.

Two D.C. motors are available — a 12V and a 6V. The 12V motor, capable of 2000 rev/min is suitable for driving the thrust propeller. It is best controlled by a double-pole switch, which will provide a reversing system in the circuit for the propeller.

The 6V motor is therefore to be used to drive the lift fan. This motor requires 0.6A to operate. Its resistance must therefore be by Ohm's Law

$$6 \div 0.6 = 10\,\Omega$$

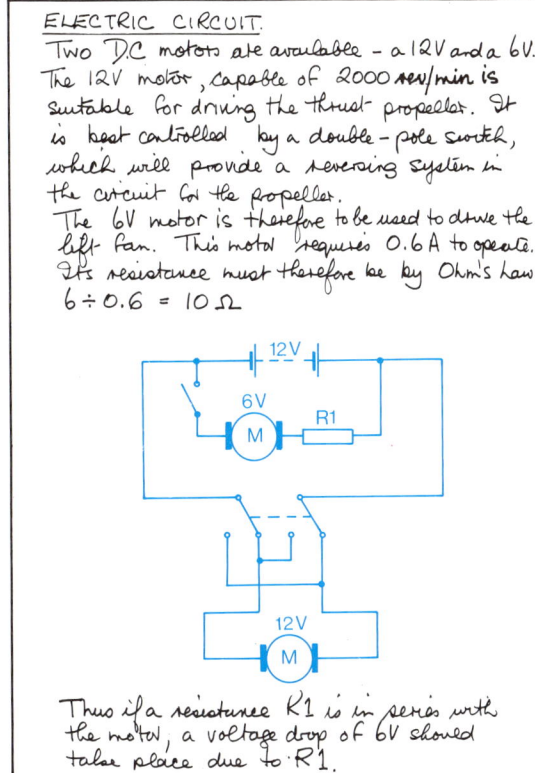

Thus if a resistance R1 is in series with the motor, a voltage drop of 6V should take place due to R1.

CONTROL BOX

A control box is to be made to house the electric circuitry

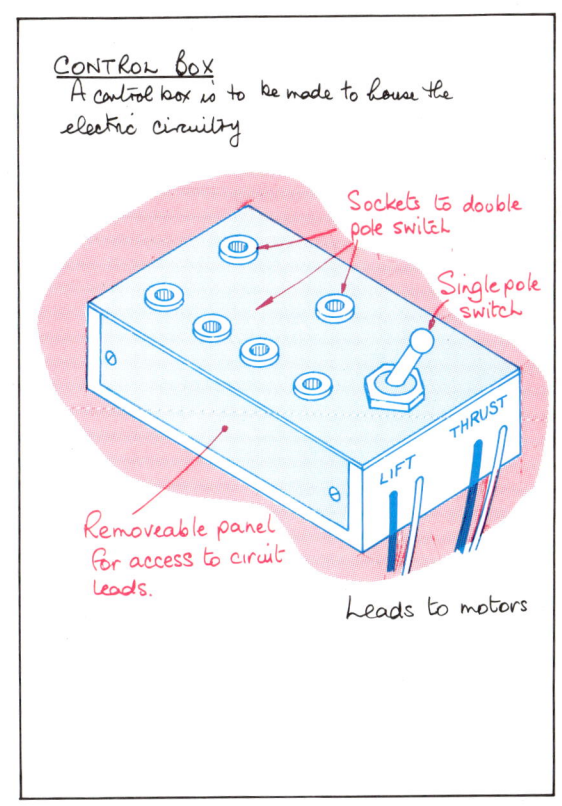

Sockets to double pole switch

Single pole switch

Removeable panel for access to circuit leads.

LIFT   THRUST

Leads to motors

HULL

Hull to be constructed from 5mm polystyrene foam tiles — sufficiently light + strong. All joints glued with pva glue — obviates nails etc.

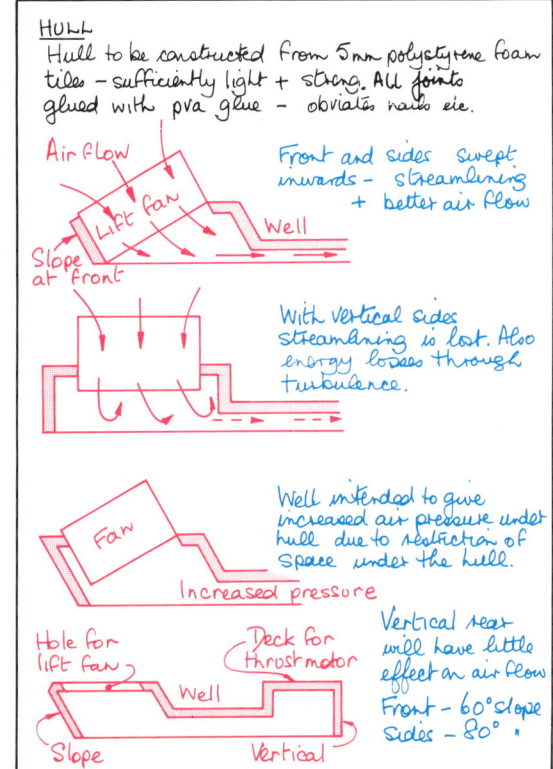

Air flow

Lift fan

Well

Slope at front

Front and sides swept inwards — streamlining + better air flow

With vertical sides streamlining is lost. Also energy losses through turbulence.

Fan

Increased pressure

Well intended to give increased air pressure under hull due to restriction of space under the hull.

Hole for lift fan

Deck for thrust motor

Well

Slope   Vertical

Vertical rear will have little effect on air flow
Front — 60° slope
Sides — 80°.

FAN MOUNTING

Assembly glued to motor and walls of ducting with "Araldite" adhesive

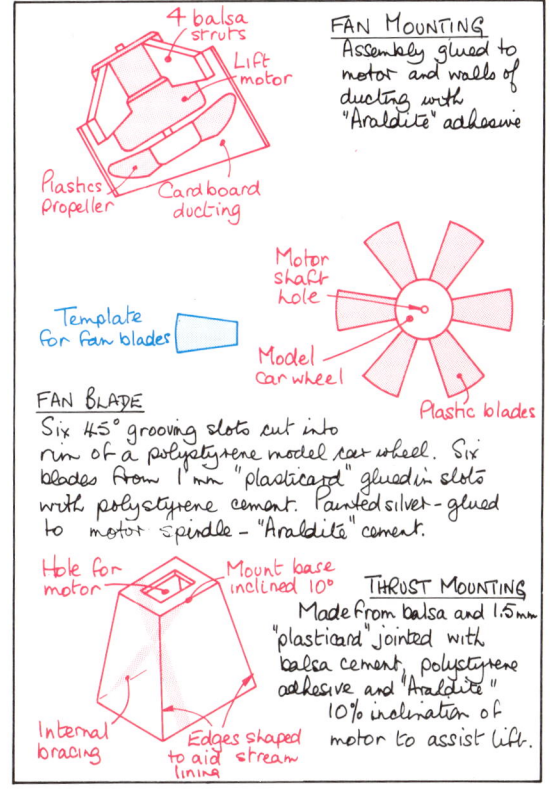

4 balsa struts

Lift motor

Plastics propeller   Cardboard ducting

Template for fan blades

Motor shaft hole

Model car wheel

Plastic blades

FAN BLADE

Six 45° grooving slots cut into rim of a polystyrene model car wheel. Six blades from 1 mm "plasticard" glued in slots with polystyrene cement. Painted silver — glued to motor spindle — "Araldite" cement.

Hole for motor

Mount base inclined 10°

Internal bracing

Edges shaped to aid stream lining

THRUST MOUNTING

Made from balsa and 1.5mm "plasticard" jointed with balsa cement, polystyrene adhesive and "Araldite" 10% inclination of motor to assist lift.

A student's design sheets for the hovercraft model

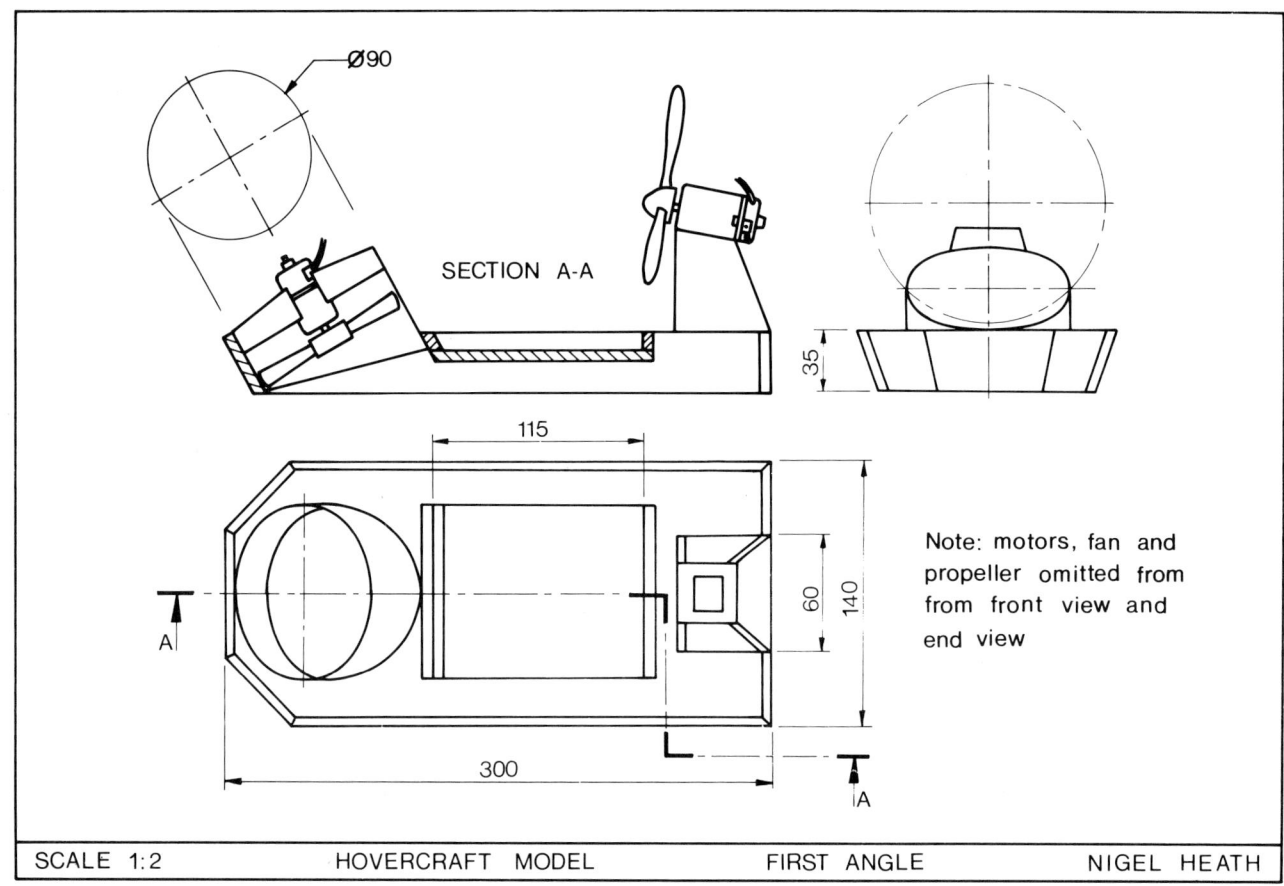

Ø90

SECTION A-A

35

115

60

140

300

A

A

Note: motors, fan and propeller omitted from from front view and end view

SCALE 1:2   HOVERCRAFT MODEL   FIRST ANGLE   NIGEL HEATH

MODIFICATIONS
Some modifications were carried out on the original hovercraft as a result of experiments. The most important of these modifications were:-
1. A larger motor and larger fan and duct was fitted to give increased and better lift.
2. The well depth was decreased to give more space underneath the hull

Modified hull, fan duct and well.

Original hull and fan duct.

3 By removing some of the fan duct a greater under-hull air pressure was obtained.

EVALUATION
After the modifications detailed above, the craft performed well. In particular the use of adhesives for joining the various parts proved to be particularly successful.

More design sheets for the hovercraft model

Project – Aero-generator

See pages 143–147.

This project has been reproduced from *Design and Technology* by A. Yarwood and A. H. Orme. Details of the project were made available by permission of the staff of the Walton High School, Stafford, where the project was designed and made. We would like to place on record here our appreciation of the generous help given, both in passing on information and allowing photographs to be taken, by the head teacher of the school, J. E. Wilkinson, Esq., M. A., and to N. A. Brittain, Esq., the teacher responsible for the project. The following pupils took part in the project: David Mellhuish, Robert Howells, Stephen McCosh, Mark Edwards, Martin Ovelton, Andrew Booth, Jeremy Robinson, Trevor Powell and David Proughton.

Problem

A greenhouse (growing unit) requires a heating system.

Design brief

Design a heating system for a greenhouse of average size (4.5 m by 2.5 m) capable of maintaining a frost-free environment at minimum energy costs.

Specification

1 The design should create and preserve heat with minimal effects upon the greenhouse facilities.
2 The shape and form of the design are not of any particular importance, but the design should occupy as little space as possible, preferably a part of the greenhouse not required for growing plants.
3 All materials employed must be durable and weather resistant. Care must be taken to avoid using materials which could damage plants.
4 All electrical/electronic circuitry must be watertight and weatherproof. Jointing should be such that accidents – such as a dropped plant container, or an accident with a watering can – will not cause the system to fail.

Possible solutions

☐ **Strength:** The design must be sufficiently strong to withstand damage caused by poor handling.

☐ **Fittings:** It is better to 'over-design' when choosing fittings. Reliability of the design is very important if plant losses are to be avoided.

☐ **Safety:** There is a slight fire risk. There is some possibility of propeller mountings and bearing becoming damaged in high winds. The tower construction for the propeller and generator system finally decided upon, must be capable of withstanding high wind speeds. The tower must also be protected against children climbing up it.

☐ **Economics:** Much of the equipment used for the design was second-hand – mainly scrap materials. Running costs are small. In fact the whole system proved to be very economic.

Power sources considered

1 Some form of waste-burning heater/boiler.
Wood: difficult to obtain in sufficient quantity.
Dustbin waste: difficult to burn.
Sump oil: possible – oildrum/water drip burner.
2 Heat generated by bio-chemical action.
Compost: decomposition.
Manure: methane gas.
3 Sun power.
Insulation: greenhouses hot during daytime.
Sun-heated storage units: bricks or water.
Solar cells: for electricity/heat conversion.
Solar panel: for water heating.
4 Water.
No natural or free standing source available.
5 Wind power.
Possible electric or rotary pumping action – site is open, with little air turbulence.

Pump control

The pump should preferably only circulate water through the system when the solar panel is heating the water – not at night or when the sunlight is inadequate during the day.

The determining factor is the difference in temperature between water entering the solar panel and water leaving the panel.

Temperature IN (A)	Temperature OUT (B)	Pump
Lower than B	Higher than A	ON
Same as B	Same as A	OFF
Higher than B	Lower than A	OFF

Solar panel control circuit

(page 146)

1 The wind-driven generator constantly charges the battery when the propeller is rotating.
2 The battery (12 V) supplies power through the circuit via a relay operated switch to the pump.
3 The circuit takes its power from two 9 V batteries.

The circuit

The circuit has two inputs – one from the 'hot' thermistor and the other from the 'cold' thermistor. The only output of the circuit is to the pump, and this is switched on when the temperature difference between the two thermistors exceeds a pre-set level.

The thermistors are mounted on each side of the solar panel, the 'cold' some distance from the input side and the 'hot' very close to the output side. The difference in the thermistor signals represents the temperature rise of the water due to solar heating.

If both thermistors were close to the solar panel, then without the pump working, no temperature difference would register and therefore the circuit would never switch the pump on. The pump is thus only on when sunlight on the solar panel heats the water by a certain amount. This amount can be varied by altering the $10\,k\Omega$ 'temperature difference adjustment' potentiometer.

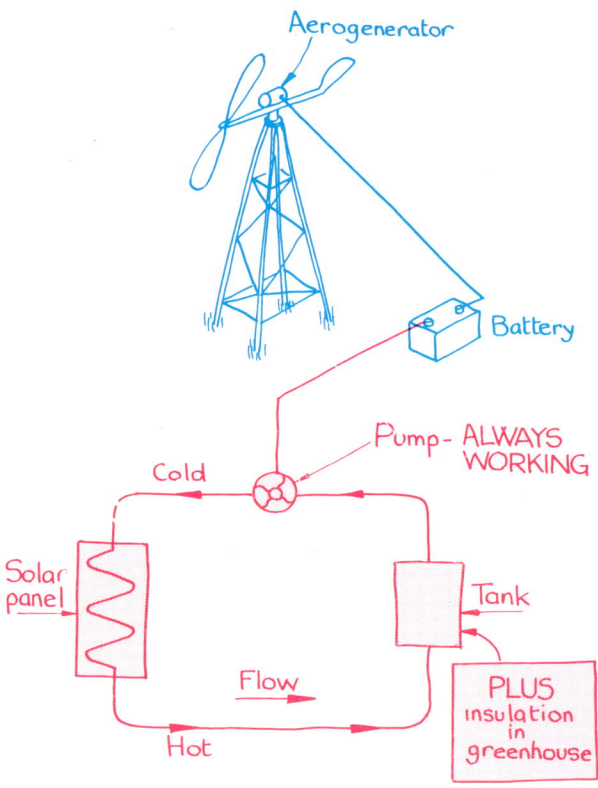

A student's idea for the aerogenerator project

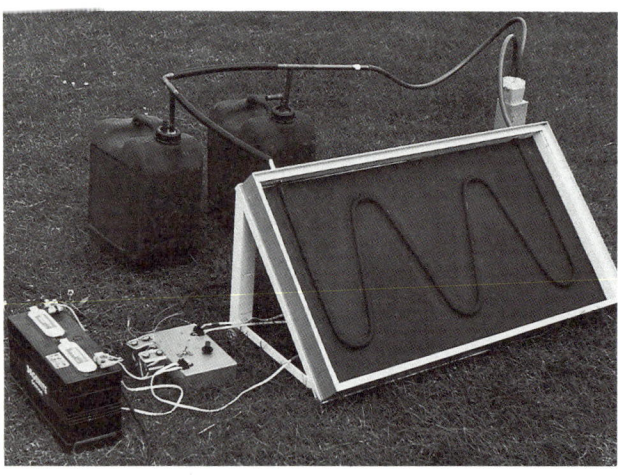

The solar panel for heating water

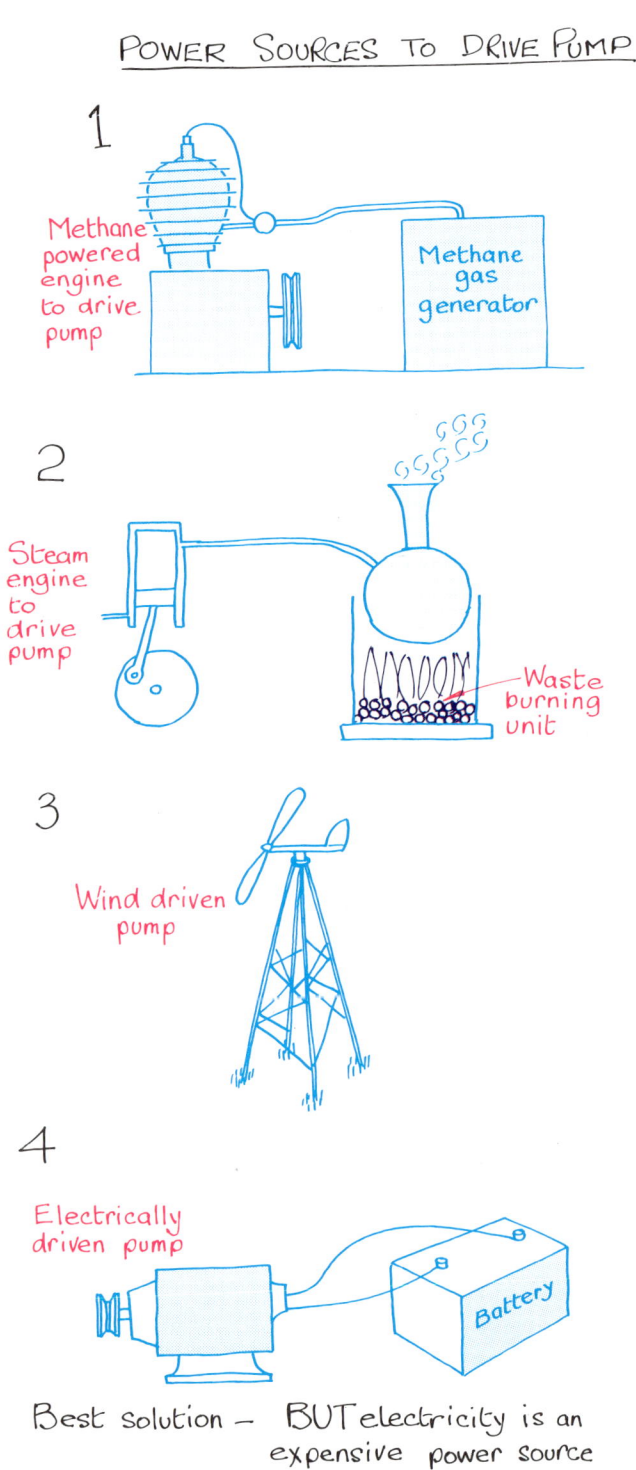

A student's notes on the aerogenerator project

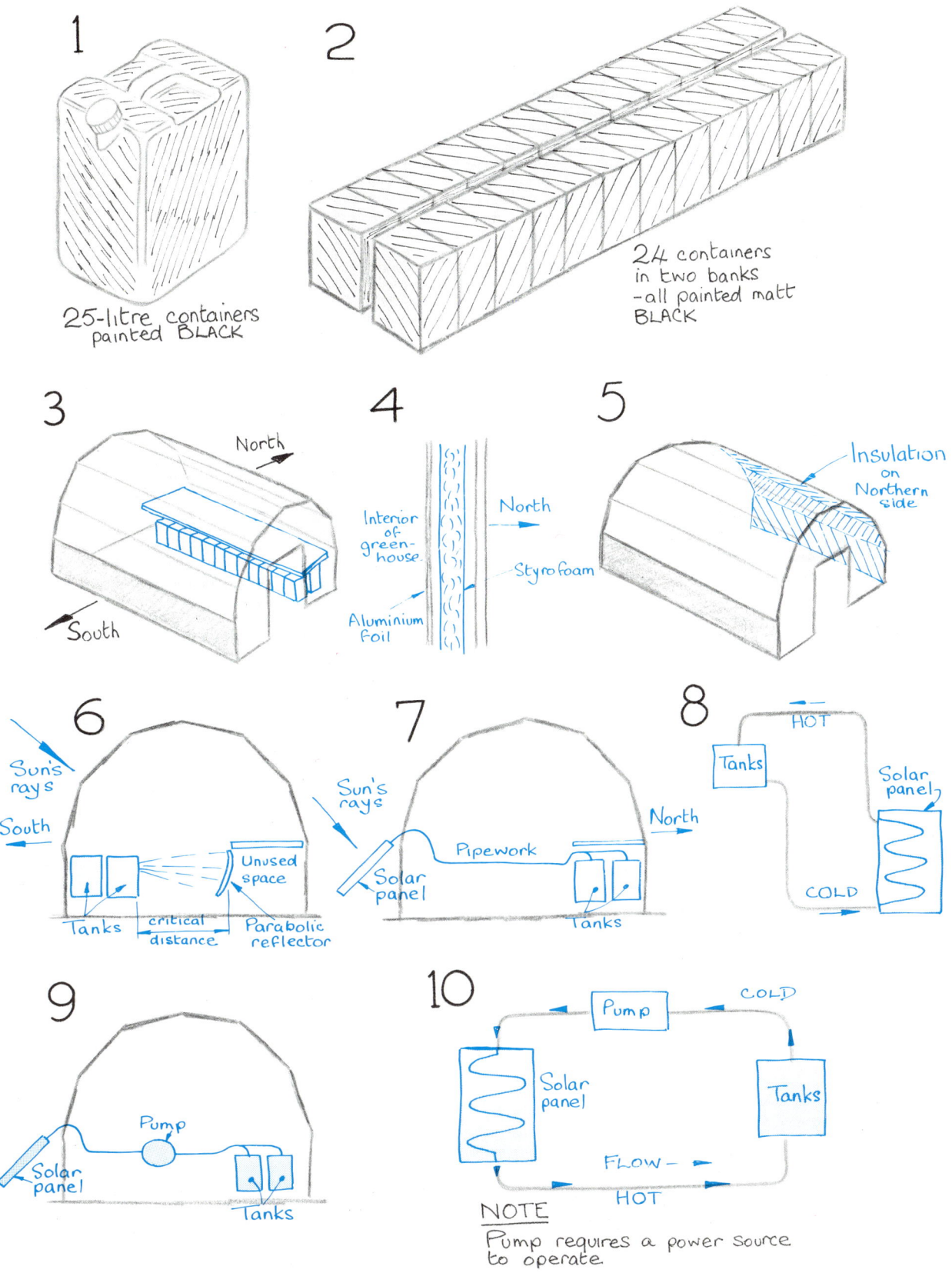

1 25-litre containers painted BLACK

2 24 containers in two banks —all painted matt BLACK

3 North / South

4 Interior of green-house / North / Styrofoam / Aluminium foil

5 Insulation on Northern side

6 Sun's rays / South / Tanks / critical distance / Unused space / Parabolic reflector

7 Sun's rays / Solar panel / Pipework / North / Tanks

8 HOT / Tanks / Solar panel / COLD

9 Pump / Solar panel / Tanks

10 Pump / COLD / Solar panel / Tanks / FLOW / HOT

NOTE
Pump requires a power source to operate.

The aerogenerator project – a page from a student's folder

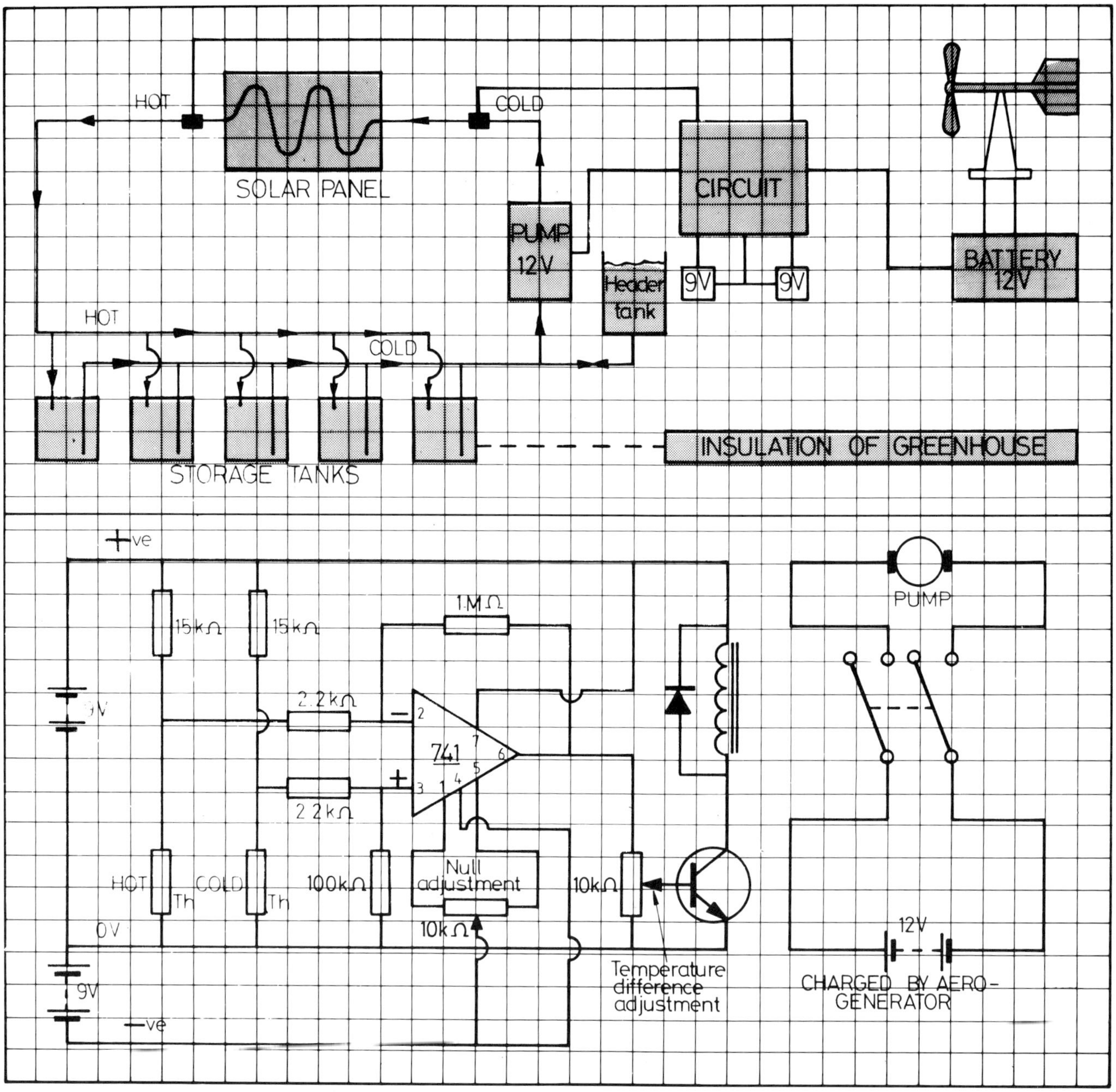

HOT
SOLAR PANEL
COLD
CIRCUIT
PUMP 12V
BATTERY 12V
Header tank
9V
9V
HOT
COLD
STORAGE TANKS
INSULATION OF GREENHOUSE

+ve
15kΩ
15kΩ
1MΩ
PUMP
9V
2.2kΩ
2
741
7 6
1 5
+
3 4
2.2kΩ
HOT Th
COLD Th
100kΩ
Null adjustment
10kΩ
0V
10kΩ
12V
9V
Temperature difference adjustment
CHARGED BY AERO-GENERATOR
−ve

A design sheet showing a possible circuit for the aerogenerator project

The 741 integrated circuit chip

The heart of the circuit above is the 741 operational amplifier integrated circuit chip. The 741 Op Amp is a high gain amplifier whch will amplify any difference between the two input voltages from the thermistors. The gain on a 741 may be 10 000 or more, but it can be reduced. The two inputs are at pins 2 and 3. Any difference in these two voltages is amplified and the resulting ouput fed to pin 6.

The final control circuitry assembled

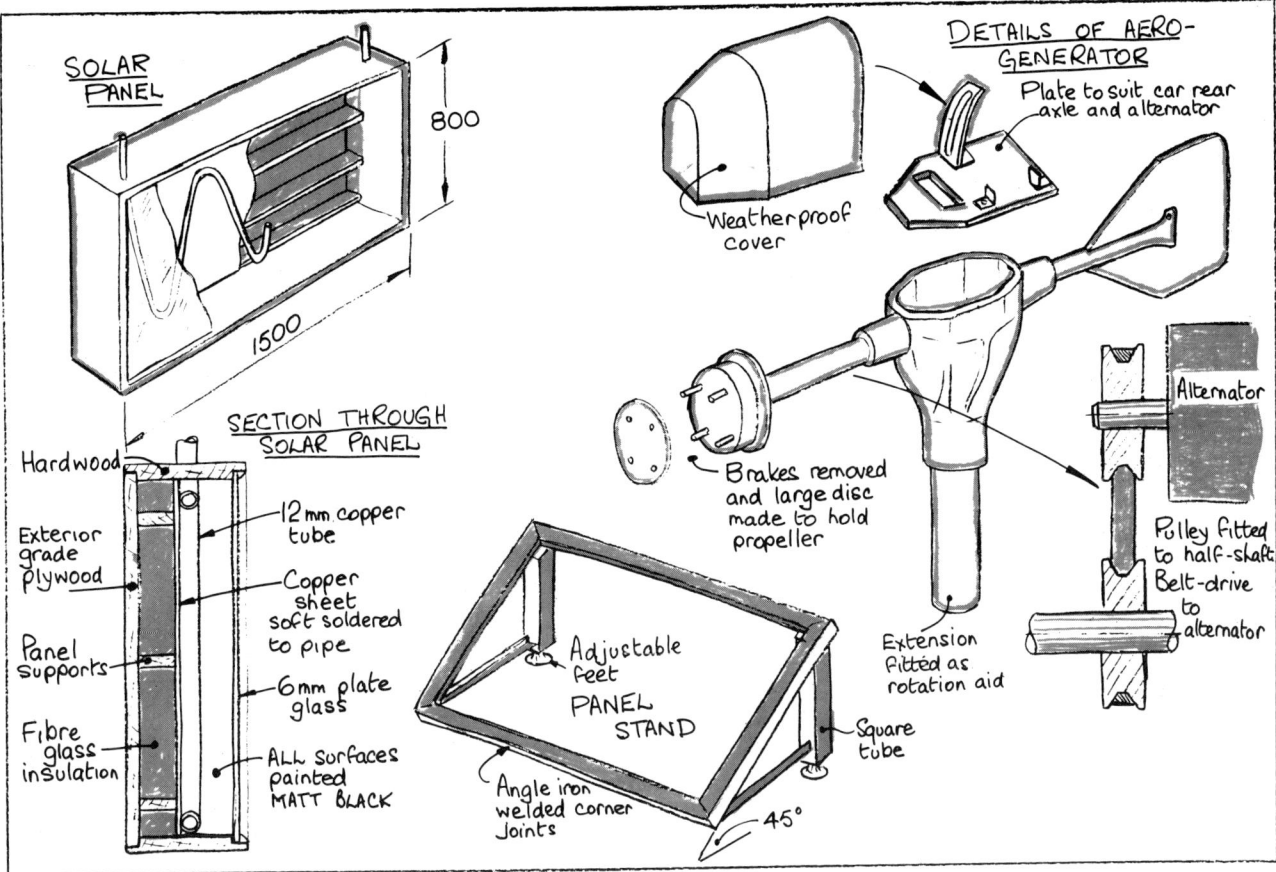

Details of a student's aerogenerator design

The complete aerogenerator project

Project – A Mechanical Robot Arm (See pages 148–149).

Project – Sweets Dispenser

Design brief

Design and make a device, operated by a pneumatic circuit, which will dispense a roll of POLO mints when a 10p coin is inserted in the device.

Problems which immediately come to mind

1 Sorting coins so that only 10p coins will work it.
2 How to make sure that a 10p coin works the device.
3 How to dispense the roll of mints when the device is worked by the coin.
4 How to make certain that only one roll is dispensed at a time.
5 How to make sure that rolls cannot be stolen from the device without putting in a 10p coin.

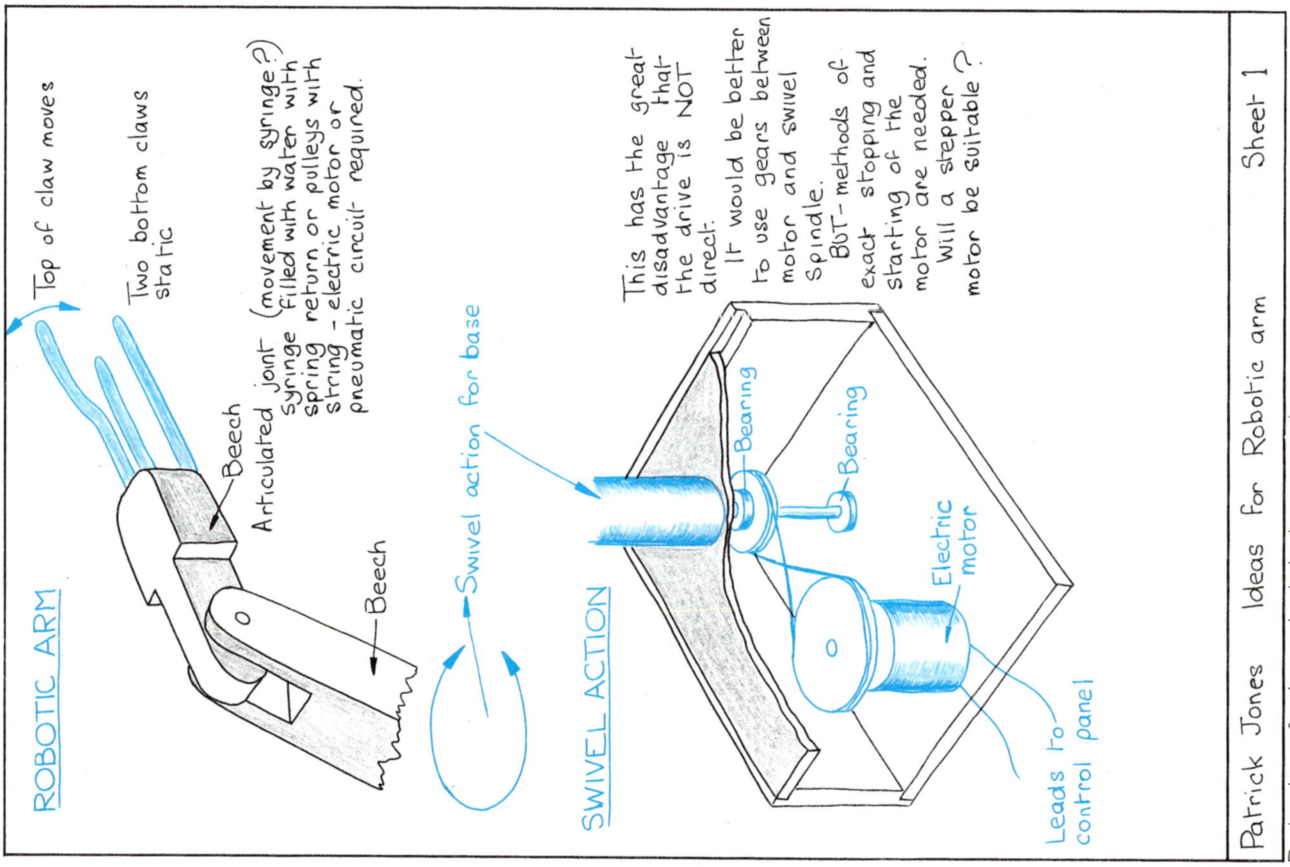

UPRIGHT OF ARM MUST BE FIRM

Any slackness between gear, washer and ply could allow movement

Gears no longer mesh

Upright

Bearing

Gear

Washer

Gear

Washer

Lower bearing

More stiffness if lower spindle supported in lower bearing

ARM MOVEMENT

Idea 1

Up and down movement of arm

Pulley

String winds and unwinds around motor spindle

Other motors control other arms if necessary

Idea 2

Piston movement

Spring return

If piston slopes would greater pressure be exerted

Water filled syringe

Patrick Jones   Ideas for Robotic arm   Sheet 2

ROBOTIC ARM

Top of claw moves

Two bottom claws static

Beech

Articulated joint (movement by syringe?) syringe filled with water with spring return or pulleys with string - electric motor or pneumatic circuit required.

Beech

Swivel action for base

SWIVEL ACTION

Swivel action

This has the great-disadvantage that the drive is NOT direct.

It would be better to use gears between motor and swivel spindle.

BUT- methods of exact stopping and starting of the motor are needed. Will a stepper motor be suitable ?

Bearing

Bearing

Electric motor

Leads to control panel

Patrick Jones   Ideas for Robotic arm   Sheet 1

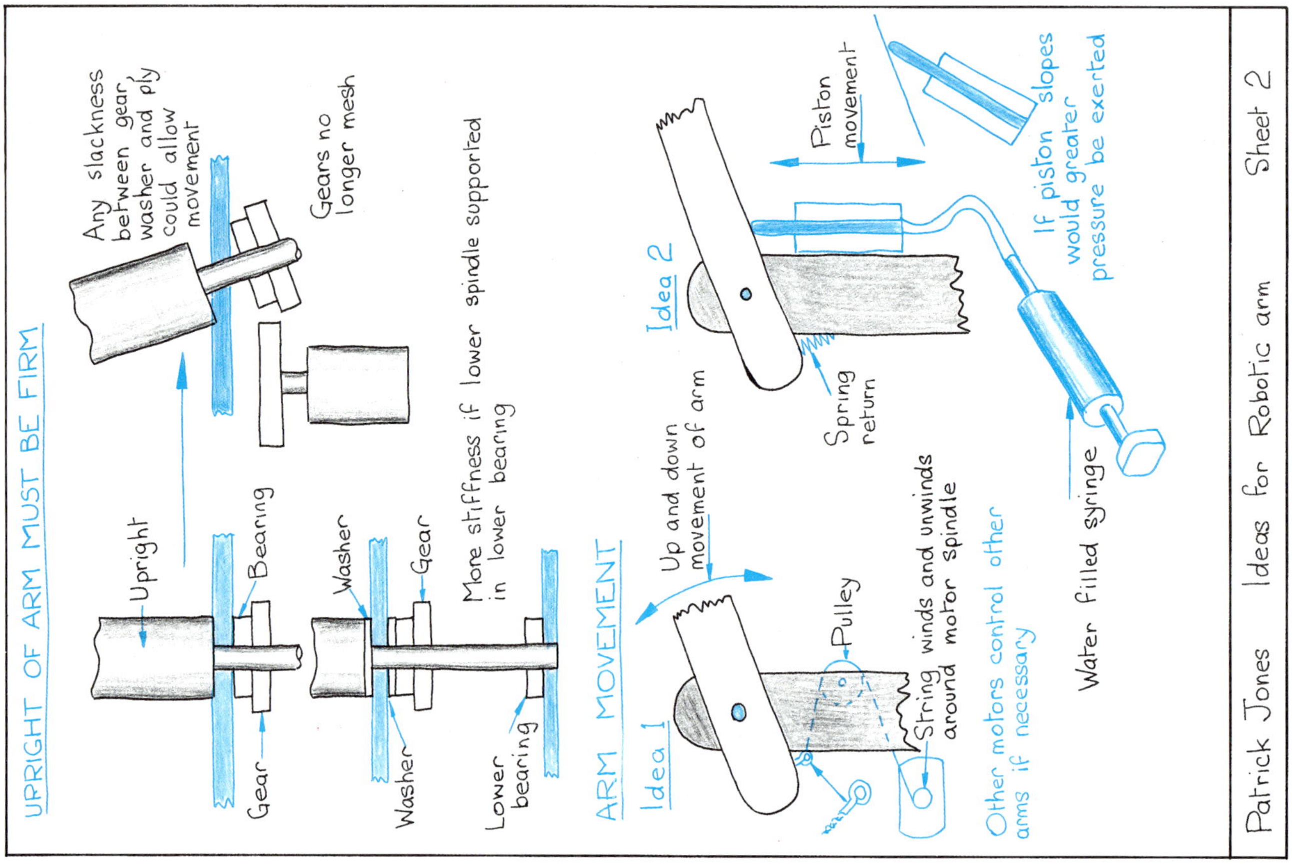

Design sheets for the mechanical robot arm project

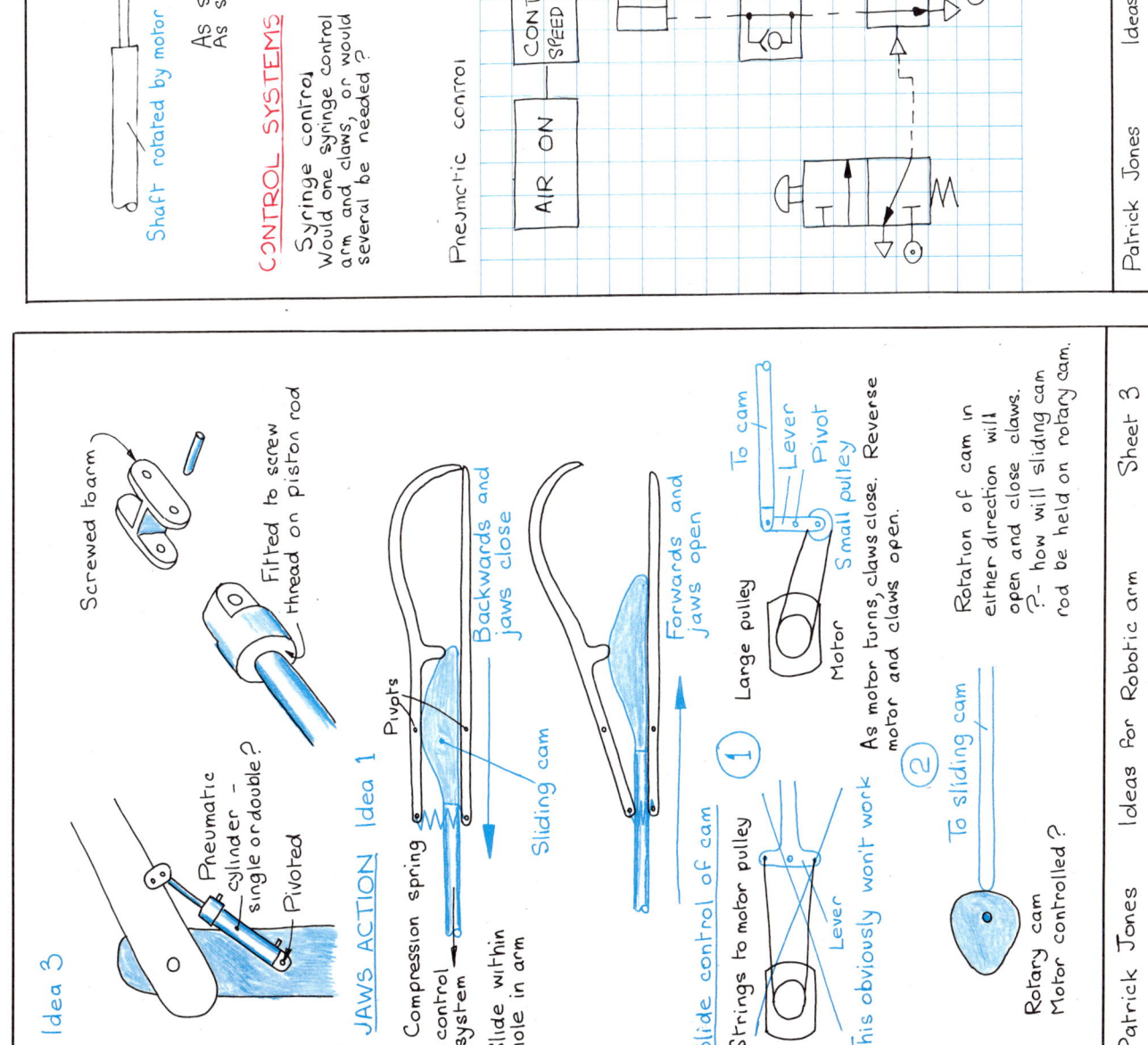

Sheet 4

Screwed bush on slider cam rod

Screw

As screw turns clockwise - jaws open
As screw turns anti-clockwise - jaws close

Shaft rotated by motor

Electric motor control
Would one electric motor control arm and claws, or would rotation, arm and claw movements? Probably not - perhaps 3 would be required -
1. - rotation of arm;
2. - up and down movement of arm;
3. - claw movement.

CONTROL SYSTEMS

Syringe control
Would one syringe control arm and claws, or would several be needed?

Pneumatic control

| AIR ON | CONTROL SPEED of AIR | MOVEMENT OF PISTON | RETURN STROKE |

Patrick Jones   Ideas for Robotic arm   Sheet 4

Idea 3

Screwed foam

Fitted to screw thread on piston rod

Pneumatic cylinder - single or double?

Pivoted

JAWS ACTION Idea 1

To control system

Compression spring

Slide within hole in arm

Pivots

Sliding cam

Backwards and jaws close

Forwards and jaws open

Slide control of cam

① Large pulley

To cam

Lever

Pivot

Small pulley

Motor

As motor turns, claws close. Reverse motor and claws open.

Strings to motor pulley

Lever

This obviously won't work

② To sliding cam

Rotation of cam in either direction will open and close claws.
?- how will sliding cam rod be held on rotary cam.

Rotary cam
Motor controlled?

Patrick Jones   Ideas for Robotic arm   Sheet 3

Design sheets for the mechanical robot arm project

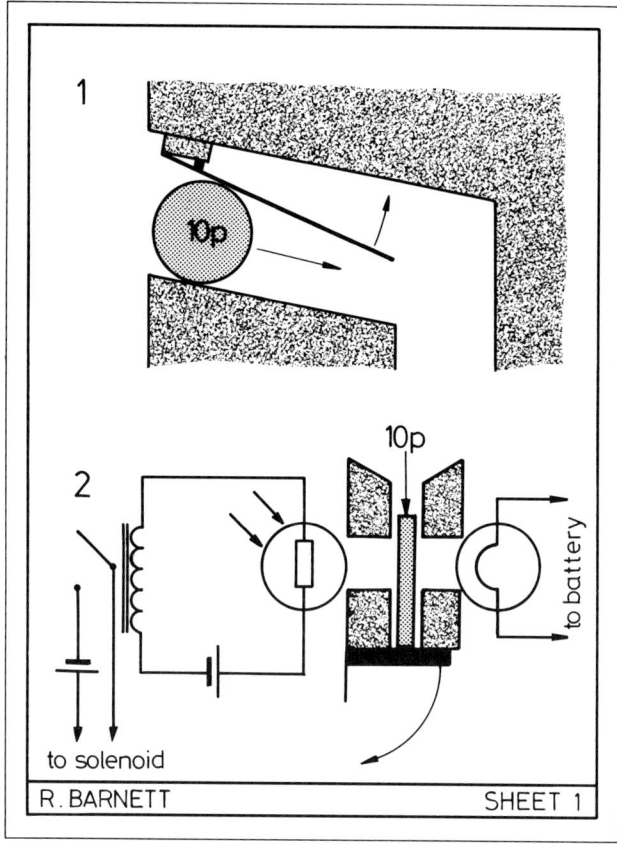

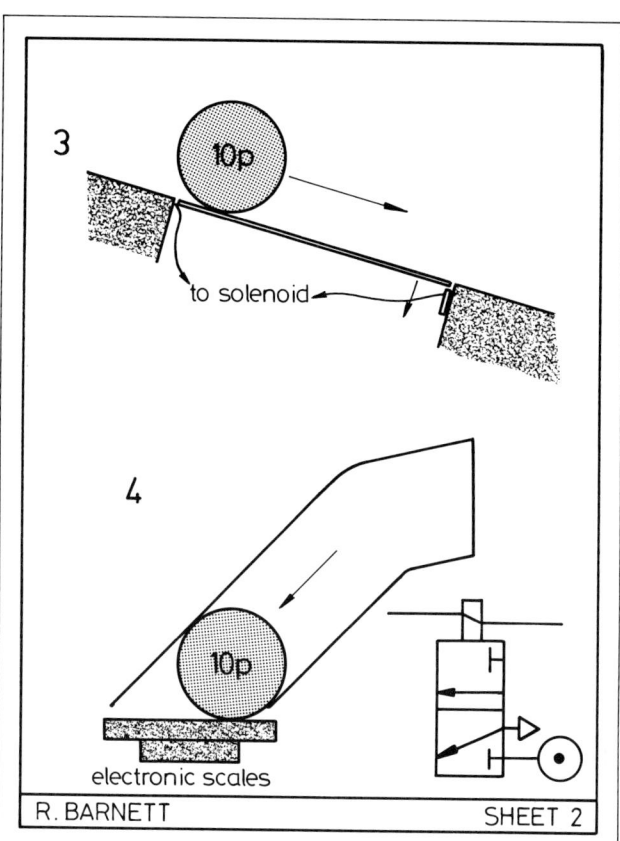

A student's ideas for the sweet dispenser project

Possible ways of solving these problems

From the start it seemed to me that the easiest way of operating the device would be by an electrically-controlled solenoid-operated pneumatics circuit. My solutions are therefore based on:
1 how the coin can be made to activate the solenoid (Sheets 1 and 2 above);
2 how the rolls of mints could be dispensed (Sheet 3, page 151);
3 a pneumatic circuit with a single-acting cylinder controlled by a solenoid actuated valve (Sheet 4, page 151).

Four ways of actuating the solenoid valve

□ **Sheet 1.**

1 With a micro-switch fitted on the end of an arm. As a coin rolls down a chute, it presses the arm up and the micro-switch is switched ON.

2 Using a light dependent resistor (LDR) and a relay circuit. As a coin falls down a chute, it blocks light from the LDR, causing the relay to de-energise and activate the solenoid. When the solenoid is activated the coin falls and light shines again on the LDR.

□ **Sheet 2.**

3 Using a pressure switch. The coin rolls over the pressure sensitive pad. The solenoid is activated as the pad switches on an electrical circuit.

4 A computer stores the weight of coins. All coins

Coin	Diameter/mm	Weight/g
1p	20	3.5
2p	25.5	7.15
5p	23	5.5
10p	28	11.2
20p	21	4.85
50p	29	13.25
£1.00	22	9.5

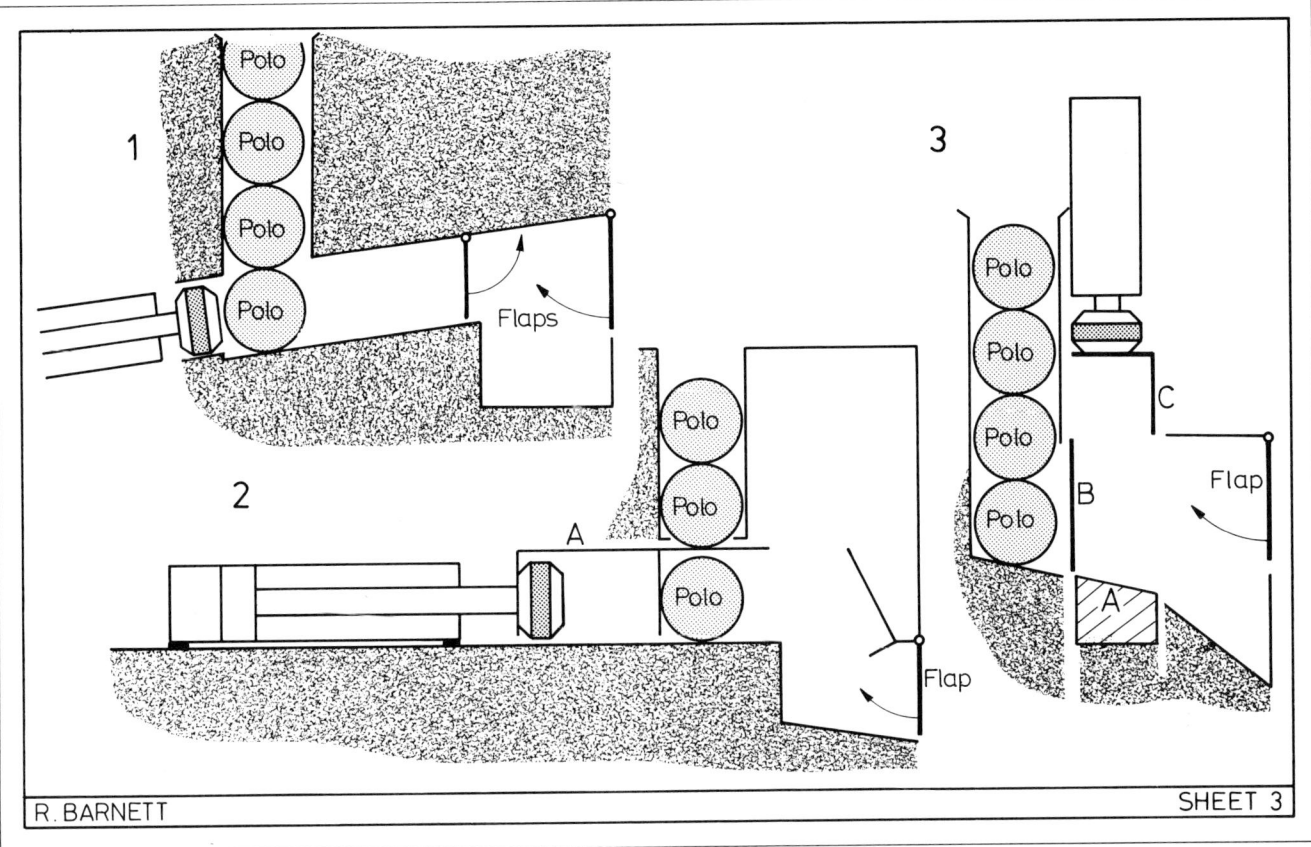

R. BARNETT   SHEET 3

Three ideas for dispensing the sweets

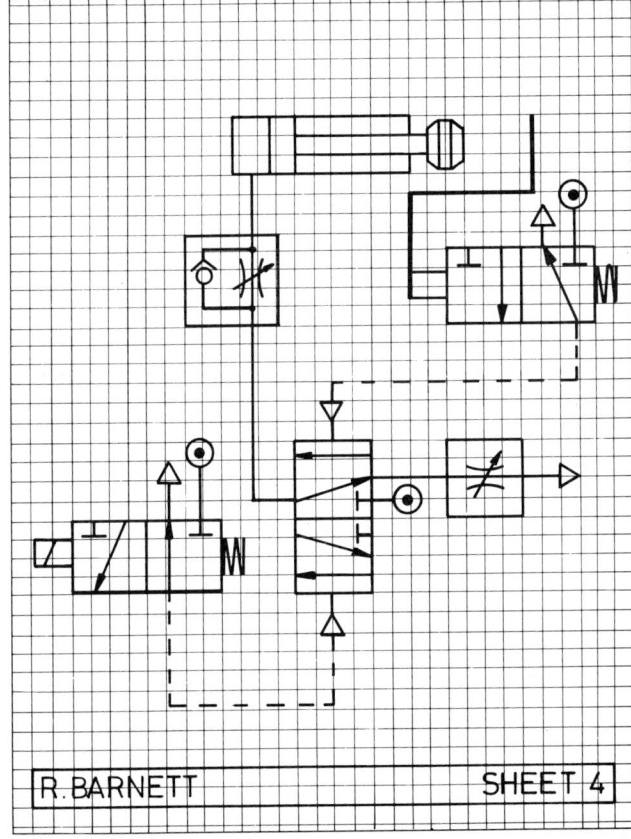

R. BARNETT   SHEET 4

A basic idea for the control circuit

fall down a chute on to an electronic weighing pad. Only if the weight agrees with the weight of a 10p coin as stored by the computer, will the solenoid in the pneumatic circuit be activated. Other weights of coins are rejected. I thought it best to weigh and measure coins in common use, in case I decided to use this system.

Ideas for dispensing the rolls of POLOs

☐ **Sheet 3** (see above).
1 As the piston goes +ve, its head pushes the roll slowly through a flap, which can only open in one direction. To take the roll from the device, a second flap is pushed forward.
2 When the piston goes +ve, plate A moves with it, pushing a roll onto a tray. When the piston moves −ve, the next roll drops down to take the place of the one pushed out.
3 A box-like device, consisting of three plates, is fixed to the piston head of the cylinder. Plate B prevents rolls falling out from a chute. When the piston head of the cylinder moves downwards, a roll falls through a gap in plate B onto plate A. The

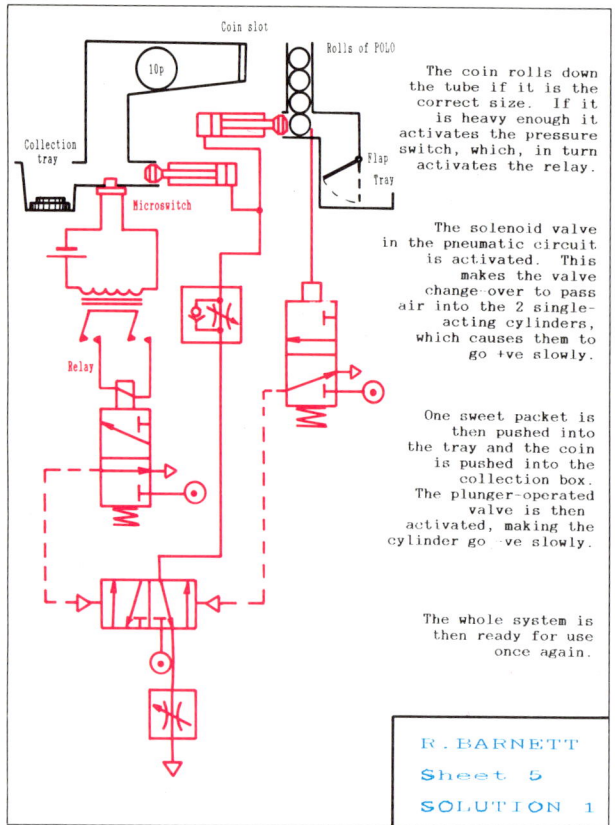

The coin rolls down the tube if it is the correct size. If it is heavy enough it activates the pressure switch, which, in turn activates the relay.

The solenoid valve in the pneumatic circuit is activated. This makes the valve change-over to pass air into the 2 single-acting cylinders, which causes them to go +ve slowly.

One sweet packet is then pushed into the tray and the coin is pushed into the collection box. The plunger-operated valve is then activated, making the cylinder go -ve slowly.

The whole system is then ready for use once again.

R. BARNETT
Sheet 5
SOLUTION 1

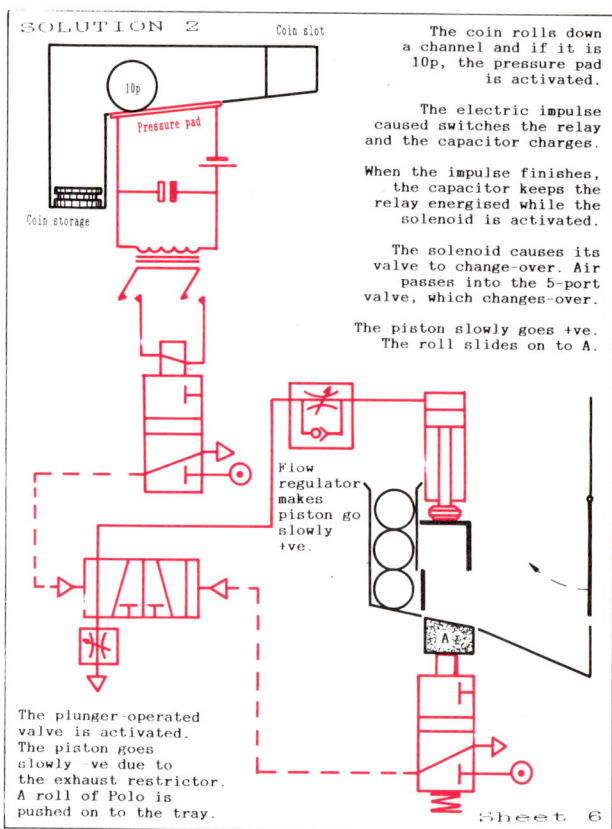

The coin rolls down a channel and if it is 10p, the pressure pad is activated.

The electric impulse caused switches the relay and the capacitor charges.

When the impulse finishes, the capacitor keeps the relay energised while the solenoid is activated.

The solenoid causes its valve to change-over. Air passes into the 5-port valve, which changes-over.

The piston slowly goes +ve. The roll slides on to A.

Flow regulator makes piston go slowly +ve.

The plunger-operated valve is activated. The piston goes slowly -ve due to the exhaust restrictor. A roll of Polo is pushed on to the tray.

Sheet 6

roll is held on plate A by plate C. When the piston goes −ve, the roll falls into a tray through a gap in plate C.

Pneumatic circuit for controlling the cylinder

☐ **Sheet 4** (see page 151).   The circuit shown will make the position move slowly +ve, so that rolls are not thrown from the dispenser, but are slowly moved out on to a tray.

☐ **Sheets 5, 6 and 7** (see above).   These three ideas for solutions to the pneumatic control circuit are based on the circuit shown in Sheet 4.

Chosen solution

The solution 2 circuit (Sheet 6) has been chosen for a final solution.
1 It is a simple system, which should be reliable.
2 Idea 1 is too complicated and the system could

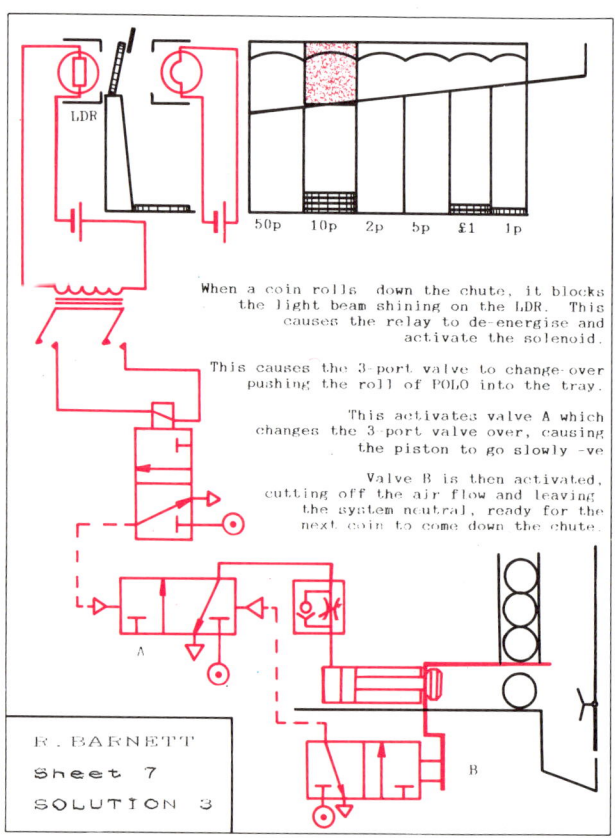

When a coin rolls down the chute, it blocks the light beam shining on the LDR. This causes the relay to de-energise and activate the solenoid.

This causes the 3-port valve to change-over pushing the roll of POLO into the tray.

This activates valve A which changes the 3-port valve over, causing the piston to go slowly -ve

Valve B is then activated, cutting off the air flow and leaving the system neutral, ready for the next coin to come down the chute.

R. BARNETT
Sheet 7
SOLUTION 3

be operated with any coin. Cylinder A may not push the required coin out of the way and into the collection tray. Also the next packet could become caught on the piston rod as the piston goes −ve, crushing the roll of mints.

3 Idea 3 was rejected. If the power supply fails or if the bulb blows (it is on all the time), a roll could be dispensed without a coin being placed in the device. Also the impulse received when the relay is de-energised could be so short that the solenoid would not charge quickly enough.

Amendments

Some changes are needed to the idea for a circuit given by solution 2.

1 A pressure pad will not be used because I cannot obtain one. Also the pad could be insufficiently sensitive to detect a 10p coin.

2 A microswitch will be used instead.

3 One problem was how to sort a 10p coin from other coins. I carried out some experiments with a Lloyd's Bank money box. This sorts coins into their various values, according to their diameters. I have adopted this method for sorting out 10p coins from any others inserted in the device.

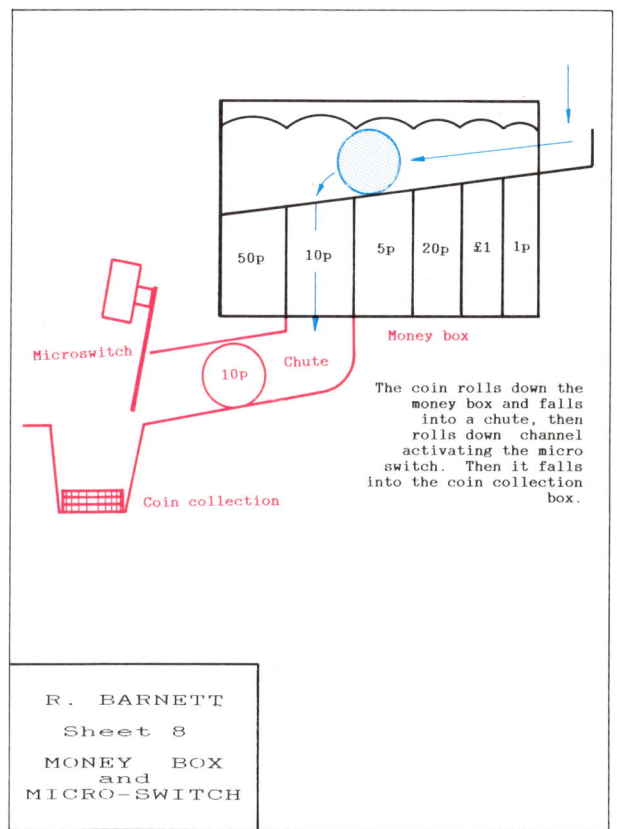

The coin rolls down the money box and falls into a chute, then rolls down channel activating the micro switch. Then it falls into the coin collection box.

R. BARNETT
Sheet 8
MONEY BOX
and
MICRO-SWITCH

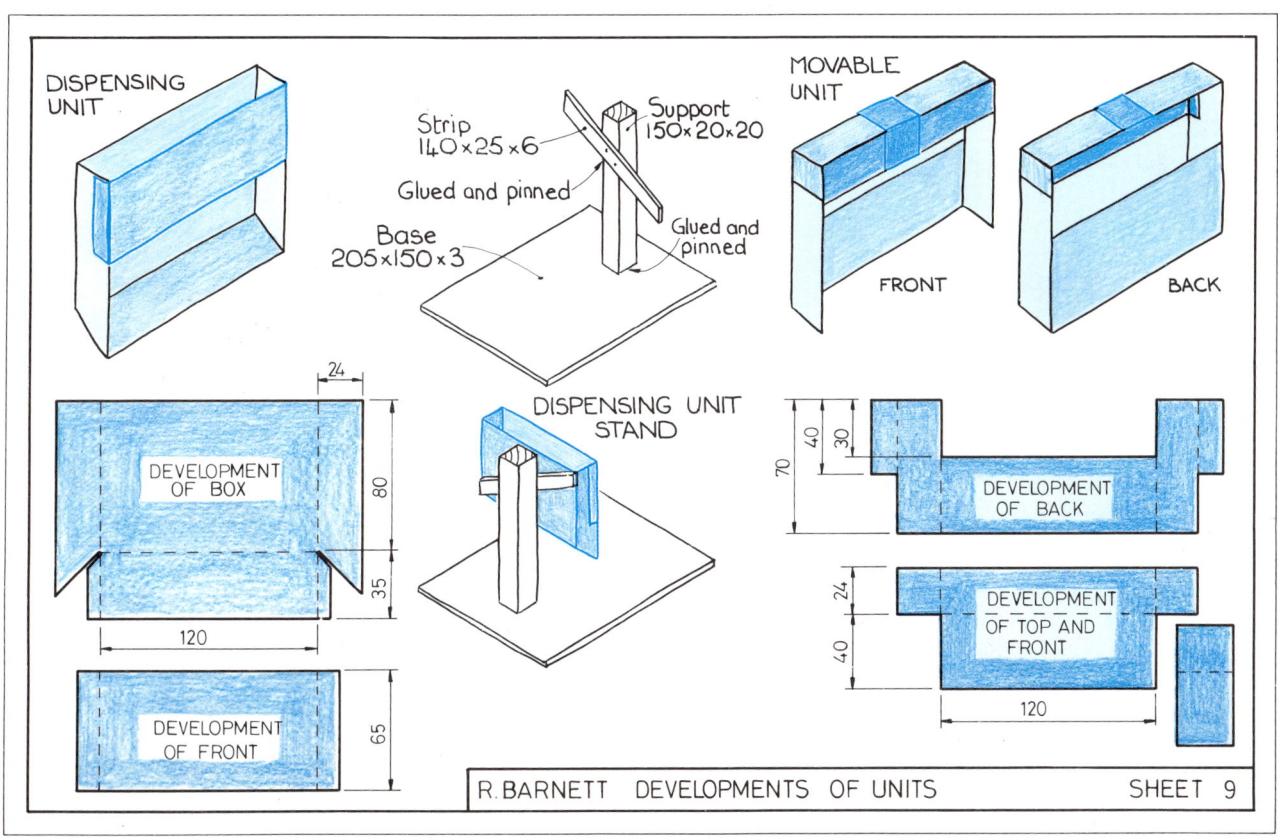

R. BARNETT   DEVELOPMENTS OF UNITS   SHEET 9

A student's realisation of the project

□ **Sheet 8** (see page 153). This drawing shows the amendments needed to the solution 2 idea for controlling which coin goes through to operate the machine.

Realisation

The design was realised by making a model.

□ **Construction of stand** Sheets 9 to 11 show the design and construction of the stand for the model of the Sweet Dispenser.

□ **Sheet 9** (see page 153). The dispensing unit, in which the rolls of mints are to be held, was made from thick (1.5 mm) card.

The movable unit which is fixed to the pneumatic cylinder head, was also made from thick card.

The base of the stand to hold the dispensing unit was made from hardboard, to which the upright was glued and pinned. The strip of wood which holds the unit was glued and pinned to the upright and the completed unit then glued to this strip.

□ **Sheet 10** (see below). This sheet shows the way in which the tray into which the mint packets roll and the door by which they are taken from the dispenser, are fitted to the model. The door is Sellotaped to the top piece of card, to hinge it to the card. The front decorative piece is Sellotaped under the base to hold it in place, as well as being glued to the tray.

Components and materials list

The following components and materials were used:

Electrical
1 microswitch
12 V power supply (from power pack)
1–2000 μF capacitor
1 D.P.D.T. relay
connecting wires

Pneumatic
1 solenoid unit
1 3-port pressure-operated valve
1 flow regulator
1 exhaust restrictor

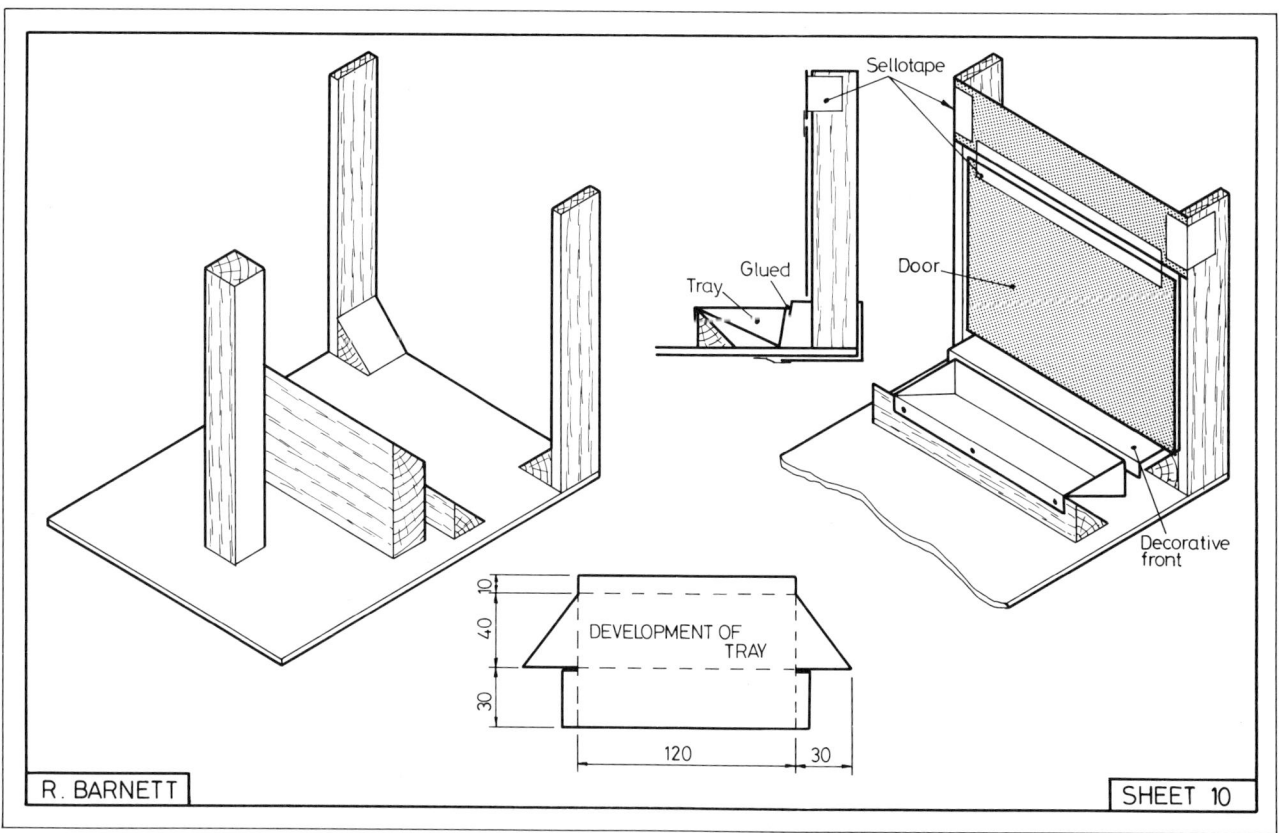

A page from the student's folder, showing the construction of the dispenser

1 single-acting spring-return cylinder
1 2-port plunger-operated valve
main air supply (from the school compressor)

Material for model
1 piece of A2 size card 1.5mm thick
2 pieces of hardboard 51.5cm × 15cm × 3mm
1 block of wood 31cm × 2cm × 2cm
1 piece of wood 29cm × 25mm × 6mm
Sellotape
1 elastic band
1 Lloyd's moneybox
1 icecream tub lid (4 litre size)
100 panel pins
1 old pen top
1 piece hard plastic 80mm × 7mm × 2mm

Evaluation

☐ **Note:** Lack of space in this book has meant that it has not been possible to include all of Ryan Barnett's notes and graphics for this project. When Ryan had completed it, he tested the working of the model and found that it needed modifications. These included:

1 changing the position of the POLO dispenser and bringing it slightly away from its supports;
2 adding a coin box to collect coins which were not 10p coins;
3 modifying the microswitch lever arm so that the switch would operate more effectively;
4 altering the position of the microswitch by mounting it on a metal plate;
5 reinforcing the dispenser, because it was found to be somewhat unsteady.

When these modifications had been made, the dispenser worked well, but a short circuit developed which was quickly cured by re-positioning the wires in the electrical circuit.

☐ **Sheet 11** (see below).
The drawings on this sheet show some of the modifications which were made as a result of the evaluation testing.

Project – Rock-a Baby-to-Sleep Device (see page 156)

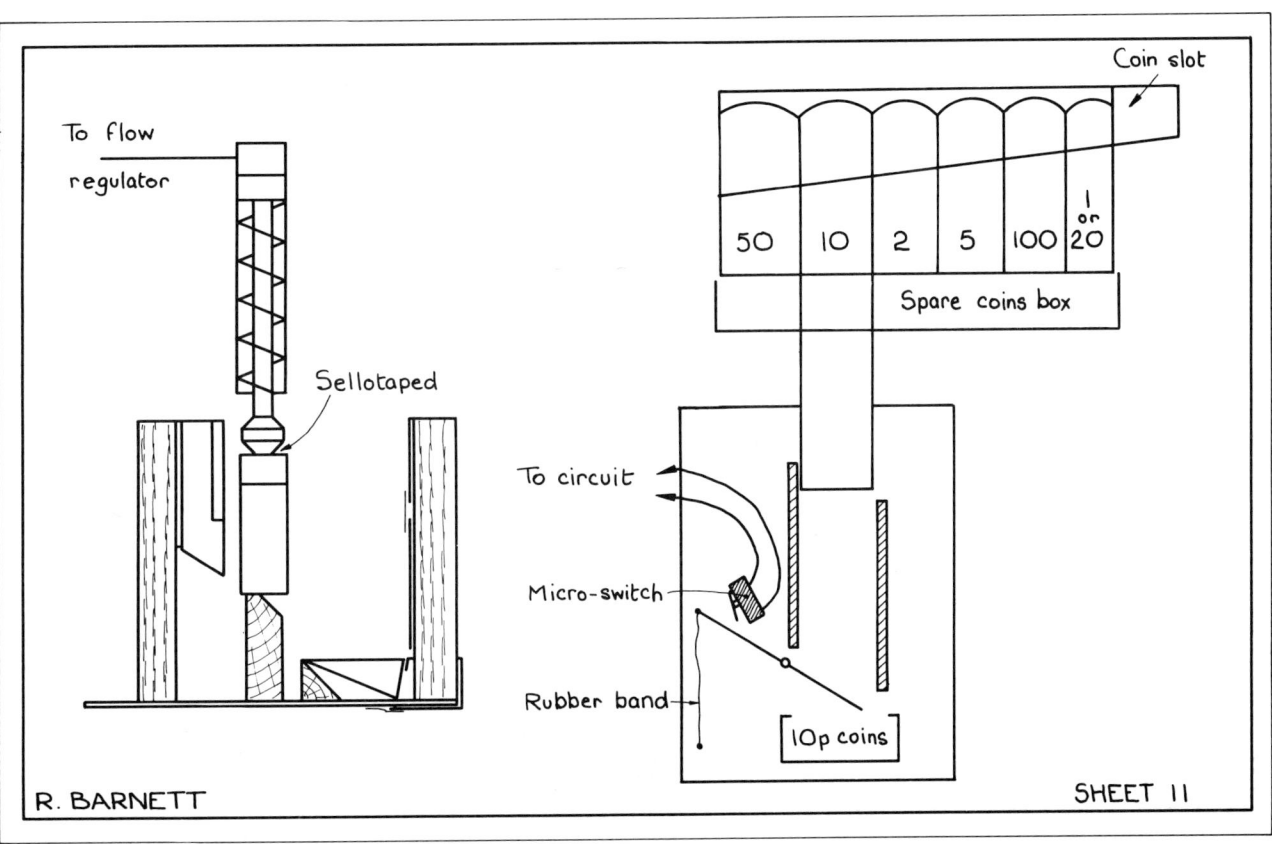

A project sheet to show modifications to the final design

ROCK-A-BABY-TO-SLEEP DEVICE

Situation

Baby wakes up and cries - the baby does not need feeding or changing, but does need to be comforted. The baby is picked up and rocked to sleep by a parent.

Design Brief

Design a device which will start a rocking motion to a cradle when the baby starts crying. The device should leave the parents free to get on with other tasks.

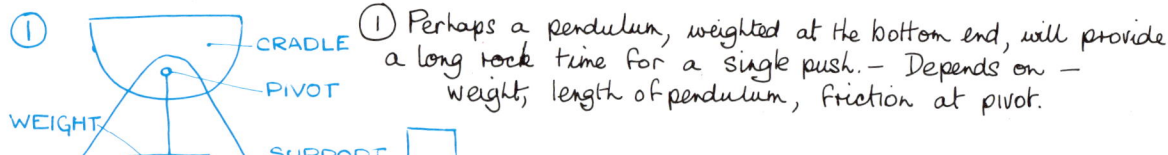

① CRADLE / PIVOT / WEIGHT / SUPPORT

① Perhaps a pendulum, weighted at the bottom end, will provide a long rock time for a single push. — Depends on — weight, length of pendulum, friction at pivot.

② SPRING

② Cheap, easy to fit, but the rocking motion would probably soon stop - BUT experiment.

③ PIVOT / CAM / SUPPORT

③ Motor turns, cams moves cradle up and down. Check on movement required. Will the cam provide the correct rocking motion? What other connections should be made from the motor spindle to the cradle. Experiment required.

④ Main problem is that a supply of compressed air is necessary. How far would cylinder have to be from the pivot? Can a cylinder long enough be purchased?

④ PNEUMATIC CYLINDER / PIVOT / SUPPORT

Queries arising so far

1. How can the speed of rocking be controlled?
2. Is rocking in one direction only sufficient? Is rocking a baby more complex?
3. If the rocking is controlled by an electric motor; is one enough or will 2 be needed? Would a variable resistor in the motor circuit vary the speed of rocking?
4. Is some method of gearing required between motor and cot-rock?
5. How can rotary motion be converted to linear motion?
6. From these queries it seems certain that from a circuit controlling an electric motor(s) a rock motion can be designed.

Block diagram of problem

1 BABY CRIES	2 CRYING HEARD	3 MOTOR ACTIVATED	4 TIME CONTROL	5 MOTOR STOPS	6 RE-SET NOISE LEVEL

So far the problem of noise activation of a circuit has not been investigated, but other problems arise before this.

Where should pivot be placed?

Is sideways rocking required?

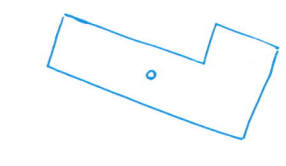

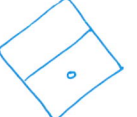

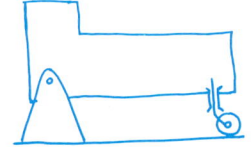

As the cradle is rocked will a slider controlled castor be needed as a safety device?

PAGE 1

A page from a student's design folder

Figure 1 The 130 kW vertical wind turbine constructed by Vertical Axis Wind-Turbine Limited on the power station site at Carmarthen Bay as part of the CEGB's 'Hosting Programme'. (Photograph by courtesy of the Central Electricity Generating Board.)

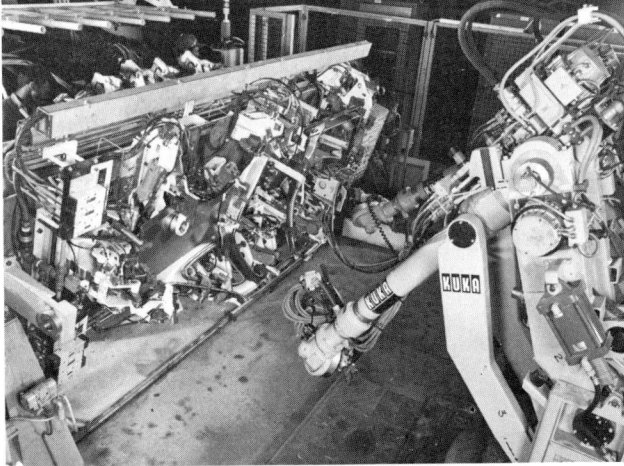

Figure 2 A robot at work constructing part of a car. (Photograph by courtesy of the Ford Motor Company Limited.)

Figure 3 A modern car. (Photograph by courtesy of the Ford Motor Company Limited.)

Actual size of the F100-L chip

Figure 4 The F100 Microprocessor Chip; an advanced 16 bit microprocessor. This is a 6 mm by 6 mm square chip containing 9000 electronic components, having a total length of 2.5 metres. (Photograph by courtesy of Ferranti Electronics Limited.)

13 Ideas for Projects

1 *Design brief:* Using sheets of newspaper and one piece of masking tape 400 mm long, construct a tower to reach the ceiling.

2 *Design brief:* Using only two pieces of card (each about equal in size to the front of an average Cornflakes packet), construct a bridge of the longest possible span to support a wire coat-hanger at its centre.

You are only allowed to used the following tools: scissors, modelling knives, straight edge or ruler, cutting board.

3 *Situation:* The Home Economics class have been asked to find out the wear rates of different materials.
Design brief: Design a pneumatic system to clamp material samples and then test their rate of wear for comparable periods of time.

4 *Design brief:* From a sheet of paper of A4 size, construct a model ship which, when floating, will carry the greatest possible quantity of dry sand for the longest period of time.

You are allowed to use the following tools: as in **2** above, plus a glue gun. (Each shot of glue will cost the user two minutes off the time which their ship holds the sand for.)

5 You have been provided with sawn softwood strips which are of 6 mm square cross-section and are of various lengths up to 220 mm. A length of 10 mm of each end of each strip has been clearly marked by dipping it in dyed water.
Design brief: Construct a stool capable of supporting any member of your group at a height of 430 mm above the ground. None of the strips of wood are to overlap another by more than the dyed length.
Tools allowed: A fine-toothed saw, glue gun, an off-cut of plywood to act as seat (not to be fixed to the stool).

6 *Design brief:* Using upholstery quality thread, design and construct a suspension bridge with a centre span of 350 mm from Meccano. Card decking of the bridge should support a small die-cast toy car.

Note: Some knowledge of the principles of *centre of gravity* (C of G) is necessary before commencing design brief 7.

7 *Situation:* The leaning tower of Pisa continues to move out of the vertical.
Design brief: Design and construct from card a tower that will lean at the greatest possible angle and yet remain standing.
Specification: Only a single thickness of card is to be used. The height of the tower must be three times its base width. No counter-weight or additional weight at the base is allowed.

8 *Situation:* Babies enjoy a rocking motion. Such motion often lulls them to sleep.
Design brief: Design an electrically driven mechanical device that could be attached to a model cot to impart a gentle rocking motion.

9 *Situation:* Some electric shavers are made with three circular cutting heads.
Design brief: Design and make a model whisk, based on the same principle.
Specification: The mechanism must drive three whisks. The beaters on the whisk must be made from card (for safety). The sweep of each beater must interlink with the sweep of the beater on either side of it, so that it can be used in confined areas.

10 *Situation:* Bathing patients in a hospital can be very time-consuming.
Design brief: Design and make a device to hang over the side of a bath which will give an alert when the correct depth of water is reached and give a broad indication of the water temperature.

11 *Situation:* Winding on the next frame of film on some cameras can be time-consuming – especially when a series of action shots is required.
Design brief: Design and make a mechanism that will cause 35 mm film to advance by one frame when the wind-on lever of a camera is moved through 45°.
Specification: A scaled-up model is permitted, but

the area occupied by the mechanism is to be kept proportionately small.

12 *Situation:* Arthritic people often have difficulty when getting into and out of chairs.
Design brief: Basing your design on the principle of levers, design and make a mechanism that will safely move the seat of a chair upwards and forwards to allow a person to stand up after being seated on a chair.

13 *Situation:* Tuning a radio may involve moving a vertical strip along the wavebands.
Design brief: Design and make a mechanism to move a vertical pointer a distance of 150 mm horizontally.
Specification: The position of the hand control is to be at right angles to the direction of horizontal movement.

14 *Situation:* A cat-flap gives a pet freedom to leave and enter a house without disturbing its owner.
Design brief: Design and make a mechanism that will allow a cat to move in and out of a house freely, yet will permit the owner to set a mechanism which will allow the cat to enter but automatically prevent exit until the mechanism is reset.

15 *Situation:* Machines which dispense parking tickets accept many different coins.
Design brief: Design and make a coin-sorting device which will store at least three different coins in separate positions.

16 *Situation:* Children who break their arm bones are encouraged by the hospital to exercise their hand.
Brief: Design and make a component that will indicate the strength of hand grip.

17 *Situation:* Fairgrounds use many forms of mechanism in their rides.
Design brief: Design and make a mechanism to be used on a fairground ride to fulfil one of the following criteria:
- a rotating arm, the end of which rises and falls four times every revolution;
- an arm which rotates vertically about its mid-point. At the end of the arm is a box, which itself rotates at twice the speed of the arm.

18 *Situation:* Industrial robots are now in use where repetitive tasks have to be performed with accuracy.
Design brief: Design and make a mechanism to be used to activate the legs on a walking robot.

19 *Situation:* Using scissors to cut many sheets of paper to size can be very time-consuming.
Design brief: Design a pneumatically controlled device which will cut A4 paper exactly in half.
Specification: The cutter is to be pneumatically controlled and driven. Safety devices must prevent the hands of an operator making contact with the cutter when it is used.

20 *Situation:* A greenhouse can have its humidity levels and temperature monitored and controlled automatically.
Design brief: Design and make a device which will automatically open and close a louvred window in order that the temperature inside a greenhouse can be maintained at a steady level.
It is suggested that an electrical/pneumatic interface could be used to operate the device.

21 *Situation:* Many charities make use of collecting tins left in shops and other public places. But a tin is not very appealing to children.
Design brief: Design and make a collecting tin which will appeal to children by giving an audible and visual response when a coin is inserted. Economy in operating costs is important.

22 *Situation:* Drilling machines in workshops are fitted with plastic safety guards. It is, however, possible to use the machine without the guard in place.
Design brief: Design and make a pneumatically controlled device which will prevent the drill chuck being lowered if the guard is not in place.

23 *Situation:* People who are badly disabled are often unable to turn the pages of a book when they are reading.
Design brief: Design and make a device which uses a short, controlled jet of air to assist a disabled person to turn the pages of a book.
Specification: The device will:
- incorporate other mechanical parts to assist in the turning over of the page;
- be designed to take several sizes of books;
- be of use to those who have lost the use of their fingers;
- be safe and economical in use.

24 *Note:* Some research into air bleed circuits is necessary before attempting this design.
Situation: A driver finds it difficult to get into and out of a car when it is parked in the family garage.
Design brief: Design and make a pneumatic circuit which will warn the driver when the nearside of the car is getting too close to the garage wall.

25 *Situation:* Young children are often frightened of the dark.
Design brief: Design and make a night light powered from a 6V battery which will automatically switch on and off according to the state of daylight.

26 *Situation:* In automated packaging systems, warning devices are built into units to alert operatives when errors occur.
Design brief: Design and make a pneumatically operated device which will automatically remove an overfull container from a conveyor.
Specification: The system must react to overfull containers only. All other containers must pass through. The system must be flexible enough to react to both slow and fast moving containers. A counter must record the number of containers rejected.

27 *Situation:* When using a glue gun, the best results are obtained if the components can be held tightly together for up to half a minute.
Design brief: Design a pneumatic press which uses one push button valve to fulfill the following sequence:
  (i) Hold glued parts together for up to half a minute.
  (ii) Then automatically release them.
  (iii) Remove the glued parts from the press.

28 A potential energy engine operates from a release of stored energy. Using fine sand (or container and winch or small bearings and chute), design and make a potential energy engine which will impart movement to the container or chute to give amusement to a seven-year-old child.

29 *Situation:* Most companies use paper which is headed with details of the company.
Design brief: Design and make a machine which will automatically print an address and a telephone number on a sheet of paper when it is positioned under the stamp head.

30 *Situation:* A hardness testing machine is used to ascertain the comparative hardnesses of a variety of materials.
Design brief: Investigate the use of a pneumatically controlled device which could be used to indent off-cuts of timber to discover their relative hardnesses.
Specification: It must be possible to hold the device positively in-stroke. An alarm must sound before the device goes out-stroke.

31 Design and make a pull-along toy for a child 3 to 4 years of age. The toy should change rotary to linear motion to provide a focus for enjoyment.

32 *Situation:* Sound, lights and sudden movement are often used either to attract attention or warn of danger.
Design briefs:
■ Design and make a system that will alert your family whenever your cat enters your house through a cat-flap.
■ Design and make a device which will tell you the front door of your house has been opened. Even if the door is shut immediately, the device must continue to function until such time as you switch it off.
■ Design and make a device to tell you it is raining on washing hanging out to dry.
■ Design and make a device to warn a person who is hard of hearing that the doorbell is ringing.
■ Design and make a device to warn a car owner audibly that someone has opened any door, bonnet or boot of the car.
Specification: The alarm must be set from inside the car. It must give the owner time to leave the car and shut the door before becoming operative. Once set, when a door is opened it must give the owner time to deactivate the alarm before it gives both visible and audible warning of entry. If a thief enters the car, the alarm must continue to sound even if the door is then closed. In case of false alarm the system must reset after ten minutes of the alarm sounding.

33 *Situation:* The effect of light on a light dependent resistor can be used to effect changes in other parts of a circuit.
Design briefs:
■ Design and construct a circuit which incorporates an LDR to count the number of people leaving a building.
■ Design and construct a circuit which will illuminate a door-bell push switch during the hours of darkness.

- Design and construct a circuit which will automatically dip the headlamp main beam of a motorcycle so that drivers approaching in the opposite direction are not dazzled.
- A business company package their product in one of three sizes of package. Design and construct a system which will automatically send the packs to one of three dispatch points. Assume the packs arrive by conveyor belt in any order, but in single file.

34 *Design brief:* Design a system which will automatically sort wine gums from Polo mints.

35 *Situation:* Museums need to protect their exhibits from theft.
Design brief: Design and make an alarm system which could be used to raise an alarm if an expensive necklace was being taken from its display case.

36 Even those with good sight may find threading a needle difficult.
Design brief: Design a device that will make threading a needle possible for non-sighted persons. Consider:
- ease of positioning the needle;
- variety of needle eye sizes;
- getting thread through the eye.

37 *Situation:* The RSPCA are often called upon to catch injured, lost and abandoned animals.
Design brief: Design and make a system which will automatically close the exit to a cage when an animal has been enticed inside. This system must be safe and humane in use.

Note: Time delay circuits often use capacitors or integrated circuits. Use this information in the design briefs **38** to **40**.

38 *Design brief:* Design and make a circuit for an egg timer which will time both a soft-boiled and a hard-boiled egg.

39 *Design brief:* Design and make a circuit to be used by a musician to help him keep to tempo. It should be adaptable to a variety of tempos.

40 *Design brief:* Design and make a device that could be used to warn of 'time-out' in a variety of game situations.

41 *Situation:* People with poor eyesight often find difficulty in reading a thermometer used to monitor room temperature.

Design brief: Design a thermometer which uses an LED readout to communicate the temperature. The thermometer must be capable of being used throughout the year.

42 *Situation:* Many motorists are unable to accurately monitor the state of important parts of the car.
Design briefs: Design and make a unit to be fixed to the dashboard of a car:
- that will indicate the state of charge of the battery;
- that will accurately show the amount of petrol left in the tank;
- that will indicate to the driver he has left the lights on (when the engine has been switched off);
- that will indicate that the water in the screenwash reservoir is dangerously low;
- that will indicate the water temperature in the radiator;
- that will allow the instrument panel light to be gradually dimmed for the convenience of the driver;
- that will indicate that the temperature outside is below freezing point;
- that will allow the driver to vary the speed of the wipers to an infinite degree;
- that will open the garage door at the push of a button.

43 Make a model from pipe cleaners of a device which would prevent a filled baby's feeding bottle being knocked over.

44 A hot iron on an ironing-board is a possible source of accidents. Design a structure which will be easy to attach and remove from the board yet hold a hot iron even when the ironing-board is moved. The iron must be easily lifted from and replaced into the design.

45 People with head injuries may need to have their heads held immobile for long periods of time. Using a school rugby ball as a model, design a structure which includes a means of holding a head securely.

46 Elderly people may have difficulty in holding a can of food while opening it. Design a device which will clip onto a work-surface and, with a simple action, grip a can of food.

47 People with arthritis may find the tuning controls of TVs or radio sets very difficult to turn.

Design a device which makes this action easier for them.

48 When carrying out a repair by torchlight there are occasions when a 'third hand' is needed to hold the torch at the correct angle. Design a structure which will:
- stand on a level though uneven surface;
- grip the torch securely;
- allow the torch to be adjusted through the greatest possible angle.

49 Using a child's large seaside bucket as a model, design a structure with associated mechanism which will allow an open-topped bucket two-thirds full of water to:
- be moved (not carried) across flat ground;
- travel (not carried) up a flight of stairs without a drop being spilt.

50 *Situation:* Hinged doors to a serving hatch are convenient to use, but sliding doors take up less space.
Design brief: Design a pneumatically operated unit to automatically open/close two sliding serving hatch doors whenever a person is close to them.

51 Exotic plants need special care.
Design brief 1: Design and make a device which will maintain temperatures in an indoor greenhouse suitable for growing healthy exotic plants.
Design brief 2: Design and make a back-up system for your design which, in the event of too high/low temperatures, will sound an alarm.
Design brief 3: Design and build a sensor which will monitor and display by means of LEDs the moisture level in the soil.

52 *Situation:* It is often difficult to find a torch in the dark when it is most needed.
Design brief: Design a circuit to be built into a torch which causes a LED to flash every 5 seconds, using a minimum of current.

53 *Design brief:* Design and make a small battery-powered vehicle to travel 3 m in the shortest possible time.

54 *Design brief:* Design and build a solar cell powered vehicle that will actively seek the brightest light source.

55 *Problem/Situation:* The local shoe repairer works at the rear of his shop. The noise of his work often prevents him from hearing customers enter his premises.
Design brief: Design and make an automatic system which will alert the shoe repairer when:
- a customer enters his shop;
- there is a great deal of noise in his workshop.

The system must operate from the moment a customer enters the shop and should continue to operate until the repairer signifies he has been alerted.

56 *Problem/Situation:* A model railway enthusiast wishes to modify one of his model road bridges to allow for the easy passage of taller carriages.
Design brief: Design and make a model bridge which lifts automatically when taller carriages approach.

The design should allow for:
- warning traffic using the road over the bridge;
- the automatic lowering of the bridge when the train has passed;
- resumption of traffic flow over the bridge.

57 *Problem/Situation:* You have become tired of a younger brother sneaking into your bedroom and leaping upon you as you are studying. Your Mum forbids the use of force.
Design brief: Design and make an early warning system that is activated by someone entering your room. The actual opening of the door should not trigger the device, nor should it remain activated for more than 30 seconds or less than 10 seconds.

58 *Problem/Situation:* The traditional door lock is useful in many situations, but it has one fault – it can be picked.
Design brief: Design an electrical lock to fulfil the following criteria:
- it can be opened at any time;
- it allows no more than three wrong methods of attempted access, before raising an alarm;
- it will continue to operate in the event of a power failure;
- it cannot be opened by random means.

59 *Problem/Situation:* House plants always seem to suffer when people go away on holidays. Invariably a kind neighbour over-waters them, or they use up the supply you stood them in within 24 hours, but either way, they die.
Design brief: Design and make a watering system which monitors the moisture level in the soil. If the soil becomes too dry, an electrical motor switches on for a short period of time to supply water through a sprinkler system.

60 *Problem/Situation:* The entrance to a warehouse is via a narrow road with a sharp bend in it. Trucks meeting on this road cannot pass and one of them must reverse.
Design brief: Design and make a traffic control system that will allow for the orderly use of the road. The use of barriers is forbidden because this would impede emergency service vehicles.

61 *Problem/Situation:* A model railway system needs to make the best possible use of a limited space.
Design brief: Design and make an automatic means of turning a model locomotive through 180° in order to return it down a track.

62 *Problem/Situation:* Many team quiz games seen on television involve the quickness of response to posed questions.
Design brief: Design and make a device which will reliably indicate which team pressed their button first.

63 *Problem/Situation:* Have you ever found yourself in the unenviable position of having to get out of a car, open a garage door and wait for dear old Dad to drive the car into the garage? It is usually raining too!
Design brief: Design and make an automatic garage door opening system. The system must fulfil the following:
- the door must open automatically as a car approaches;
- the door must remain open until the car is fully in the garage;
- the door must close automatically;
- when you want to leave the garage, the door must open and then close again when the car has left.

Other suggestions

1 A table-tennis ball server
2 A pet feeder which will dispense food twice daily at stated times
3 A soccer ball force and accuracy indicator
4 A burglar containment device
5 A factory sorting system
6 A mobile robot smoke detector
7 A television control device for the severely disabled
8 A bird feeder and counter
9 A toothpaste dispenser
10 An electric self-bailer which will keep the level of water in a dinghy at a given level
11 A mechanism which will lift an egg from boiling water after a set cooking time

Appendix 1
SI Units

Quantity	Unit	Symbol
length	metre	m
velocity	metre per second	m/s
acceleration	metre per second per second	m/s^2
mass	gram	g
	kilogram	kg
density	grams per cubic centimetre	g/cm^3
force	newton	N
moment	newton metre	Nm
	newton millimetre	Nmm
electrical current	ampere	A
potential difference	volt	V
resistance	ohm	Ω
capacitance	farad	F
work	joule	J
power	watt	W
angle	degree	°
	minute	′
	second	″
	radian	rad

Table 1 SI units

Prefix	Value	Symbol
mega	one million	M
kilo	one thousand	k
hecto	one hundred	h
deca	ten	da
deci	one-tenth	d
centi	one-hundredth	c
milli	one-thousandth	m
micro	one-millionth	μ
nano	one-thousand-millionth	n
pico	one-million-millionth	p

Table 2 Prefixes to SI units

Appendix 2 Pneumatics

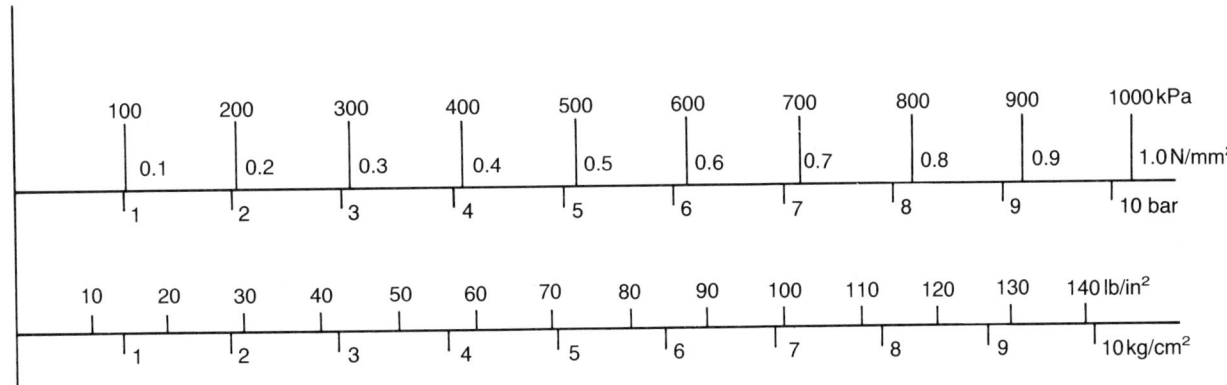

Figure 1 A scale comparing units of pressure

Units of measurement of pressure

The pressure of compressed air is measured in a number of different units. Figure 1 gives a comparison between the four units commonly used – kilopascals (kPa), bars, pounds per square inch (lb/in^2) and kilograms per square centimetre (kg/cm^2).

Actuator symbols

Name of actuator	Conventional use
Lock-down lever	Hand-operated. With a 3-port type, permanent pressure when in ON position and will not go OFF until lever is re-positioned. With a 5-port type, permanent pressure in both ON and OFF positions.

Name of actuator	Conventional use
Key	Hand-operated. Permanent pressure while key is ON.
Push button	Hand-operated. Supplies air only when pressed down. Immediately allows air to exhaust when released.
Foot pedal	Foot-operated. Supplies air only when down.
Plunger and roller	Machine operated, e.g. by the +ve or −ve stroke of a piston.
Uni-directional roller	Machine operated in one direction only.
Diaphragm	Operated by pressure of 0.2 bar or greater.
Pilot pressure	Operated by pressure of 3 bar or greater.
Solenoid	Operates when an electric current flows through the solenoid.

Note: All actuators may be returned by a spring.
Table 1

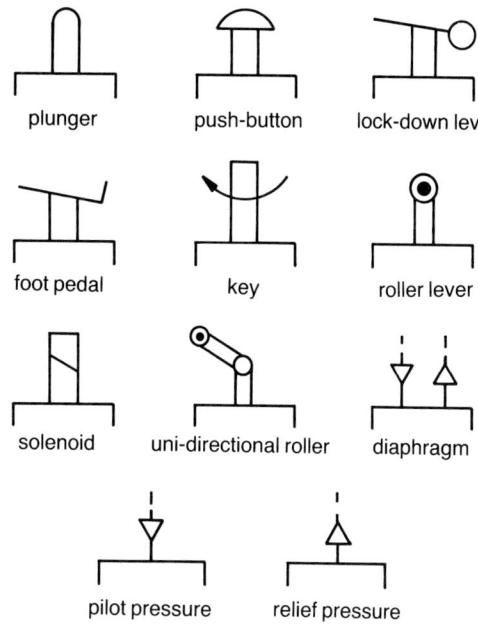

Figure 2 BS symbols for actuators

Valve port identification

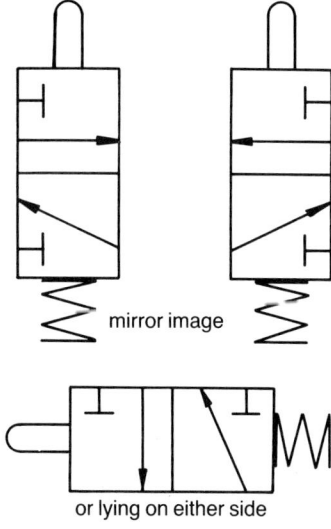

Figure 3 BS symbols for spring operated 3-port plunger valve

In order to connect up circuits correctly, you must know the position of each port in a valve. If you examine a 3-port valve, you will see that each port is:

1 numbered, or

2 marked with a letter, or

3 marked with an abbreviated word. See Table 2.

Port	Description	Action
1 or A or none	inlet port	Connect main air supply.
2 or B or CYL	cylinder port	Air delivery. Connect to valve/cylinder.
3 or C or EXH	exhaust port	Do not blank off.

Note: The symbol for a pneumatic valve bears little resemblance to its actual shape.

Table 2 3-port valves

Figure 4 A 3-port valve

3-port and 5-port double pilot pressure-operated valves

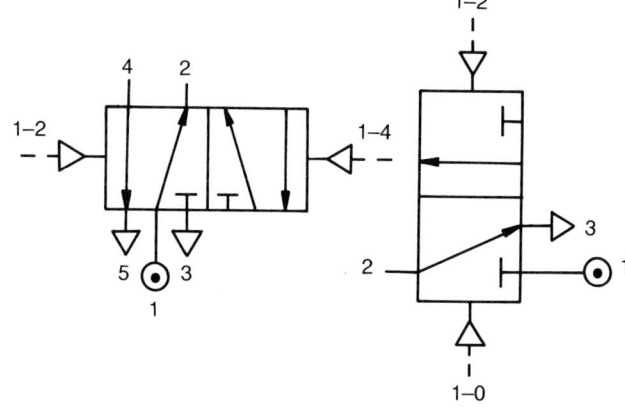

Figure 5 BS symbols for 3-port and 5-port valves with identification

Port	Description	Action
I	inlet port	Connect main air supply.
2 and 4	cylinder port	Air delivery. Connect to valve/cylinder.
3 and 5	exhaust ports	Do not blank off.
I–0, I–2 and I–4	signal ports	Receive compressed air from a valve to effect a change-over of the pilot valve.

Table 3

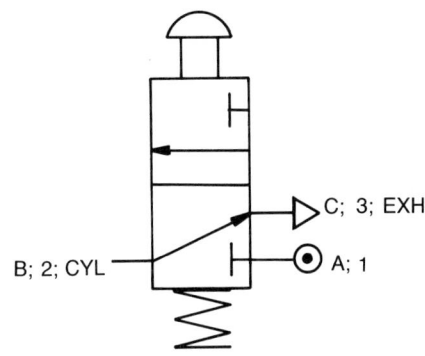

Figure 6 BS symbol for 3-port valve with port identification

Spool valves

Note: the details given in Tables 2 and 3 apply equally well to spool valves, but with spool valves it is permissible to connect the main air to the cylinder or exhaust ports without damaging the valve. DO NOT attempt this interchanging with poppet valves.

Connect main air to:	Result
Cylinder port	Air pressure flows through the valve and out through the exhaust port. When the valve is actuated, the air flows out through the inlet port.
Exhaust port	Air flows through the valve and out through the cylinder port. When the valve is actuated, air ceases to flow.

Table 4

Other BS Pneumatics symbols

Symbol	Meaning of symbol
Main air feed line	Continuous line showing a connection between a control valve and a cylinder.
Pilot signal line	A broken line showing a control signal line.
Crossing lines	Pipes are NOT joined together.
Connecting lines	Pipes ARE joined together.
Reservoir	Stores compressed air. Often used in series with a flow control valve to put a delay into a circuit.
Exhaust restrictor	Restricts the flow of air from the exhaust of a valve and so controls the speed of a piston movement.
Flow control valve	Adjustable so as to restrict the flow of air in one direction only. Full flow always allowed in opposite direction.
Air bleed	A pipe continuously leaks a controlled amount of air. When an object passes close to the pipe or stops the leak, pressure builds up in the circuit.

Symbol	Meaning of symbol
Tee connector	A fitting enabling a pipe to supply or receive air in two directions. Two tee connectors joined together with a short length of pipe produce a 4-way connector
Blanking-off piece	Prevents the exit of air in a circuit. Do NOT blank off the exhaust ports of valves.

Table 5

Appendix 3 Resources

Briefs and design ideas

1 Egg Race Books
The British Association for the
Advancement of Science,
Fortran House,
23 Savile Row,
London WW1X 4SU

2 *Designing* magazine
The Design Council,
23 Haymarket,
London SW1 4SU

3 *School Technology* magazine
TICST,
Trent Polytechnic,
Burton St,
Nottingham NG1 4BU

4 *Electronic Systems News* magazine
IEE,
Station House,
Nightingale Road,
Hitchin,
Hertfordshire SG5 1PJ

Robotics and NC machines

1 Clwyd Technics Ltd,
Antelope Industrial Estate,
Rhydymwyn,
Mold,
Clwyd CH7 5JH

2 Commotion,
241 Green Street,
Enfield EN3 7TD

3 Shesto-Tech Ltd,
Unit 2,
Sapcote Trading Centre,
374 High St, Willesden,
London NW10 2DH

Engineering plastics

1 A.B.G. (Industrial) Ltd,
Galowhill Road,
Brackmills Industrial Estate,
Northampton NN4 OEE

2 Proops Distributors Ltd,
Heybridge Estate,
Castle Road,
London NW1 8TD

3 Trylon Ltd,
Thrift St,
Wollaston,
Northants NN9 7QJ

4 Ema Model Supplies,
56–60 The Centre,
Feltham,
Middlesex TW13 4BH

Electronic components

1 Rapid Electronics Ltd,
Hill Farm Industrial Estate,
Boxted,
Colchester,
Essex CO4 5RD

2 Cirkit Distribution Ltd,
Park Lane,
Broxbourne,
Herts EN10 7NQ

3 Kelan (Hobbyboard),
North Works,
Hookstone Park,
Harrogate

4 JPR Electronics, Unit M,
Kingsway Industrial Estate,
Kingsway,
Luton LU1 1LP

5 Maplin Electronic Supplies Ltd,
PO Box 3,
Rayleigh,
Essex SS6 8LR

Electronic systems

1 Testbed Technology Ltd,
PO Box 70,
Clarendon Road,
Blackburn BB1 9TD

2 E & L Instruments Ltd,
Whitegate Industrial Estate,
Wrexham,
Clwyd LL13 8UG

3 Omega,
12 Oxhill Road,
Middle Tysoe,
Warwicks CV35 0SX

Pneumatic equipment

1 Tech-Air Ltd,
26 Britannia Court,
Burnt Mills,
Basildon,
Essex SS13 1EU

2 Vento Solenoids Ltd,
43 Burners Lane,
Kiln Farm,
Milton Keynes,
Buckinghamshire MK11 3HA

3 Festo Didactic,
7 High Street,
Teddington TW11 8EH

Technology equipment suppliers

1 Economatics (Education) Ltd,
Epic House,
Orgreave Road,
Handsworth,
Sheffield S13 1EU

2 Technology Teaching Systems Ltd,
Penmore House,
Hasland Road,
Chesterfield S41 0SJ

3 Technology Supplies,
6 Stoke Court,
Market Drayton,
Shropshire TF9 2DY

Technology equipment

1 Energy Facilities Management Ltd,
Pendlebury Industrial Estate,
Manchester M27 1FJ

2 Environmental Improvements Ltd,
Unit 34,
Engineer Park,
Sandycroft,
Clwyd CH5 2QD

Gears, pulleys, motors

1 Durr-Technik,
7 West Rd,
Woolston,
Southampton SO2 9AH

Technological information (including software)

1 BP Education Service,
Britannic House,
Moor Lane,
London EC2Y 9BU

2 British Gas,
Education Service,
Room 707A,
326 High Holborn,
London WC1V 7PT

3 British Nuclear Fuels Ltd,
Information Services Directorate,
Risley,
Warrington WA3 6AS

4 British Standards Institute Sales,
Linford Wood,
Milton Keynes MK14 6LE

5 British Steel Corporation,
Information Services,
9 Albert Embankment,
London SE1 7SN

6 British Telecom Education Service,
PO Box 10,
Wetherby,
W. Yorks LS23 7EL

7 Central Electricity Generating Board,
Department of Information,
Sudbury House,
15 Newgate Street,
London EC1A 7AU

8 Centre for Alternative Technology,
Machynlleth,
Powys,
Wales

9 Education Service of the Plastics Industry,
University of Technology,
Loughborough,
Leics LE11 3TU

10 Electricity Council,
30 Millbank,
London SW1P 4RD

11 Friends of the Earth Trust Ltd,
377 City Road,
London EC1V 1NA

12 Glass Manufacturers Federation,
Information Officer,
19 Portland Place,
London W1N 4BH

13 Institution of Metallurgists,
Education Officer,
PO Box 471,
1 Carlton House Terrace,
London SW1Y 5BE

14 United Kingdom Atomic Energy Authority,
Information Services Branch,
11 Charles II St,
London SW17 4QP

The Dewey decimal system of book classification

Non-fiction books in libraries are classified by subject in this system. The Dewey system is very complex and as a guide to those looking for information when designing a technology project, the following short list from the system is given. The whole system will be found in all public libraries.

Subject	Dewey subject number
Aeroplanes	629.1
Bridges	624
Cameras	771.3
Chemistry	540
Computing	001.64
	510
Design (Craft)	745.4
Electric circuits	537.61
	621.31
Electronic circuits	621.381
Energy conservation	333.72
Fluidics	629
Graphic design	741.6
Hydraulics	621.8
Hovercraft	629.3
Materials	620.1
Mathematical logic	511.3
Mechanics	620.1
Microscopes	535.332
Plastics	668.4
	745.57
Rockets	623.45
Structures	624.1
Technical Drawing	604.2
Technology	607.2
Telephones	621.385
Telescopes	621.38
Television	621.388

14 Exam questions

Core questions

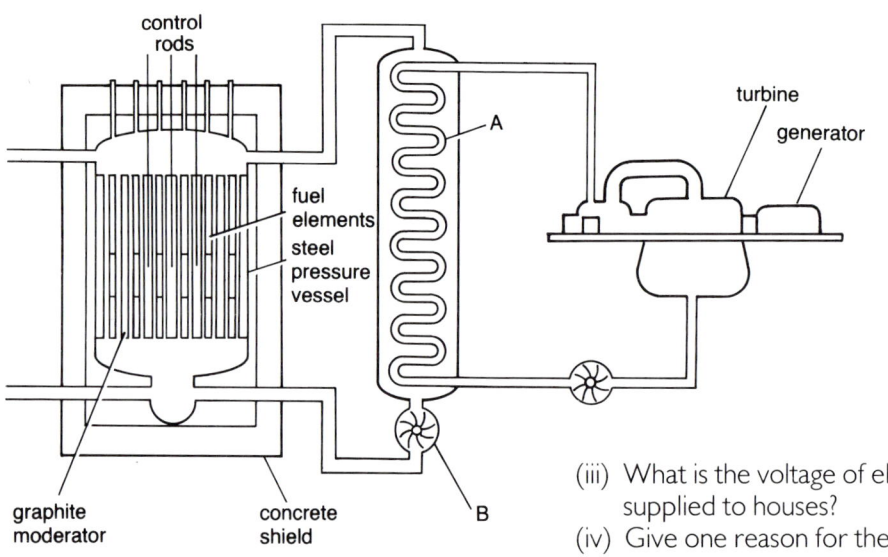

1 The drawing above represents the main parts of a 'Magnox' nuclear reactor which is sometimes used as the first stage in the generation of electricity.

■ Name the input and output of the energy conversion process which takes place inside the pressure vessel. (2)

■ The turbine is an energy converter.
What is produced in the part labelled A in order to drive the turbine? (1)

■ It is important that the operators know of any changes in the volume of cooling gas flowing round the fuel elements.
Explain why an analogue meter is more suitable than a digital meter for giving this information. (2)

■ (i) The pump marked B may fail in use. What effect would this have on the pressure vessel? (1)

(ii) Describe a suitable safety system which would operate should this pump fail. (3)

■ The generator produces electricity for the National Grid.

(i) What is the purpose of the National Grid?

(ii) At approximately what voltage is electric current carried by the National Grid system?

(iii) What is the voltage of electric current supplied to houses?

(iv) Give one reason for the differences between these two voltages. (4)

NEA

2 Explain, with the aid of a sketch, one method by which the energy of the wind may be converted into electrical energy. (8)

WJEC

3 This question is based on the common core topic, acid rain, as specified in the syllabus. You may need to read the following passage more than once before answering.

As long ago as 1872, the term acid rain was used by Robert Angus Smith, a British chemist, in a book he wrote. It was not until 1926, however, that the Inspector of Freshwater Fisheries in Norway noted that the sudden death of newly hatched salmon in Norwegian waters seemed to be linked to water acidity. The diagram on the next page shows the production of acid rain.

Damage from acid rain used to be only in areas close to the source of pollution. With the building of much taller chimneys to reduce local pollution, the problem was merely moved farther away, often to a country foreign to the source of the pollution. The Tyneside area of the United Kingdom is possibly the main source of acid rain pollution in Sweden, but the movement of pollutants is a two way process. The "produced" and "deposited"

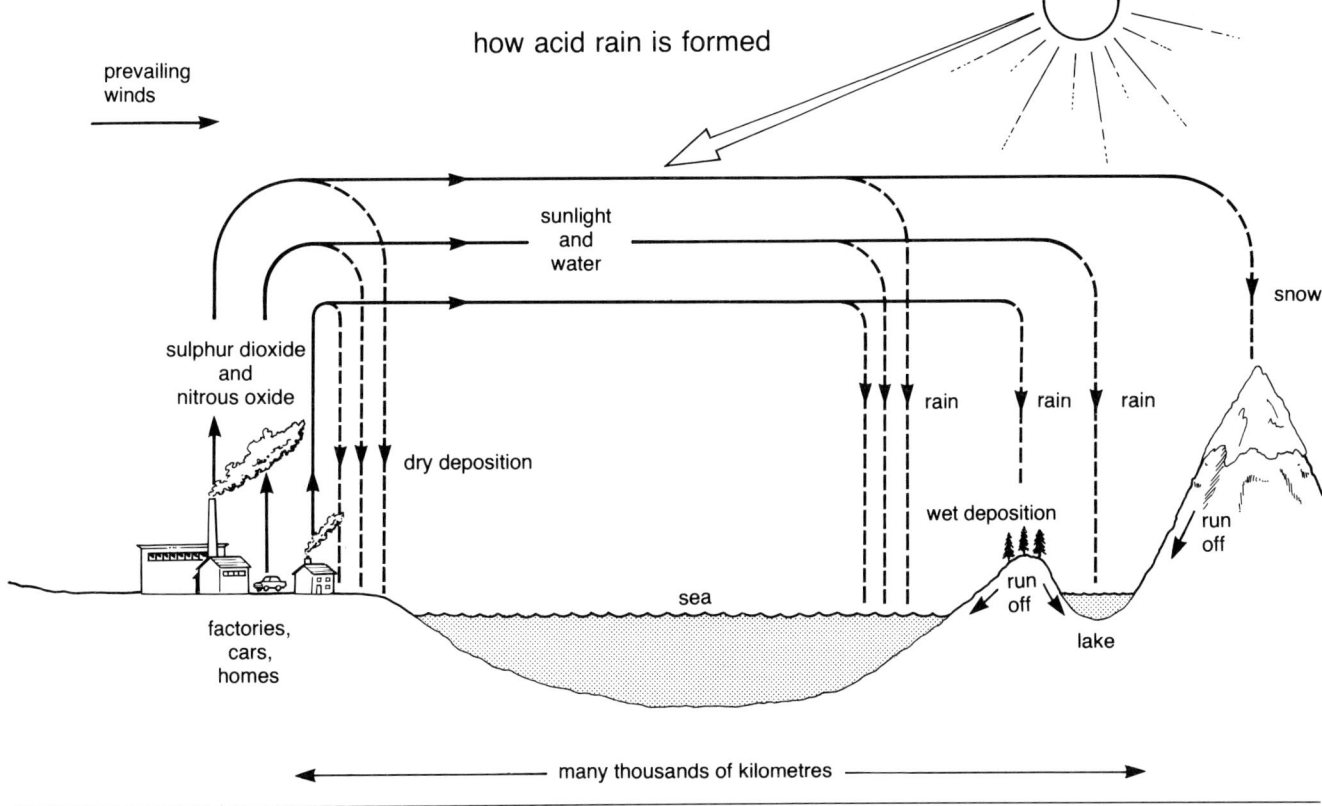

how acid rain is formed

figures for sulphur pollution in some European countries are given in the following table.

COUNTRY	tonnes of sulphur produced (× 1000)	tonnes of sulphur deposited (× 1000)
Belgium	404	161
France	1800	1212
Germany	1815	1158
Iceland	6	74
Italy	2200	1132
Poland	2150	1330
Sweden	275	472
United Kingdom	2560	847

SULPHUR POLLUTION TABLE

Attempts are now being made to neutralize acidified lakes, but the process was costing Sweden 1.2 million pounds per year by the end of the 1970s. Sweden now expects to spend large sums of money on this problem every year, as it will be some time before the emission of acidifying substances is significantly reduced.

■ State the year in which the term acid rain was first used. (1)
■ Name the country which produces the largest amount of sulphur pollution. (1)
■ Briefly explain why the introduction of taller chimneys led to the "exportation" of pollution. (1)
■ Briefly explain how snow falling on a mountain can affect the acidity of a lake in the valley. (1)
■ Rain is naturally acidic to a small degree, without the addition of pollutants. Name the acid which is present in "unpolluted" rain. (1)

MEG

4 Technology in Society.
"In society we rely more and more on technology."
■ Explain in **one paragraph** what you understand by this statement. (4)
■ Describe briefly **one** situation in our day to day lives where you feel the effects of technology are extremely good. (4)
■ Describe briefly **one** other situation in our day to day lives where you feel the effects of technology are bad. (4)
■ Explain why you think 'western technology' is not always suitable in third world countries. (8)

WJEC

5 ■ Complete the truth table for the three input OR gate shown in the diagram. (2)

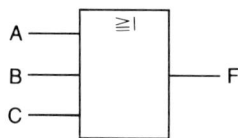

Truth table			
A	B	C	F
0	0	0	
0	0	I	
0	I	0	
0	I	I	
I	0	0	
I	0	I	
I	I	0	
I	I	I	

■ Draw a logic circuit which will give the output $F = AB + C$. (2)

WJEC

6 ■ For each of the circuits shown below, name the logic gate which can act in the same way and draw its symbol. (6)

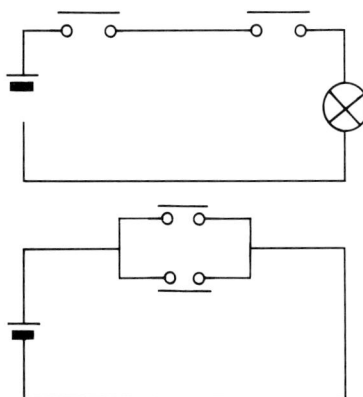

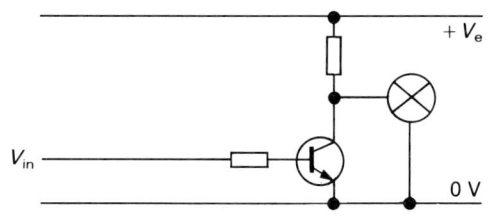

■ During the operation of a photocopying machine many sub systems need to be monitored. Of these, the motor which drives the paper feed and the motor driving the scanning lamp's movement are vital and have sensors fitted to them.
Should either motor fail, then the main power must be switched off.

(i) Complete the table showing all the possible output logic states of the two motor sensors. Assume that motor running = logic I. (4)

Paper Feed Motor	Scanning Lamp Motor

(ii) The machine also has a heater fitted. If the heater fails, the power supply must be switched off.
If either of the motors or the heater fails, name a single logic gate which will change state when any sensor signals fail. (2)

(iii) Photocopying machines require a supply of toner, a black powder which is used to make the print on the photocopies.
The tank holding the toner is fitted with a sensor which gives a logic state 0 until the tank is empty.
Draw clearly a logic system which would sense any failure condition. The final output logic should be 0 if one of the motors or the heater fails, or if the toner tank becomes empty. (5)

NEA

7 ■ Write down which of the following materials are metals:
aluminium; brass; carbon; copper; glass; low carbon steel; paper; p.v.c. (4)

- Explain, with the aid of a sketch, the construction of **one** of the following:
  blockboard; plywood; chipboard. (5)
- State **one** practical application for the material chosen above. (1)

WJEC

8 ■ Complete the table below by choosing, from the following list, the correct property for the definitions given:
hardness; plasticity; conductivity; toughness; strength. (4)

Property	Definition
	Deforms under load and does not return to its original shape when the load is removed.
	Resists abrasion and penetration.
	Withstands shock loads without fracture.
	Withstands loads which tend to make it increase in length.

- Complete the table by stating a definition for the given property. (4)

Property	Definition
Ductility	
Malleability	
Elasticity	
Brittleness	

- Explain what is meant by the term thermal conductivity. (2)

WJEC

9 The following diagram shows a simple lever. The load, effort and fulcrum are also shown. Practical examples of levers are shown (i) – (iv).
On **each** of the diagrams (i) – (iv), indicate the **load, effort** and **fulcrum**. (6)

WJEC

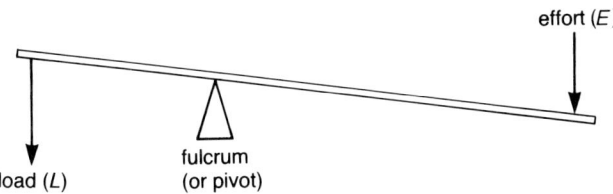

effort (E)

load (L)   fulcrum (or pivot)

(i) **Crow bar**

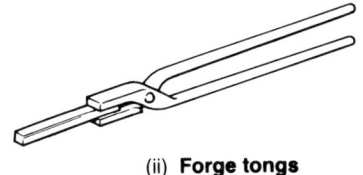

(ii) **Forge tongs**

(iii) **Pair of tweezers**

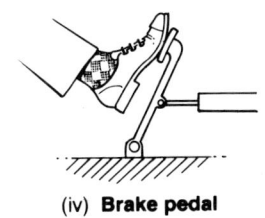

(iv) **Brake pedal**

10 Below is a plan view of a toddler's kart.

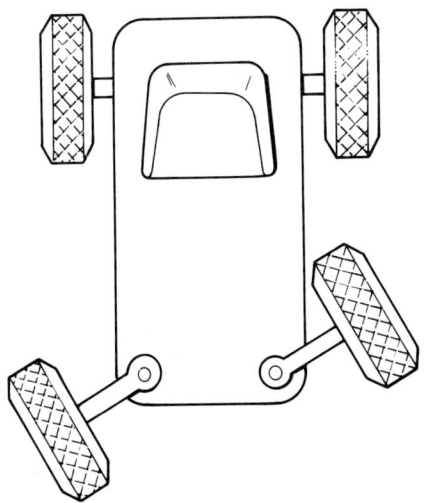

■ Mark on your drawing a way of linking the two front wheels so that they steer together. (2)
■ Sketch clearly a mechanism which links the movement of a steering wheel to the movement of the front wheels. (4)
SEG

11 Here is part of a car's headlight circuit.

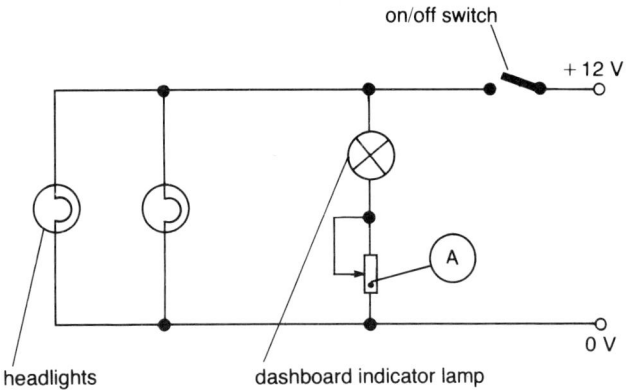

■ State why the two headlamp bulbs are connected in parallel. (1)
■ (i) Name component A. (1)
(ii) State the purpose of component A in this circuit. (1)
■ (i) Name a safety device which can be included in the circuit to restrict the current to a safe maximum value.
(ii) Draw the symbol for this device.
(iii) Mark on your diagram with a cross where this device should be connected. (3)
■ Complete a copy of the following diagram so that the component S switches on either the main or the dipped beam. *SEG*

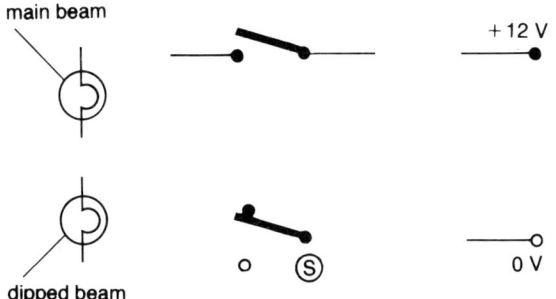

12 Many households have a step-ladder; these are commonly made from wood, aluminium or steel.
■ Three factors which affect the choice of material for this application are cost, weight and resistance to corrosion; for each of these factors, state the order of merit (low, medium or high).

Material	Untreated Pine	Painted Steel	Aluminium
Cost			
Weight			
Resistance to corrosion and decay			

(6)

■ Below are four possible cross-sections for the step of a step-ladder.

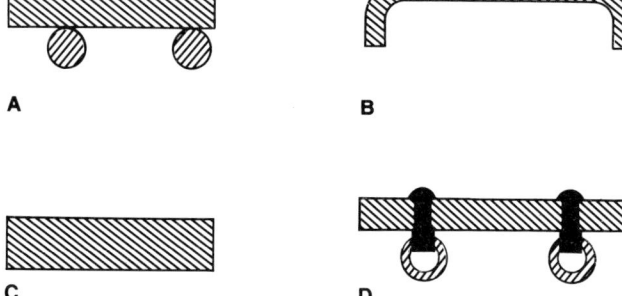

State the two section(s) suitable for manufacture from aluminium. (2)
SEG

Modular questions
Electronics

1 ■ By reference to the colour code table state what colour bands you should find on resistors with the following values:
(i) 47k; (1)
(ii) 2k; (1)
(iii) 180R; (1)
(iv) 5k6; (1)
(v) 1M. (1)

number	colour
0	black
1	brown
2	red
3	orange
4	yellow
5	green
6	blue
7	violet
8	grey
9	white

■ Consider the circuit shown below.
  (i) Name component A. (1)
  (ii) Name the configuration formed by
       components A and B. (1)
  (iii) If component B is rated at $6\,V - 0.36$ Watts,
        state the reading required at V_1. (1)
  (iv) Calculate the value setting of component A
       which would give the required reading V_1. (2)

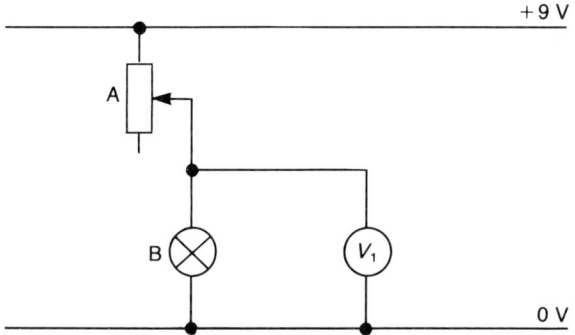

■ An ammeter has a scale of $0 - 100\,mA$, but it is to
  be used to measure a current of $500\,mA$, from a
  supply of $6\,V$. Given that the resistance of the
  meter is 0R2:
  (i) draw a circuit diagram to show a suitable
      circuit you could use to do this; (2)
  (ii) state the full specification of any components
       used in your circuit. (2)
■ The following diagram shows two waveforms
  displayed on an oscilloscope. Waveform V_i is the
  input to an amplifier, and waveform V_o is the
  simultaneous output from the amplifier.
  (i) State the peak to peak value of the waveform
      V_i. (2)

  (ii) State the gain of the amplifier. (2)
  (iii) State the frequency of the waveform V_o. (2)

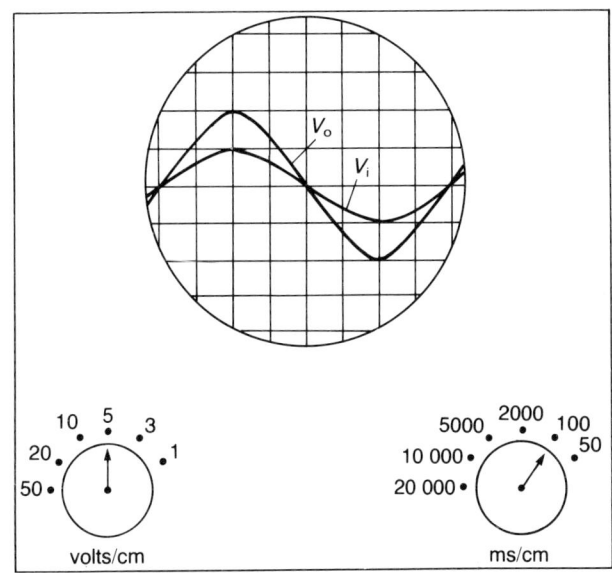

■ Consider the circuit shown below.
  (i) Give a suitable use for such a circuit. (2)
  (ii) Name, and give the circuit symbols for, two
       components which could be connected to the
       circuit to make it complete. (4)
  (iii) Calculate the value of I_e. (2)
  (iv) Calculate the current gain of the transistor as
       shown below. (3)
  (v) Calculate the value of R_4. (3)
  (vi) Explain clearly the purpose of components C_1
       and C_3. (2)

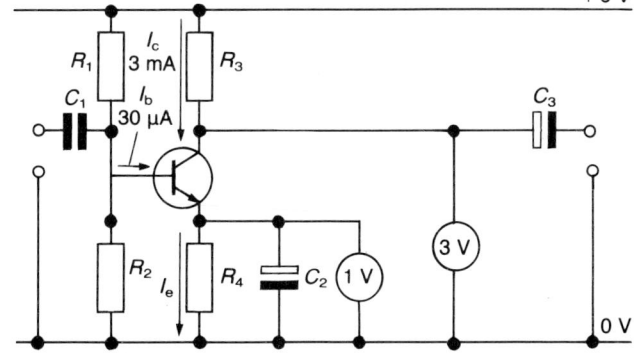

■ In an alarm system, a small buzzer is sounded for
  about 15 seconds when a switch is operated.
  (i) Draw a circuit diagram of a two transistor
      circuit which could perform this function. (9)

(ii) Give the values of any components which are critical to the timing of the sound output, showing calculations to justify the values chosen. (5)

MEG

2 Examine carefully the circuit diagram shown below.

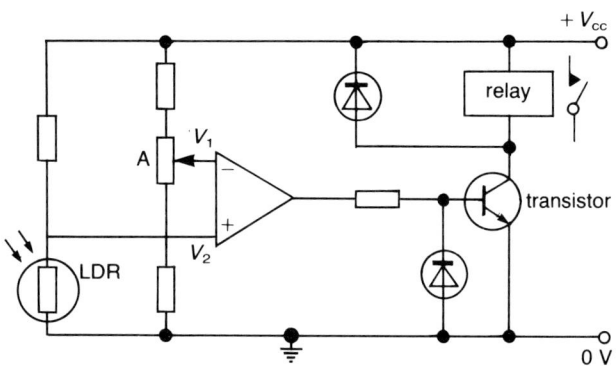

- Name the component marked A. (1)
- State whether the transistor is a pnp or an npn type. (1)
- If the circuit is being used as a light operated switch, state the function of the operational amplifier. (2)
- As darkness falls what happens
  (i) to the resistance of the LDR, (2)
  (ii) to the voltage at V_2 compared with the voltage at V_1, (2)
  (iii) to the output of the operational amplifier, (2)
  (iv) to the transistor, (2)
  (v) to the relay? (2)
- Redesign the circuit pictured so that it may be used as a burglar alarm which is activated when a light source falls on the LDR. (6)

WJEC

3 A pupil wishes to build an audio amplifier for a baby alarm. He finds an 'amplifier circuit' in an electronics book. It is an inverting amplifier using a 741 operational amplifier. The circuit is shown.

- What does *inverting* mean? (1)
- The voltage gain of the amplifier is 10. What controls the voltage gain of this circuit? (1)
- The pupil connects an 8Ω loudspeaker to the output and a small microphone to the input. Give two reasons why this circuit will not work as an audio amplifier. (2)
- The pupil decides that it would be a good idea to

be able automatically to switch the amplifier ON when it is dark and OFF when it is light. Design and draw a circuit that would be able to do this. (5)

NEA

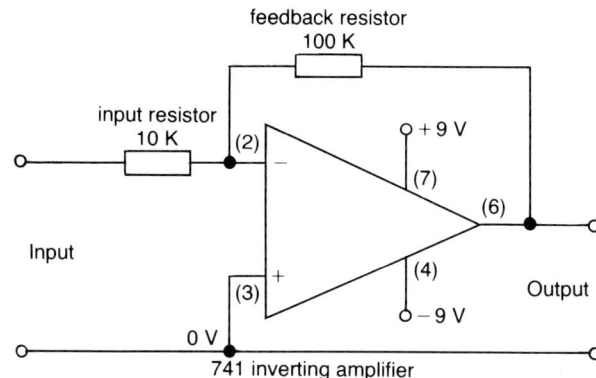

4 The diagram below shows an incomplete circuit which is designed to operate a 240 V a.c. motor to open a garage door. The circuit switches the motor on when the car drives over the pressure switch.
- Name the component marked A on the circuit. (1)
- In the space below complete the circuit so that it will operate correctly and safely. (5)

NEA

5 When constructing the circuit shown, only the following resistors are available.

130 Ω, 470 Ω, 680 Ω, 1.1 kΩ, 1.6 kΩ, 1.8 kΩ.

- (i) Show, by using resistors in series, how you would achieve a resistance of approximately 2.2 kΩ at R_1.
  (ii) Show, by using resistors in parallel, how you would achieve a resistance of approximately 270 Ω at R_2. (4)

(iii) Briefly explain why the resistance obtained at R_2 would not be exactly $270\,\Omega$. (4)

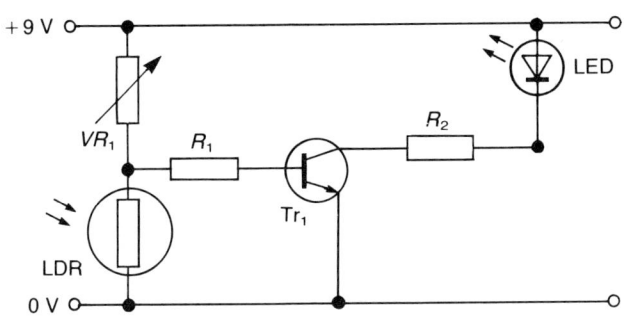

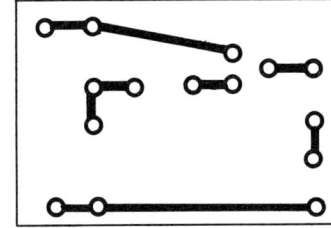

■ Assuming that all the resistors you have used were of 10 per cent tolerance, give the colour code for the highest value of resistor used.

Use the colour code table below to help. (4)

LEAG

1st Colour Band 1st Digit		2nd Colour Band 2nd Digit		3rd Colour Band Number of Zeros		4th Colour Band Tolerance	
Black	0	Black	0	Orange	0	Gold	5%
Brown	1	Brown	1	Brown	1	Silver	10%
Red	2	Red	2	Red	2		
Orange	3	Orange	3	Orange	3		
Yellow	4	Yellow	4	Yellow	4		
Green	5	Green	5	Green	5		
Blue	6	Blue	6	Blue	6		
Violet	7	Violet	7	Violet	7		
Grey	8	Grey	8	Grey	8		
White	9	White	9	White	9		

Pneumatics

1 ■ Identify the circuit symbols shown below. (3)

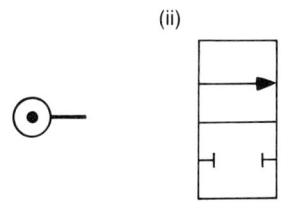

A builders merchant has a pneumatically operated sand bagging system, as illustrated on the next page. Two outlets are provided for the sand, so that the operator has time to change bags on one side while the bag on the other side is being filled. Cylinder **A** controls the quantity of sand delivered, and cylinder **B** controls which bag is to be filled. Cylinder **A** is a cushioned cylinder, and cylinder **B** is a non-cushioned type.

■ (i) Explain the difference between cushioned and non-cushioned types of cylinder. (6)
  (ii) State the type of use for which a cushioned cylinder should be chosen. (2)
  (iii) State the type of use for which a non-cushioned cylinder would be suitable. (2)
  (iv) Name a component which can be used to control the speed of the piston in a cylinder. (2)

The bagging system operates in the following sequence:
– the operator attaches a bag to side **X**, and switches a lever operated, lever return 3-port valve to supply air to the system;
– the operator presses a button operated, spring return 3-port valve at side **X**;
– cylinder **B** outstrokes to allow sand to pass to bag **X**, operating a roller operated, spring return, one way trip 3-port valve when it nears the end of its stroke;
– cylinder **A** goes negative, slowly, and then positive, quickly, after a time delay;
– while bag **X** is being filled, the operator attaches a bag to side **Y**;
– the operator presses a button operated, spring return 3-port valve at side **Y** (after cylinder **A** has gone positive);
– cylinder **B** instrokes to allow sand to pass to bag **Y**, operating a roller operated, spring return, one way trip 3-port valve when it nears the end of its stroke;
– cylinder **A** is operated as before;
– the operator changes the bag at **X**, and so the process continues.

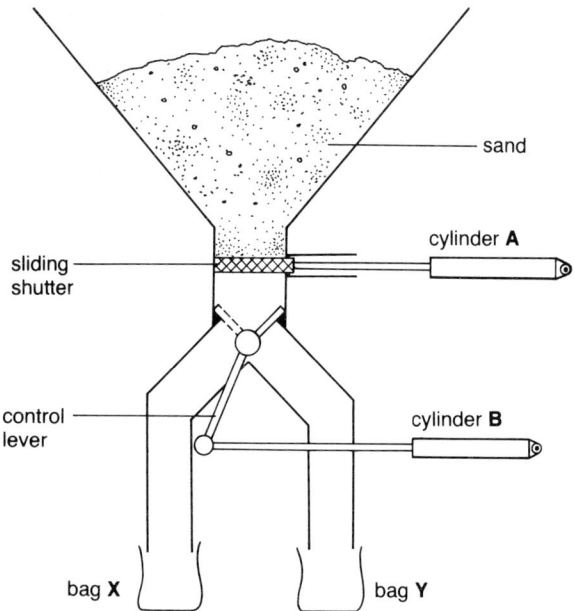

- Draw a flow chart to show the sequence as described on the previous page. (8)
- On your drawing of the following diagram show the circuitry necessary for the system to operate as described. Some components have been outlined, these are to be completed, and you must add any other components which are necessary. (18)
- When the equipment was first installed, the operation of the time delay on cylinder **A** was unsatisfactory. State what problem might have occurred, and why. (5)
- (i) State why the one way trip valves are not positioned at the ends of the strokes of cylinder **B**. (2)
  (ii) Give a reason for the use of one way trip valves in conjunction with cylinder **B**. (2)

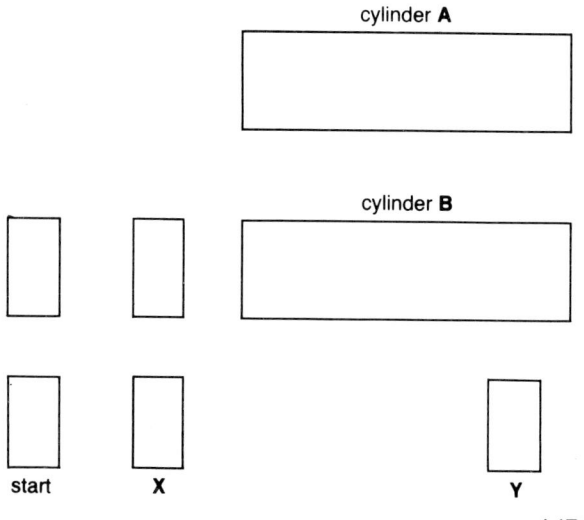

MEG

2 A block diagram pneumatic time delay circuit for the opening and closing of a car park barrier is shown.

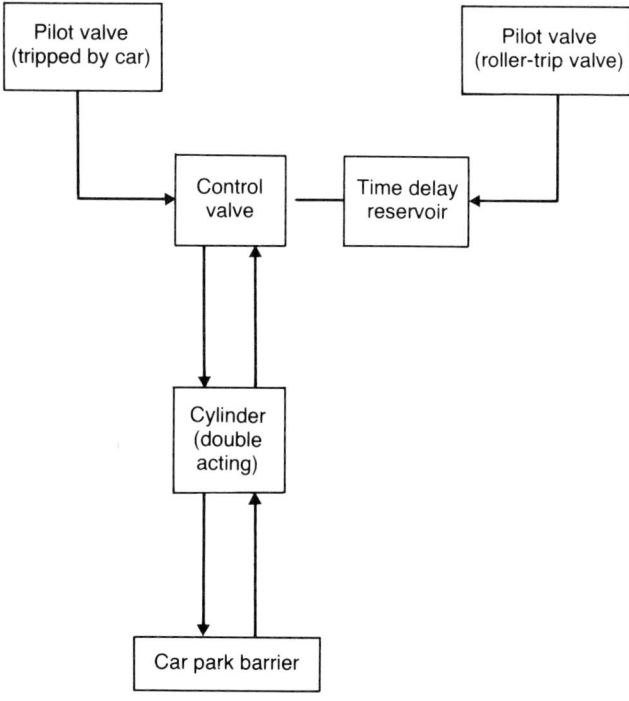

Using British Standard graphical symbols, draw a pneumatic circuit diagram for the car park barrier. (8)
WJEC

3 ■ Explain what is meant by the term *pneumatic*. (2)
  ■ State three examples of devices which produce non-linear motion when using compressed air. (3)
  ■ What is the difference between the motion of a pneumatic road drill and a dentist's drill? (2)
  ■ Which stroke of a single acting cylinder normally will give the greater force; that caused by compressed air or that caused by the spring? (1)

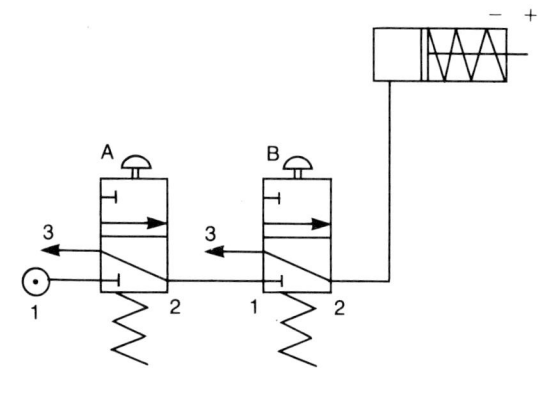

■ Inspect the circuit given above and then complete the truth table after drawing it out yourself. (4)

AND *Truth Table*		
Valve state. input		*Cylinder state.* output
A	*B*	*+ve or − ve*
off	off	
off	on	
on	off	
on	on	

■ Design a pneumatic circuit which would operate the cylinder when valve A or valve B is operated. Use one extra component to those shown in the diagram. (8)

WJEC

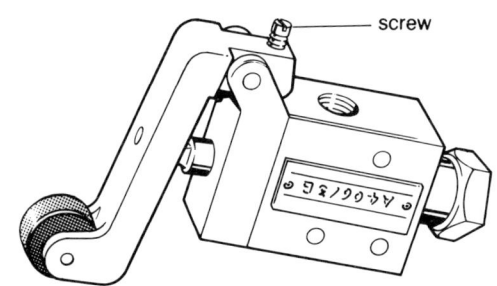
screw

■ Name the component. (1)
■ What is the purpose of the screw? (1)
■ Give one example where this component would normally be used. (1)

NEA

5 A furniture manufacturer wishes to test a new range of drawer guides. The guide must operate 1 000 000 times without failure. A double-acting cylinder is used to move the drawer in and out.

■ Give one advantage of using a double-acting cylinder rather than a single-acting cylinder for this purpose. (1)
■ State two factors you would need to consider before choosing a suitable double-acting cylinder for this purpose. (2)
■ Complete the piping of the circuit shown which is to be used to push and pull the drawer. (3)

4 A faulty component has been removed from a pneumatic system. A sketch of the component is shown below.

■ Give two safety precautions when using pneumatic equipment. (4)

NEA

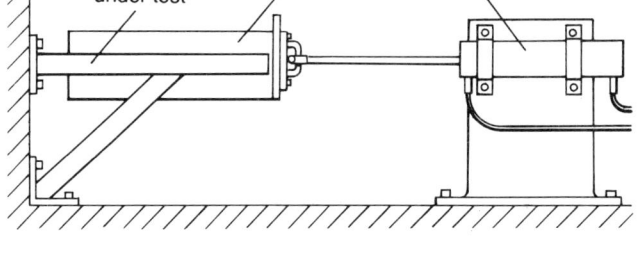
drawer guide under test · drawer · double-acting cylinder

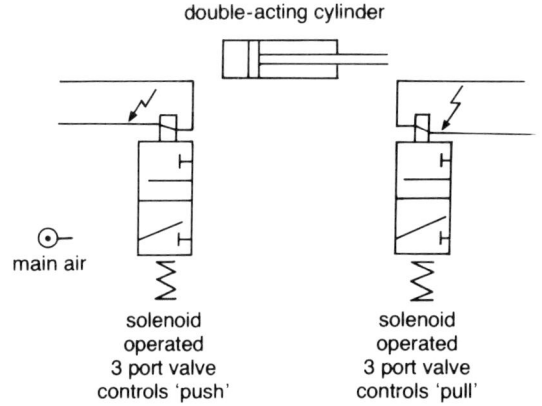
double-acting cylinder

main air

solenoid operated 3 port valve controls 'push'

solenoid operated 3 port valve controls 'pull'

Computing and Digital Micro-Electronics

1 ■ Explain fully the meaning of the following terms:
  (i) RAM; (2)
  (ii) ROM; (2)
  (iii) EPROM. (2)
■ Name two devices whose electrical characteristics change considerably with temperature. (2)
■ The diagram is of a central heating system. The electrical devices in the system operate from a 240 V mains supply.

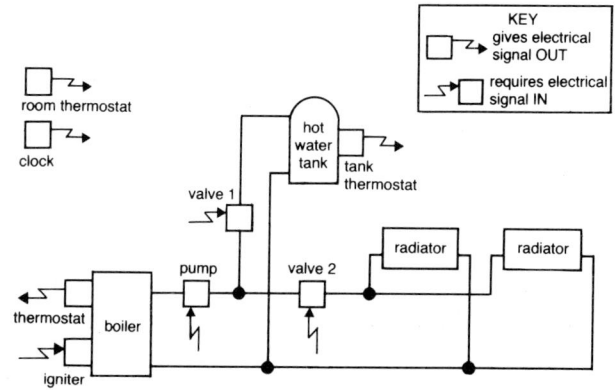

KEY
☐ gives electrical signal OUT
☐ requires electrical signal IN

room thermostat

clock

valve 1

hot water tank

tank thermostat

pump

valve 2

radiator

radiator

thermostat

boiler

igniter

(i) Draw the circuit diagram of a suitable interface which could be used to control any one of the devices from an output port of a microcomputer. Assume that a logic I out will turn the device ON. (10)

(ii) Draw the circuit diagram of a suitable interface circuit which could be used to send a logic I signal to the microcomputer when the boiler reached the required temperature. (6)

- Control of the central heating system must conform to the following criteria:
  - if the clock provides a logic 0, the boiler is to be off;
  - the pump must be on with the boiler;
  - either the room thermostat or the water tank thermostat must indicate a need for heat (logic I) for the boiler to be turned on.

  (i) Draw a flow chart to show the inter-relationship of all the elements of the system. (8)

  (ii) Draw a circuit diagram, using logic gates, to show how the signals from the clock, tank thermostat, and room thermostat could be processed to achieve the correct state at the boiler and pump. (6)

- The devices in the system are connected to a computer in the following way:
  bit 0 – pump;
  bit I – boiler;
  bit 2 – valve I;
  bit 3 – valve 2.

  (i) Write a computer program which would give the following pattern of control of the system:
  - boiler, pump and valve 2 switched on for five minutes;
  - valve I is switched on for three minutes, starting at the same time as the boiler, pump and valve 2;
  - all switched off. (8)

  (ii) Alongside your program, indicate the purpose of each line, and at the end, state the model of computer that your program is written for. (4)
  MEG

2 ■ Explain what is meant by the term binary. (2)
  ■ Convert the denary (decimal) number 12 to a binary number. (1)
  ■ Convert the binary number 1111 to a denary number. (1)
  ■ What will the hexadecimal number 5D be in denary? (2)
  ■ Change the denary number 103 to its equivalent hexadecimal number. (2)
  WJEC

3 ■ Name *three* devices in the home which could be controlled by a microprocessor. (3)
  ■ Select *one* of the devices you listed above and then discuss three advantages of using microprocessor control as compared with traditional methods. (5)
  WJEC

4 When connecting mechanical switches to a microprocessor, contact bounce is often a problem.
  ■ What is meant by the term contact bounce? (2)
  ■ Explain why contact bounce is undesirable in a microprocessor system. (2)
  ■ Suggest a device that can overcome the problem of contact bounce. (2)
  ■ Explain how your device overcomes the problem. (2)
  WJEC

5 ■ Explain briefly the need for an analogue to digital converter when measuring changing voltages with a microprocessor. (3)
  ■ A four bit digital to analogue converter has a maximum output of 4.5 V. During a particular conversion the digital input is set at 0011. What is the analogue output of the converter? (4)
  ■ The following block diagram is for a microprocessor controlled greenhouse system. Part of the diagram is completed for you.

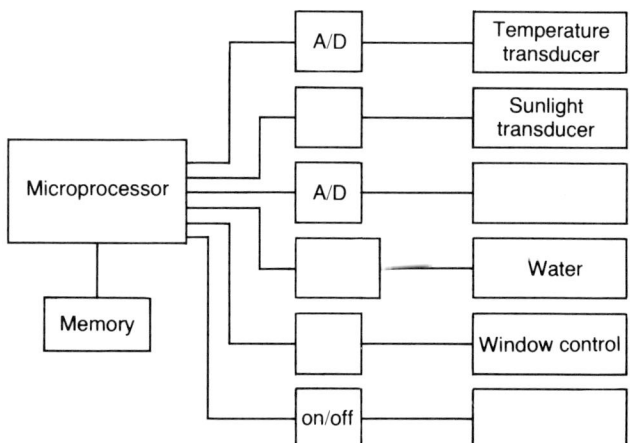

  (i) Complete the block diagram showing the sequence of control using items from the following list:
  moisture transducer; blinds; A/D; D/A; ON/ OFF.

  (ii) Indicate on the block diagram the direction of data flow in *each* case. (5)

  ■ When linking a microprocessor control based system to devices using high voltages, opto isolators are often used. Describe with the aid of a sketch the action and construction of an opto isolator. (8)
  WJEC

6 The sketch below shows part of the block diagram of a microprocessor controller.

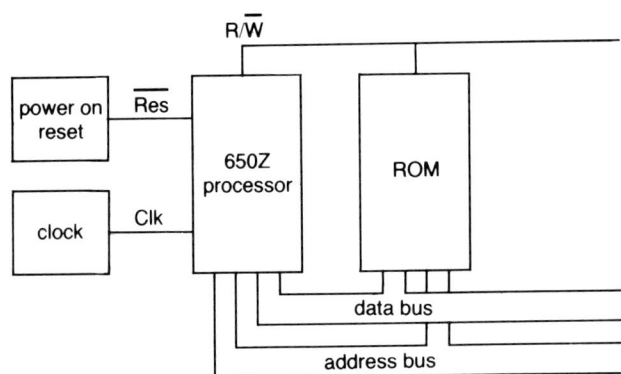

- What does ROM stand for? (1)
- What semiconductor material is the processor made from? (1)
- Explain briefly why a **POWER ON RESET** signal is needed by the processor. (1)
  NEA

7 The fuel of a car is controlled by a microprocessor. Fuel consumption is measured from a flow meter. A simplified sketch of a fuel flow meter is shown below.

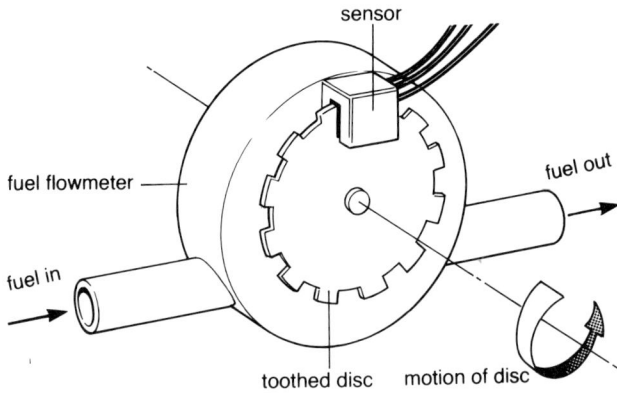

- Name a suitable sensor that could be used to detect the rotation of the toothed disc. (1)
- With a fuel flow of 6 litres per hour, the disc turns at 300r/m. The disc has 20 teeth. Calculate how many pulses per second will be produced by the sensor assuming that it produces one pulse every time a tooth passes it. (2)
- Fuel flow is controlled by a voltage which depends upon the position of an accelerator pedal. An electric motor drives a fuel pump. The faster the motor turns the more fuel is delivered. Signals from the flow meter are used to monitor how much fuel is used. Fuel is injected into the correct cylinder through one of four electrically operated valves which can be switched on or off by the

processor. A switch produces a single pulse for one complete rotation of the engine.

Design and draw a block diagram of a suitable fuel control system. Show clearly any interfaces and feedback control loops. (6)
NEA

8 ■ Here you are shown the circuit symbol for a two input NOR gate. Complete the truth table for the gate. (4)

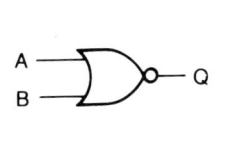

B	A	Q
0	0	
0	1	
1	0	
1	1	

- Complete a copy of the circuit diagram shown below, showing how you would use two NOR gates to make a simple quiz game switch. The switch should always show which contestant was ready to answer first. (8)

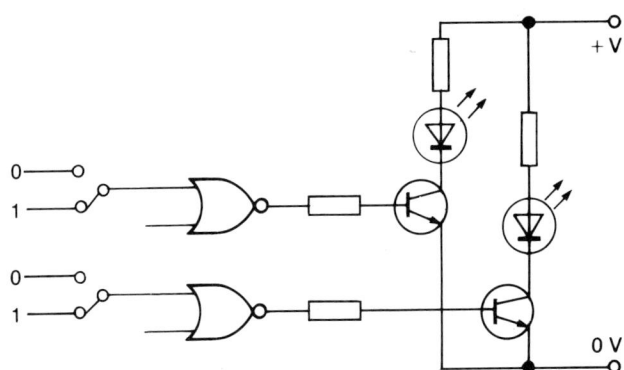

- Name the circuit you have made from the two NOR gates. (4)
  LEAG

Structures

1 This simple framework is used when loading or unloading cargo. The members of the framework are in tension or compression.

- (i) Give the term used for a member which is in tension. (1)

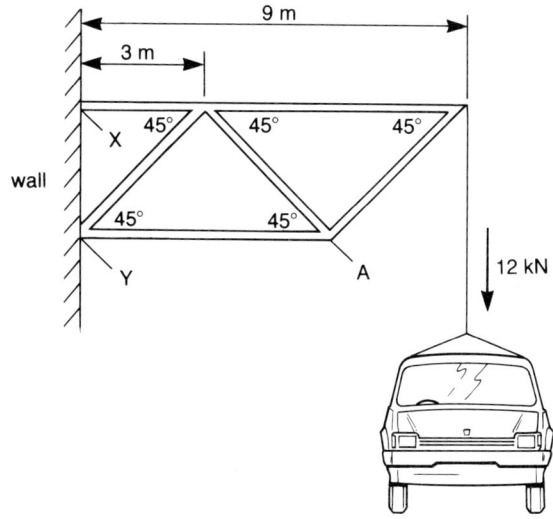

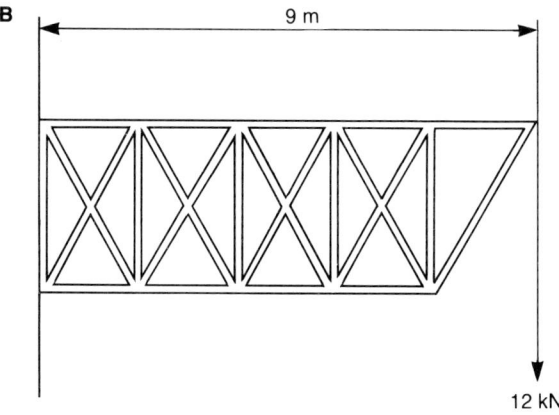

(ii) Give the term used for a member which is in compression. (1)

(iii) Name the property of a triangle which makes it suitable for use in the construction of frameworks. (1)

(iv) Give the name used for a structure which is supported at one end only. (2)

■ The framework shown is made from angle iron. (1)

(i) Illustrate what is meant by the term angle iron. (1)

(ii) Using notes and diagrams, show *three* ways in which the frame could be joined at point A. (6)

■ Before the framework shown was selected, two other designs of a similar size, A and B as shown below, were considered.

(i) Give one advantage and one disadvantage of the framework A, when compared with that chosen. (2)

(ii) Give one advantage and disadvantage of the framework B, when compared with that chosen. (2)

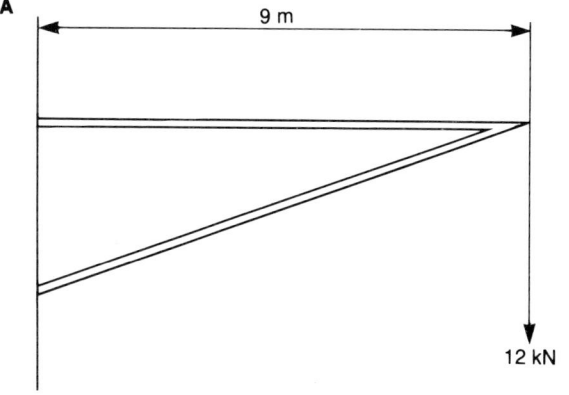

■ (i) By drawing or calculation, determine the values of the forces in *all* of the members of the framework shown in the first diagram when the car is lifted clear of the ground, clearly stating whether each member is in tension or compression. (18)

(ii) Draw a diagram to show clearly the direction and nature of the forces acting on the wall at points **X** and **Y**. (4)

(iii) Describe the effect that each of the forces identified in part (ii) above will have on the wall at points **X** and **Y**. (2)

(iv) Illustrate *one* suitable method of attaching the framework shown in the first diagram to the wall at points **X** and **Y**. (3)

■ Strain gauges are an important aid in monitoring surface movement in structures.

(i) Explain why a rule is not suitable for this type of measurement. (1)

(ii) Draw a large clear diagram to illustrate the appearance of a strain gauge. (3)

(iii) Describe the principle of operation of strain gauges. (3)

MEG

2 A load of 1000 N is supported by two wires, one attached to the roof and one attached to a wall, as shown below. Determine the tension in each wire. (8)

WJEC

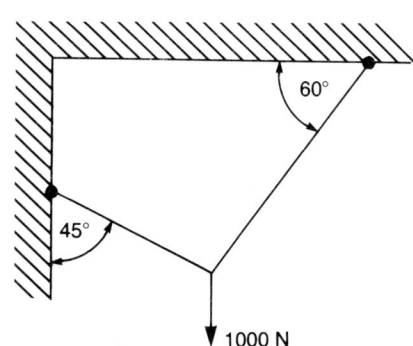

3 Describe
  (i) how a strain gauge is used to measure the strain in a material,
  (ii) the use of a dial test indicator on a structure. Give a practical example of where *each* technique may be used. (8)
WJEC

4 ■ State why it is necessary to have bridges. (2)
  ■ Draw a simple example of *each* of the following types of bridge:
    (i) beam;
    (ii) cantilever;
    (iii) arch;
    (iv) suspension. (8)
  ■ Faced with the problem of designing a bridge over a river, list six factors that need to be considered in the design. (10)
WJEC

5 A large supplier of home furniture has become concerned over the number of complaints about a kitchen stool which it sells. Several customers have been hurt when the stool collapsed. Details of the stool are shown below.

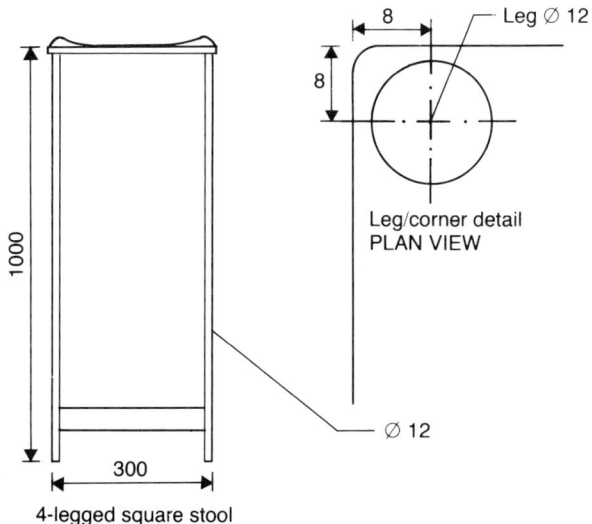

Leg ∅ 12
8
8
Leg/corner detail
PLAN VIEW

1000
300
∅ 12
4-legged square stool
FRONT VIEW

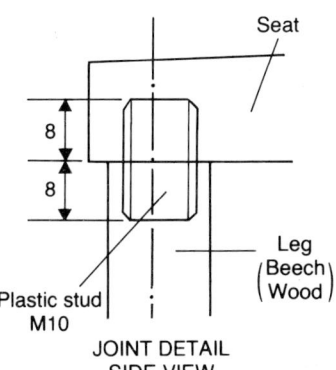

Seat
8
8
Plastic stud
M10
Leg
(Beech
Wood)
JOINT DETAIL
SIDE VIEW

■ What principle of structures has *not* been applied which would have made the structure rigid? (1)
■ Clearly identify *two different* weaknesses in the design/construction that could lead to the collapse of the stool in normal use. (2)
■ In the space below, design and sketch *one* modification that could be made to the stool. Clearly state the problem that your modification solves. (3)
NEA

Mechanisms

1 The diagram shows an old fashioned treadle type sewing machine.

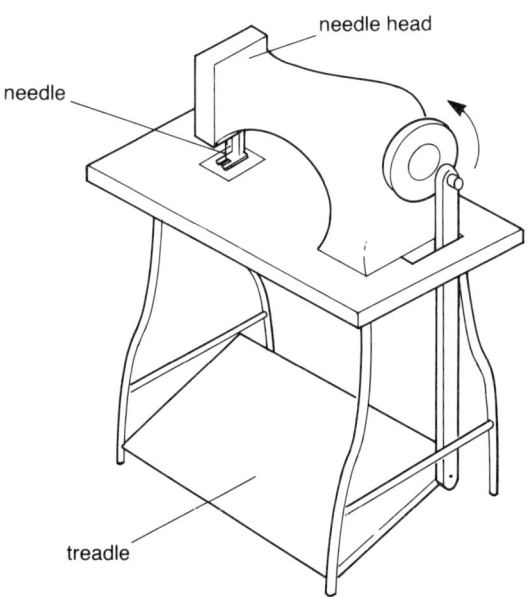

needle head
needle
treadle

■ Name the type of motion on the sewing machine at:
  (i) the needle; (1)
  (ii) the treadle. (1)
■ Draw and label clear diagrams to show **two** methods by which the motion at the wheel could be converted to the motion at the needle. (8)
■ On the next page is an enlarged view of the needle head, looking from a different angle. The foot is moved down, by use of the lever, to exert pressure on the cloth being sewn, by the use of a toggle linkage. Draw a clear diagram to illustrate a toggle linkage. (4)
■ Teeth under the foot move the cloth forward to be sewn. This is a ratchet type of mechanism. Draw and label a diagram of a pawl and ratchet, and give two examples to show other ways in which it can be used. (5)

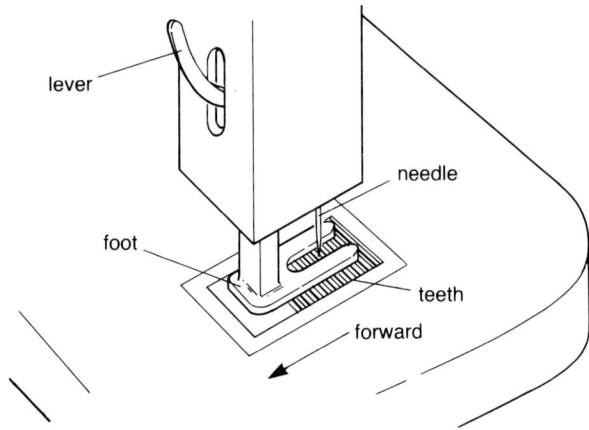

- The sewing machine shown in the first diagram is to be modernised by the addition of a motor with a belt and pulley drive.
  - (i) Draw clear diagrams to show both the belt and pulley type which would be suitable for the modification of the sewing machine, naming the type chosen. (6)
  - (ii) Show how the tension on the belt you have chosen could be adjusted and maintained, as it stretched with use. (4)
  - (iii) Draw a diagram to illustrate a suitable type of brake which could be fitted to the machine. (4)
- The sewing machine makes one stitch for each revolution of the machine pulley, and the motor runs at 780 r.p.m. If the motor is fitted with a pulley of 20 mm diameter, and the machine is fitted with a pulley of 65 mm diameter:
  - (i) state the velocity ratio; (2)
  - (ii) calculate the length which could be sewn in 30 seconds, with a stitch size of 2.5 mm. (4)
- Draw clear diagrams to show two methods by which the pulley could be attached to the motor, labelling the significant features in each case. (6)
  - (i) Name **two** types of bearing which could be used to support the motor shaft in the motor casing. (2)
  - (ii) Choose **one** of the bearings you have named in answer to part (i) above, stating which you have selected, and draw a large clear diagram to illustrate its main features. (3)

MEG

2 Explain, with the aid of a sketch, how a pawl and ratchet can be used on a wheel and axle to stop a cable from unwinding when it has been wound in a clockwise direction. (8)

WJEC

3 ■ Below is a diagram of a mechanically operated brake drum. Identify the parts numbered 1 to 7. (7)
  ■ The direction of the frictional force (F) is given. State the direction of the drum's rotation. (1)

WJEC

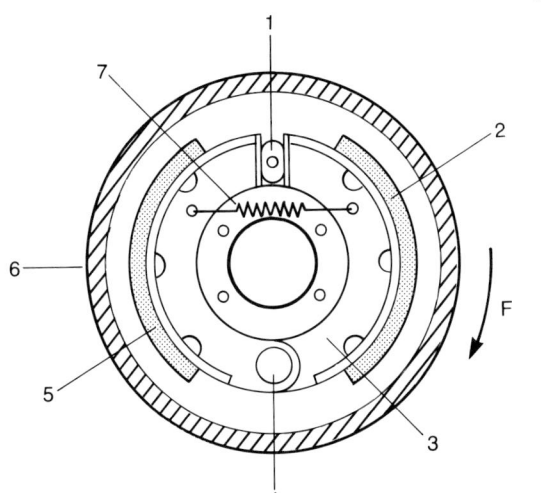

4 ■ Draw the British Standard graphical symbol for a set of meshed spur gears. (2)
  ■ Describe, with the aid of a sketch, what is meant by a *gear train, wheel and pinion, driver and driven gear.* (6)
  ■ State **two** practical engineering applications of a compound gear train. (2)
  ■ The diagram shows a small boat with a tiller operated rudder.
    Design a mechanism to operate the rudder by means of a steering wheel situated at the front of the boat. The tiller may be removed if required. (10)

WJEC

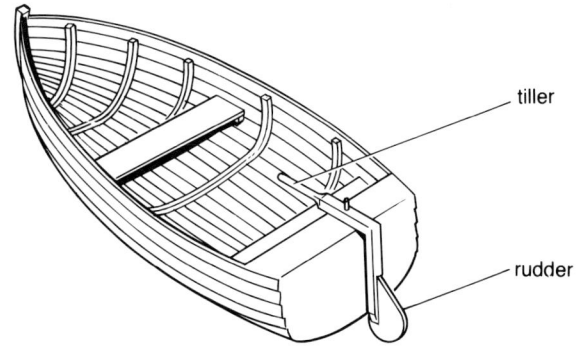

Index